VOLUME NINETY

COMPREHENSIVE
ANALYTICAL CHEMISTRY
Analysis of Cannabis

Advisory Board

Hendrik Emons
Joint Research Centre, Geel, Belgium

Gary Hieftje
Indiana University, Bloomington, IN, USA

Kiyokatsu Jinno
Toyohashi University of Technology, Toyohashi, Japan

Uwe Karst
University of Münster, Münster, Germany

Gyrögy Marko-Varga
Lund University, Lund, Sweden

Janusz Pawliszyn
University of Waterloo, Waterloo, Ont., Canada

Susan Richardson
University of South Carolina, Columbia, SC, USA

VOLUME NINETY

COMPREHENSIVE
ANALYTICAL CHEMISTRY
Analysis of Cannabis

Edited by

IMMA FERRER
Center for Environmental Mass Spectrometry,
Department of Environmental Engineering,
University of Colorado, Boulder, CO, United States

E. MICHAEL THURMAN
Center for Environmental Mass Spectrometry,
Department of Environmental Engineering,
University of Colorado, Boulder, CO, United States

ELSEVIER

Elsevier
Radarweg 29, PO Box 211, 1000 AE Amsterdam, Netherlands
The Boulevard, Langford Lane, Kidlington, Oxford OX5 1GB, United Kingdom
50 Hampshire Street, 5th Floor, Cambridge, MA 02139, United States

Copyright © 2020 Elsevier B.V. All rights reserved.

No part of this publication may be reproduced or transmitted in any form or by any means, electronic or mechanical, including photocopying, recording, or any information storage and retrieval system, without permission in writing from the publisher. Details on how to seek permission, further information about the Publisher's permissions policies and our arrangements with organizations such as the Copyright Clearance Center and the Copyright Licensing Agency, can be found at our website: www.elsevier.com/permissions.

This book and the individual contributions contained in it are protected under copyright by the Publisher (other than as may be noted herein).

Notices
Knowledge and best practice in this field are constantly changing. As new research and experience broaden our understanding, changes in research methods, professional practices, or medical treatment may become necessary.

Practitioners and researchers must always rely on their own experience and knowledge in evaluating and using any information, methods, compounds, or experiments described herein. In using such information or methods they should be mindful of their own safety and the safety of others, including parties for whom they have a professional responsibility.

To the fullest extent of the law, neither the Publisher nor the authors, contributors, or editors, assume any liability for any injury and/or damage to persons or property as a matter of products liability, negligence or otherwise, or from any use or operation of any methods, products, instructions, or ideas contained in the material herein.

ISBN: 978-0-444-64341-4
ISSN: 0166-526X

For information on all Elsevier publications
visit our website at https://www.elsevier.com/books-and-journals

Publisher: Zoe Kruze
Acquisitions Editor: Jason Mitchell
Editorial Project Manager: Leticia M. Lima
Production Project Manager: Denny Mansingh
Cover Designer: Mark Rogers

Typeset by SPi Global, India

Contents

Contributors to volume 90   *xi*
Preface   *xv*
Series editor preface   *xix*

Part 1
Current overview of methods for cannabis analysis

1. Comprehensive analytical testing of cannabis and hemp   3

Anthony Macherone

1. Introduction and background   3
2. Biosynthesis of phytocannabinoids   5
3. Cannabinoid profiling, total THC, and total CBD quantitation   7
4. Analysis of residual pesticides and mycotoxins   18
5. Residual solvent analysis   24
6. Summary   27
Acknowledgements   27
Disclaimers   27
References   27

2. Medicinal cannabis: Pharmaceutical forms and recent analytical methodologies   31

María Alejandra Fanovich, María Sandra Churio,
and Cristina Luján Ramirez

1. Introduction   31
2. The endocannabinoid system   32
3. *Cannabis sativa* constituents   37
4. Pharmaceutical forms of cannabinoids and novel delivery systems   41
5. Pharmacokinetics of cannabis   45
6. Analysis of cannabinoids and cannabinoid metabolites in organic matrices   47
7. Conclusions   52
Acknowledgements   52
References   54

vi

3. Analysis of cannabinoids in plants, marijuana products and biological tissues   **65**

Markel San Nicolas, Aitor Villate, Mara Gallastegi,

Oier Aizpurua-Olaizola, Maitane Olivares, Nestor Etxebarria,

and Aresatz Usobiaga

1. Introduction	65
2. Analysis of cannabinoids in plants	68
3. Analysis of cannabinoids in marijuana products	72
4. Analysis in biological fluids and tissues	78
5. Conclusions	96
References	96

Part 2
Applications of GC-MS techniques for cannabis analysis

4. State of the art solventless sample preparation alternatives for analytical evaluation of the volatile constituents of different cannabis based products   **105**

Gyorgy Vas

1. Introduction	105
2. Static head-space based cannabis testing	107
3. Solid phase microextraction	116
4. Dynamic head space (DHS)	127
5. Twister	129
6. Vacuum assisted sorbent extraction (VASE)	130
7. Cannabis and electronic delivery devices	132
8. Summary	134
References	134

5. Quantitating terpenes/terpenoids and nicotine in plant materials and vaping products using high-temperature headspace gas chromatography–mass spectrometry   **139**

Trinh-Don Nguyen, Seamus Riordan-Short, Thu-Thuy T. Dang,

Rob O'Brien, and Matthew Noestheden

1. Introduction	140
2. Materials and methods	143
3. Results and discussion	150

| Contents | vii |

4. Conclusion — 165
Acknowledgement — 166
References — 166

6. Improving cannabis differentiation by expanding coverage of the chemical profile with GCxGC-TOFMS — 169

Elizabeth M. Humston-Fulmer, David E. Alonso, and Joseph E. Binkley

1. Classifications of cannabis and the need for nontargeted chemical analyses — 169
2. Analytical tools and methods provide nontargeted chemical profiling for chemovar classifications — 172
3. Application of GCxGC-TOFMS for nontargeted chemical analysis and chemovar classifications — 184
4. Methods — 194
References — 196

7. Analysis of terpenes in hemp (*Cannabis sativa*) by gas chromatography/mass spectrometry: Isomer identification analysis — 197

E. Michael Thurman

1. Introduction — 198
2. Major terpenes and terpenoids — 198
3. Methods and materials — 202
4. Results and discussion — 203
5. Conclusions and major takeaway ideas — 232
Acknowledgements — 232
References — 232

8. Gas chromatography/electron ionization mass spectrometry (GC/EI-MS) for the characterization of phytocannabinoids in *Cannabis sativa* — 235

Jodie V. Johnson, Adam Christensen, Daniel Morgan, and Kari B. Basso

1. Introduction — 235
2. Experimental — 239
3. Results and discussion — 241
4. Concluding remarks — 272
References — 272

viii · Contents

Part 3
Applications of LC-MS techniques for cannabis analysis

9. The analysis of pesticides and cannabinoids in cannabis using LC-MS/MS   **277**

Paul Winkler

1. Introduction	277
2. LC-MS/MS instrumentation	279
3. Method development	286
4. Sample analysis	304
5. Summary	313
References	313

10. Cannabis and hemp analyzers for improved cannabinoid potency accuracy and reproducibility   **315**

Masayuki Nishimura, Tairo Ogura, Yohei Arao, Taka Iriki, Craig Young,

Andy Sasaki, Bob Clifford, A.J. Harmon-Glaus, Raz Volz,

Niloufar Pezeshk, Jeff Dahl, Will Bankert, Paul Winkler, Max Wang,

Jordan Frost, Sandy Mangan, John Easterling, and Scott Kuzdzal

1. A new era of medical cannabis and cannabis quality control testing emerges	316
2. The importance of sample homogenization	319
3. Potency testing instrumentation considerations	320
4. The development of cannabis and hemp analyzers	320
5. Development of easy to use, fit for purpose overlay software	323
6. Ability to operate in 21 CFR 11 compliance mode	324
7. Sample preparation and performance data	325
8. Cannabis flower sample preparation	325
9. Important notes regarding cannabis flower sample filtration	325
10. Proper handling of cannabinoid standards	326
11. Methods section	326
12. Calibration of the HPLC system by use of a standard solution	328
13. Standard curves	329
14. Total THC potency formula	330
15. Application of the cannabis analyzer for the quantitative determination of cannabinoids in cannabis flower and edible products	331

16. The need for a hemp analyzer		333
17. Future directions of cannabis and hemp analyzer development		333
18. Expansion to a 15-cannabinoid standards analysis		334
19. 'Full Spectrum' cannabis and automated analysis of tinctures/oils		334
20. Summary		336
Disclaimers		336
References		337

11. Marihuana safety: Potency of cannabinoids, pesticide residues, and mycotoxin in one analysis by LC/MS/MS   339

Jerry Zweigenbaum and Agustin Pierri

1. Introduction		340
2. Method		343
3. Results and discussion		355
4. Summary		363
Acknowledgements		363
References		363

12. Using sesame seed oil to preserve and concentrate cannabinoids for paper spray mass spectrometry   367

Brandon J. Bills and Nicholas E. Manicke

1. Introduction		367
2. Methods		371
3. Results and discussion		379
4. Suitability for cannabinoid testing		391
5. Conclusion		392
Acknowledgements		393
References		393

13. Quantitating cannabinoids in edible chocolates using heated ultrasonic-assisted extraction   397

James W. Favell, Ryan Hayward, Emily O'Brien, Seamus Riordan-Short, Nahanni Sagar, Rob O'Brien, and Matthew Noestheden

1. Introduction		397
2. Materials and methods		400
3. Results		406
4. Conclusions		412
References		413

14. Analyses of cannabinoids in hemp oils by LC/Q-TOF-MS 415

Imma Ferrer

1. Introduction	415
2. Methods	417
3. Results and discussion	420
4. Conclusions	450
Acknowledgements	450
References	451

15. The estimation of cannabis consumption through wastewater analysis 453

Lubertus Bijlsma, Daniel A. Burgard, Frederic Been, Christoph Ort,
João Matias, and Viviane Yargeau

1. Wastewater-based epidemiology	453
2. Cannabis biomarkers	457
3. Analytical methodology	461
4. Wide-scale use of WBE for cannabis estimates in Canada and US.	472
5. Future research and current asset	476
Acknowledgements	477
References	477

Index   *483*

Contributors to volume 90

Oier Aizpurua-Olaizola
Dinafem Seeds (Pot Sistemak S.L.), San Sebastian, Spain

David E. Alonso
LECO Corporation, Saint Joseph, MI, United States

Yohei Arao
Shimadzu Scientific Instruments, Columbia, MD, United States

Will Bankert
Shimadzu Scientific Instruments, Columbia, MD, United States

Kari B. Basso
Department of Chemistry, University of Florida, Gainesville, FL, United States

Frederic Been
KWR Water Research Institute, Nieuwegein, The Netherlands

Lubertus Bijlsma
Research Institute for Pesticides and Water, University Jaume I, Castellón, Spain

Brandon J. Bills
Department of Chemistry and Chemical Biology, Indiana University Purdue University at Indianapolis, Indianapolis, IN, United States

Joseph E. Binkley
LECO Corporation, Saint Joseph, MI, United States

Daniel A. Burgard
Department of Chemistry, University of Puget Sound, Tacoma, WA, United States

Adam Christensen
Essential Validation Services (EVS); Botanica Testing, Inc., Gainesville, FL, United States

María Sandra Churio
Departamento de Química y Bioquímica, FCEyN, Universidad Nacional de Mar del Plata; Instituto de Investigaciones Físicas de Mar del Plata (IFIMAR) CONICET—Universidad Nacional de Mar del Plata, Mar del Plata, Argentina

Bob Clifford
Shimadzu Scientific Instruments, Columbia, MD, United States

Jeff Dahl
Shimadzu Scientific Instruments, Columbia, MD, United States

Thu-Thuy T. Dang
Department of Chemistry, I.K. Barber School of Arts and Sciences, University of British Columbia, Kelowna, BC, Canada

John Easterling
Happy Tree Microbes, Los Angeles, CA; Laughing Dog Farms, Portland, OR, United States

Nestor Etxebarria
Department of Analytical Chemistry, Faculty of Science and Technology, University of the Basque Country (UPV/EHU), Leioa; Research Centre for Experimental Marine Biology and Biotechnology (PIE), University of the Basque Country (UPV/EHU), Plentzia, Spain

María Alejandra Fanovich
Instituto de Investigaciones en Ciencia y Tecnología de Materiales (INTEMA) CONICET—Universidad Nacional de Mar del Plata, Mar del Plata, Argentina

James W. Favell
Supra Research and Development; Department of Chemistry, I.K. Barber School of Arts and Sciences, University of British Columbia, Kelowna, BC, Canada

Imma Ferrer
Center for Environmental Mass Spectrometry, Department of Environmental Engineering, University of Colorado, Boulder, CO, United States

Jordan Frost
Shimadzu Scientific Instruments, Columbia, MD, United States

Mara Gallastegi
Dinafem Seeds (Pot Sistemak S.L.), San Sebastian, Spain

A.J. Harmon-Glaus
Shimadzu Scientific Instruments, Columbia, MD, United States

Ryan Hayward
Supra Research and Development, Kelowna, BC, Canada

Elizabeth M. Humston-Fulmer
LECO Corporation, Saint Joseph, MI, United States

Taka Iriki
Shimadzu Scientific Instruments, Columbia, MD, United States

Jodie V. Johnson
Department of Chemistry, University of Florida, Gainesville, FL, United States

Scott Kuzdzal
Shimadzu Scientific Instruments, Columbia, MD, United States

Anthony Macherone
Agilent Technologies, Inc., Wilmington, DE; Johns Hopkins University School of Medicine, Baltimore, MD, United states

Sandy Mangan
SPEX SamplePrep LLC, Metuchen, NJ, United States

Nicholas E. Manicke
Department of Chemistry and Chemical Biology, Indiana University Purdue University at Indianapolis, Indianapolis, IN, United States

João Matias
European Monitoring Centre for Drugs and Drug Addiction, Lisbon, Portugal

Contributors to volume 90

Daniel Morgan
Botanica Testing, Inc., Gainesville, FL, United States

Trinh-Don Nguyen
Department of Chemistry, I.K. Barber School of Arts and Sciences, University of British Columbia; Supra Research and Development, Kelowna, BC, Canada

Masayuki Nishimura
Shimadzu Scientific Instruments, Columbia, MD, United States

Matthew Noestheden
Supra Research and Development; Department of Chemistry, I.K. Barber School of Arts and Sciences, University of British Columbia, Kelowna, BC, Canada

Emily O'Brien
Supra Research and Development, Kelowna, BC, Canada

Rob O'Brien
Supra Research and Development; Department of Biology, I.K. Barber School of Arts and Sciences, University of British Columbia, Kelowna, BC, Canada

Tairo Ogura
Shimadzu Scientific Instruments, Columbia, MD, United States

Maitane Olivares
Department of Analytical Chemistry, Faculty of Science and Technology, University of the Basque Country (UPV/EHU), Leioa; Research Centre for Experimental Marine Biology and Biotechnology (PIE), University of the Basque Country (UPV/EHU), Plentzia, Spain

Christoph Ort
Eawag, Urban Water Management, Swiss Federal Institute of Aquatic Science and Technology, Dübendorf, Switzerland

Niloufar Pezeshk
Shimadzu Scientific Instruments, Columbia, MD, United States

Agustin Pierri
Weck Laboratories, City of Industry, CA, United States

Cristina Luján Ramirez
Departamento de Química y Bioquímica, FCEyN, Universidad Nacional de Mar del Plata, Mar del Plata, Argentina

Seamus Riordan-Short
Supra Research and Development, Kelowna, BC, Canada

Nahanni Sagar
Supra Research and Development, Kelowna, BC, Canada

Markel San Nicolas
Dinafem Seeds (Pot Sistemak S.L.), San Sebastian; Department of Analytical Chemistry, Faculty of Science and Technology, University of the Basque Country (UPV/EHU), Leioa, Spain

Andy Sasaki
Shimadzu Scientific Instruments, Columbia, MD, United States

E. Michael Thurman
Center for Environmental Mass Spectrometry, Department of Environmental Engineering, University of Colorado, Boulder, CO, United States

Aresatz Usobiaga
Department of Analytical Chemistry, Faculty of Science and Technology, University of the Basque Country (UPV/EHU), Leioa; Research Centre for Experimental Marine Biology and Biotechnology (PIE), University of the Basque Country (UPV/EHU), Plentzia, Spain

Gyorgy Vas
VasAnalytical, Flemington, NJ, United States

Aitor Villate
Department of Analytical Chemistry, Faculty of Science and Technology, University of the Basque Country (UPV/EHU), Leioa, Spain

Raz Volz
Shimadzu Scientific Instruments, Columbia, MD, United States

Max Wang
Shimadzu Scientific Instruments, Columbia, MD, United States

Paul Winkler
SCIEX LLC, Framingham, MA, United States

Paul Winkler
Shimadzu Scientific Instruments, Columbia, MD, United States

Viviane Yargeau
Department of Chemical Engineering, McGill University, Montreal, QC, Canada

Craig Young
Shimadzu Scientific Instruments, Columbia, MD, United States

Jerry Zweigenbaum
Agilent Technologies, Wilmington, DE, United States

Preface

Cannabis has been known for several millennia, even the ancient Roman and Egyptian civilizations used it for medical and religious ceremonies. Since then it has been used widely in countries around the world. It was not till the early 20th century that it became an illegal drug. Most recently, in 2012, in the states of Colorado and Washington, marihuana became legal for recreational purposes in spite of the fact that it is still federally illegal in the United States. However, the farm bill of 2018 legalized hemp (also called 'industrial hemp') at the federal level. Hemp is a variety of *Cannabis sativa* high on cannabidiol (CBD) and containing less than 0.3% tetrahydrocannabinol (THC). Because of this, during these last few years we have seen a tremendous increase in products in health food stores containing CBD, which claim to have remarkable health properties. As we write this preface in the middle of the COVID-19 crisis, we wonder about the future of our world and the importance of understanding the potential health benefits of the many compounds present in the cannabis plant.

With the massive existence of products on the market (oils, creams, lotions, dietary supplements, etc.) it is highly important to develop instrumental methods for the chemical characterization of all these products. This includes analysis of solvents, trace metals, pesticides, mycotoxins, active ingredients (cannabinoids and terpenes) to name but a few. There are no standard methods for these types of analyses at this time. Each individual lab usually develops its own methodologies. For these reasons, we were anxious to prepare a volume on cannabis analysis.

Why did we get involved in cannabis analysis? That is a good question, since we have been mainly environmental chemists focused on water, soil and food analyses. However, we watched the cannabis industry grow right in front of us, especially with regard to edibles containing CBD and THC. Because of the federal laws prohibiting the study of cannabis in the early days, the University would not allow us to work on this type of research. But research topics are pursued not by accident, but by the crossing of several paths. In our case, it was the combination of three paths. First, we heard a talk by one of our colleagues from the Chemistry Department, Dr. Bob Sievers, who was very enthusiastic about the use of CBD for medical purposes (treating epilepsy, seizures, for example), and that enthusiasm encouraged us to follow the topic closely. Second, it was the coming of the Farm Bill in 2018, which allowed the University of Colorado to pursue research on hemp products, so now we had an opportunity to apply our analytical chemistry skills to this field. This, together with the challenge of unravelling the complexity of the cannabis plants, presented a very good example for our type of research. Finally, it was not until we got one of those products in our hands (a pet oil containing CBD) that we felt the urgency to test the material for its safety and unknown products. Yes, our Beloved Kitty was the final "catalyzer" for this research effort, as we treated him for his serious illness, and to him this book is dedicated. He was always part of our silent writing in the computer and he is missed a lot.

This book is divided into three main sections. Section 1 gives a current overview of the methods used for cannabis analysis. It comprises three chapters: Chapter 1 gives an overview of the analytical testing methods needed for some of the parameters included in regulations for cannabis and hemp. Chapter 2 describes the pharmacokinetics of cannabinoids in biological systems, novel pharmaceutical formulations, as well as their methodologies of analyses in biological matrices. Finally, Chapter 3, in this first section, reviews all the recent analytical methodologies in the literature for the analysis of cannabinoids in plants, cannabis products and biological tissues. Section 2 is focused on GC–MS techniques. Chapter 4 gives a comprehensive review of all the solventless and head space sample preparation techniques for the analysis of volatile constituents from cannabis products and their hyphenation with mass spectrometry techniques. Chapter 5 describes the quantitation of an important group of compounds in cannabis, terpenes and terpenoids, in plant and vaping products by headspace and GC–MS. Chapter 6 describes in a unique way the combination of GC, MS and two-dimensional GC (GCxGC) for the comprehensive chemical profiling of cannabis samples and nontargeted analyses. Chapter 7 takes a deep look at the mass spectral information obtained for terpenes and terpenoids by GC–MS and establishes fragmentation patterns. Chapter 8 describes the GC–MS spectral information for major and minor cannabinoids and the identification of new cannabinoids for the characterization of *Cannabis sativa*. Section 3 is focused on LC–MS techniques. Chapter 9 provides a description for the analysis of pesticides and cannabinoids using LC–MS–MS techniques in a way that is very understandable for a person not expert in this field of analysis, as well as a discussion on how to interpret the data reported by validation laboratories. Chapter 10 describes the use of a commercial hemp analyser for the analysis of cannabis products, including sample preparation, potency testing, software tools and quality control data, as well as examples on the analysis of cannabis flower and edible products. Chapter 11 presents a one analysis methodology using LC–MS–MS and isotope transitions for the simultaneous analysis of potency, pesticides and mycotoxins. Chapter 12 describes the use of sesame oil for the preservation and concentration of cannabinoids and applies it to a methodology using paper spray mass spectrometry for the detection of THC. Chapter 13 describes an important methodology for the quantitation of cannabinoids in edible chocolates, very much needed for product safety and quality control purposes. Chapter 14 deals with the application of high resolution time-of-flight mass spectrometry for the analysis of cannabinoids and it takes a deep look at fragmentation patterns using accurate mass. Finally, Chapter 15 proposes and demonstrates the use of wastewater analysis for the estimation of cannabis consumption in a given population environment.

This is our fourth edited book with our various Publishers, including Elsevier, which is part of a series dedicated to chemical analysis. In each of previous books we have predicted what we think will be the future of the book and its topics. In the case of *Cannabis sativa*, we have the following thoughts. First, cannabis will continue to grow, so to speak, as an important crop, especially in regard to industrial hemp. We think that hemp will become a crop, perhaps not as abundant as corn, but will definitely be a major crop over the next decade. The reason for this expansion of growth is the valuable isomers of CBD, their acids, terpenes and terpenoids that will be useful for health and medicine. Second, the analysis of contaminants in the hemp market will also grow accordingly, especially in regard to pesticides, metals and solvents. Third, new compounds will be discovered in the hemp plant and become important nutraceuticals or medicines in the near future. Finally, we think that Big Pharma will become an important player in this field as the profitability of hemp grows.

Lastly, we would like to acknowledge our two colleagues from CU Boulder, Dr. Bob Sievers and Dr. Randy Shearer, who have been inspirational to us with the study of cannabis and their dedication to this topic (from growing hemp to harvesting, extracting, purifying and analysing the products). We also would like to thank our friend and Series Editor, Dr. Damià Barceló, again one more time, for offering us this chance to edit this book together, the first on this topic from this Series of *Comprehensive Analytical Chemistry*. Our appreciation also goes to the Editorial staff of Elsevier for their support through the steps of this Book preparation. And finally and most importantly, to all our individual authors who contributed to this book and by doing so helped to tremendously expand the knowledge on this important topic, the analysis of cannabis.

IMMA FERRER
E. MICHAEL THURMAN
The Writers Cabin
Jamestown, CO, USA

Series editor preface

It was not so difficult to convince to my old friends and colleagues, Drs Imma Ferrer and E.M. Thurman, to edit a book on cannabis analysis. As soon as I was aware that they started to work on this plant I was sure that they would be able to organize the right team to compile a book on the different methods of analysis of cannabis and related chemicals and impurities. As most of the readers already know, during the last few years in US and Canadian conferences on analytical chemistry, i.e. ACS or ASMS sessions devoted to cannabis analysis were included in their programs. This is a consequence of the growing interest on this new commercial product in the United States and Canada that is being sold in shops as medicinal plant. In Europe, we are still far from its approval for commercialization and use as medicinal plant. With the exception of the Netherlands where cannabis derivatives have been sold in their coffee shops for more than 30 years. But the thing is that this Dutch tradition is related to the acceptance of cannabis as illicit drug by the people of the Netherlands.

The book that you have now in your hands contains 15 chapters. The first three chapters present general information on analytical methods of cannabis, pharmaceutical forms, impurities like residual pesticides and analysis of cannabis in plants and marijuana. The second group of chapters from 4 to 8 are devoted to gas chromatography–mass spectrometry (GC-MS) methods applied to cannabis analysis, including time-of-flight (TOF) applications. This is not a surprise since Imma and Mike did edit another book in this series on TOF and high-resolution methods. The last group of chapters, form 9 to 15 report liquid chromatography–mass spectrometry (LC-MS) applications for cannabis analysis. There are plenty of applications to determine the quality of cannabis. Among the list of compounds analyzed there are terpenes, phytocannabiods, pesticide residues and micotoxins. Analysis of cannabis in other matrices like chocolate and hemp oils is also reported. The last chapter of the book on wastewater-based epidemiology (WBE) needs attention because looks to cannabis analysis from a different point of view. WBE is a way to determine cannabis consumption either as prescribed or illicit drug in cities and uses LC-high resolution or tandem MS.

To this end, the book contains a comprehensive list of GC–MS and LC–MS methods to analyze cannabis and other chemicals and impurities

xix

in plants and related products. Most importantly, it is written by well-known researchers with great expertise on cannabis analysis. I would like to thank to all of them as well to both the editors working very hard under the Covid-19 outbreak in their Writers Cabin in Jamestown, Co.

That being said, I strongly believe that the book is timely and fits 100% the scope of CAC series. I am pretty sure that it will be of interest to many different types of readers, from expert analytical chemists who want to know more details on the quality control side of cannabis, to the newcomers and to the general public interested to search information on this very popular medicinal plant. It will be as well of great help in the near future to the analytical chemists of other parts of the world, like Europe or Asia, when cannabis will be allowed to be sold as medicine.

Damia Barceló
ICRA and IDAEA
Girona and Barcelona, Spain

PART 1

Current overview of methods for cannabis analysis

CHAPTER ONE

Comprehensive analytical testing of cannabis and hemp

Anthony Macherone*

Agilent Technologies, Inc., Wilmington, DE, United States
Johns Hopkins University School of Medicine, Baltimore, MD, United States
*Corresponding author: e-mail addresses: anthony_macherone@agilent.com; amacher1@jhmi.edu

Contents

1. Introduction and background — 3
2. Biosynthesis of phytocannabinoids — 5
3. Cannabinoid profiling, total THC, and total CBD quantitation — 7
   3.1 Sample preparation — 7
   3.2 Sample preparation procedures for flower and oils — 8
   3.3 Cannabinoid analyses with HPLC-UV — 8
   3.4 Cannabinoid analyses with liquid chromatography: Mass spectrometry (LC-MS) — 14
   3.5 Analytical challenges to cannabinoid analyses using HPLC-UV and LC-MS — 15
   3.6 Best practices — 18
4. Analysis of residual pesticides and mycotoxins — 18
   4.1 Background — 18
   4.2 Sample preparation — 19
   4.3 Liquid chromatography-tandem mass spectrometry (LC-MS/MS) — 21
   4.4 Gas chromatography-tandem mass spectrometry (GC-MS/MS) — 23
5. Residual solvent analysis — 24
   5.1 Reagent and sample preparation — 24
   5.2 Headspace GC-MS analysis (HS GC-MS) — 25
6. Summary — 27
Acknowledgements — 27
Disclaimers — 27
References — 27

1. Introduction and background

Traditionally, cannabis has been viewed from the post-consumption toxicology perspective, criminalistic identification of confiscated bulk seizures, and research and development for potential new drugs that may modulate the endocannabinoid system. More recently in the United States and Canada, legislation has been passed to legalize the recreational use of cannabis and cannabinoid products for adults over the age of 21. Canada has passed their legislation at the federal level to define the available consumer products, how the products will be dispensed, and regulatory requirements to ensure the safety and quality of the products prior to retail distribution. At the federal level in the United States, marijuana (cannabis) continues to be a Schedule I narcotic meaning it has no "acceptable medical use and has a high potential for abuse" [1]. In this context, marijuana is defined as any *Cannabis* spp. with a Δ^9-tetrahyrocannabinol (THC) or total THC (Eq. 1) content greater than 0.3% by dry weight ($\leq$13% moisture content).

$$Total\ THC = 0.877*[THCA] + [THC] \tag{1}$$

where, [THCA] is the concentration of Δ^9-tetrahyrocannabinolic acid (THCA), [THC] is the THC concentration, and 0.877 is the molecular weight ratio THC/THCA (314.469/358.478). As seen in Eq. (2), total CBD is calculated through substitution of [THCA] with [CBDA], and [THC] with [CBD].

$$Total\ CBD = 0.877*[CBDA] + [CBD] \tag{2}$$

According to Leafly.com [2], 11 states and the District of Columbia have fully legalized adult recreation use programs. Four states remain fully illegal. The remaining states have a defined medicinal or medical CBD program. This state-by-state variability translates to regulatory testing requirements with nearly every state stipulating a different target list and action (not to exceed) limits for each test.

Hemp in United States is an altogether a different story. In 2016, U.S.D.A., U.S. D.E.A., and U.S. F.D.A. defined industrial hemp in the Federal Register (FR 53365) as any part of *Cannabis sativa L.* with a dry weight concentration not greater than 0.3% (wt./wt.) [3]. In December 2018, the Agriculture Improvement Act, known colloquially as the Farm Bill, was signed into law and legalized hemp as an industrial crop. Therefore, hemp legislation *is* defined at the federal level and the governance of hemp in the U.S. falls under the auspices of U.S.D.A. In October 2019, U.S.D.A. entered "Establishment of a Domestic Hemp Production Program" into the Federal Register [4]. Therein, they defined sampling and testing protocols. Of particular note was that they positioned both high performance liquid chromatography (HPLC), and gas phase chromatography (GC) as the preferred testing equipment to determine THC content. U.S.D.A. further stipulated that all testing must occur in a DEA registered lab but in early 2020, a moratorium was place on this stipulation until October 2020.

The dialogue above points out a primary reason for confusion when it comes to the various regions, the laws, and the regulated safety and testing requirements. Even though state regulation varies greatly, there is a commonality to the testing needs. All regions including Canada, require total THC (potency) and other cannabinoids to be quantified in the legalized products. Residual pesticide and mycotoxin quantitation are generally required as is testing for heavy metals such as lead, cadmium, mercury, and arsenic. Residual solvent analysis of processed cannabinoid products and terpenes profiling are typically required. Lastly, screening for potentially harmful bacteria, moulds, and yeast is a commonly regulated test.

2. Biosynthesis of phytocannabinoids

Through two polyketide pathways, fatty acids and coenzymes synthesize olivetolic acid and divarinolic acid. These acids react with the substrate geranyl pyrophosphate and geranyl-diphosphate:olivetolate geranyltransferase synthesize cannabigerovarinic acid (CBGVA) and cannabigerolic acid (CBGA), respectively. The *Cannabis* spp. genome further encodes for THCA synthase, cannabidiolic acid (CBDA) synthase, and cannabichromenic acid CBCA synthase which react with CBGVA and CBGA to synthesize six acid phytocannabinoids as shown in Fig. 1 [5].

The acid phytocannabinoids decarboxylate to the neutral compounds after harvest and upon exposure to heat and light. An example of this is shown in Fig. 2 for THCA to $THC + CO_2$ decarboxylation. Therefore, in the living plant, THC, CBD, cannabigerol (CBG), or the other decarboxylated neutrals are not present in large concentrations. Other processes such as photo-irradiation converts CBCA and cannabichromene (CBC) to cannabicyclolic acid (CBLA) and cannabicyclol (CBL), respectively, and like other acid phytocannabinoids, CBCA decarboxylates to create a secondary pathway to CBL. Isomerization of THC to Δ^8-tetrahyrocannabinol (Δ^8-THC) can also occur, and oxidative aromatization transforms THC into cannabinol (CBN) which can be photochemically rearranged into cannabinodiol (CBND). Photo-oxidation and exposure to heat of CBDA transform it into cannabielsolic acid A (CBEA-A) which

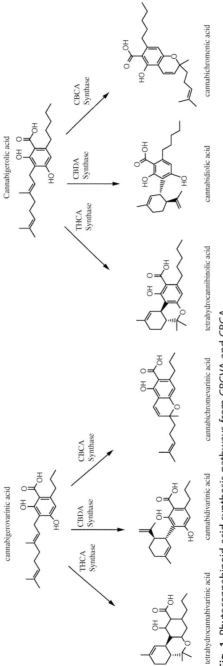

Fig. 1 Phytocannabinoid acid synthesis pathways from CBGVA and CBGA.

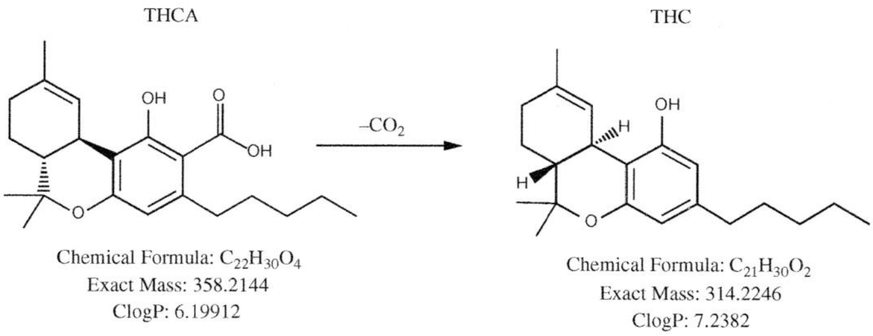

THCA → THC

Chemical Formula: $C_{22}H_{30}O_4$
Exact Mass: 358.2144
ClogP: 6.19912

Chemical Formula: $C_{21}H_{30}O_2$
Exact Mass: 314.2246
ClogP: 7.2382

Fig. 2 THCA decarboxylation to neutral THC.

decarboxylates into cannabielsoin (CBE). Photo-oxidation of CBD also transforms into CBE with exposure to heat [6]. This is just a fraction of the potential number of cannabinoids that can be found in cannabis plants.

3. Cannabinoid profiling, total THC, and total CBD quantitation

3.1 Sample preparation

The determination of cannabinoid content in flower, gummies, oils, tinctures, distillates, concentrates, chocolates, waxes, butters, topicals—the list of products goes on and on—requires different sample preparation procedures. It's out of scope to attempt to address all of these here but there are some commonalities:

- A homogeneous representation of the product is critical. For solid samples, this is typically performed through pulverization or grinding.
- The mass of the sample, the initial volume of the extract, and the dilution factor must be accurately recorded. These are required for the conversion from µg/mL determined via the analysis to % wt. as is commonly performed for potency determinations.
- Usually, cannabinoid profiling and quantitation regulations are focused on the accurate determination of total THC and total CBD content. These are defined by Eqs (1) and (2). Most laboratories however target more cannabinoids than regulated in their region.
- Due to the high THC or CBD content in many cannabinoid products, a typical sample is significantly diluted. For hemp authentication, the upper cut-off is 0.3% (wt./wt.). This represents 3 mg/g or 3 parts per thousand, which contemporary technologies can easily measure. In fact,

a sample containing 3 mg/g THC still requires a 100-fold dilution to get the concentration within the range of a typical calibration curve.

3.2 Sample preparation procedures for flower and oils [7]

3.2.1 Sample preparation for homogenized hemp oils, CBD oils, concentrates, tinctures, or resins

Pipette a 100 μL aliquot of sample into a tared 10 mL volumetric flask. Accurately record the weight of the product. Add 8 mL high-purity ethanol, cap and mix well. Bring to volume with ethanol. Using a glass syringe fitted with a 0.45 μm regenerated cellulose syringe filter, transfer 2 mL of the solution into a clean glass vessel. Perform an additional 10-fold dilution into an amber glass 2 mL auto-sampler vial by adding 900 μL high-purity ethanol to 100 μL filtered sample. Cap and vortex briefly to mix. Depending on the product, higher dilution factors may be required. In this example, M is the mass of the sample, the initial volume V, of the extract is 10 mL, the dilution factor D, is 100, and the total dilution factor is 1000.

3.2.2 Sample preparation for flower or hemp plant material

Accurately weigh 200 mg flower/leaf cutting into a 50-mL centrifuge tube. Homogenize using ceramic homogenizers, and a commercial grinder. Add 20 mL of ethanol and shake for 10 min to extract. Centrifuge at 5000 rpm for 5 min. Filter 1 mL of supernatant with a 0.45 μm regenerated cellulose syringe filter into a new vial. Transfer 50 μL of the filtered solution into a 2 mL amber glass auto-sampler vial. Add 950 μL ethanol and vortex to mix. In this example, M is the actual mass weighed into the 50-mL tube, the initial volume V, of the extract is 20 mL, the dilution factor D, is 100, and the total dilution factor is 2000.

3.3 Cannabinoid analyses with HPLC-UV

All regions including the United States and Canada require the determination of the THC content for several reasons: (1) a measure of the psychoactive potential of the product, (2) ensuring compliance with labelled THC content per unit measure of the product, and (3) to ensure a hemp product has a THC content $<0.3\%$ by weight. In most cannabis regulatory testing laboratories, this determination is performed on a HPLC system with ultra-violet (UV) detection. Typical cannabinoids that are regulated include THC, THCA, CBD, and CBDA. The state of California further regulates CBN and CBG. Δ^8-THC is also measured to ensure it is chromatographically resolved from THC and does not contribute to the quantitative THC content. Many laboratories may also include up to 18 more cannabinoids in their testing services and these include cannabinoid acids like THCVA, decarboxylated cannabinoids like CBC, and cannabinoid degradation products like CBL.

HPLC-UV systems can vary in their capabilities and maximum pressure limits. At a minimum, the system must include a binary or quaternary pump, a heated column compartment, a chilled auto-sampler, an analytical column (stationary phase), a detector, and a solvent compartment with a de-gassing unit for the mobile phases. Generally, the mobile system consists of an aqueous channel and an organic channel. The aqueous and organic mobile phases may be buffered with a low percentage (0.1% v/v) formic acid and/or millimolar concentrations (5 mM) of ammonium formate. The mobile phase composition can be operated in isocratic mode, as a gradient, or as an iso-gradient. In an isocratic method, the percentage of the mobile phase remains constant throughout the analysis. Gradient methods generally have a brief isocratic period followed by a change in the percent ratio of the aqueous and organic phases over time. In a reverse-phase method, this is typically biased to the aqueous phase early in the analysis and ramped to more organic phase towards the end of the analysis. For the analyses of cannabinoids, the iso-gradient approach may be most useful. Iso-gradient methods have a longer hold time of the initial conditions followed by a fast ramp to high organic phase. This methodology can result in excellent chromatographic resolution with the added benefit of flushing the analytical column of unwanted matrix materials at the end of the analysis.

Over recent decades, analytical column technology for HPLC and ultra-high-performance liquid chromatography (UHPLC) analyses has evolved dramatically in the range of chemical compositions that make up the stationary phase, and the particle size and porosity. For the analysis of cannabinoids, C_{18} column chemistries are most common. Sub 2-μm solid core particles enhance the resolving power of the analytical column at the expense of increased pressure requirements. Methods using these columns generally need expensive (>$125,000 USD) UHPLC systems capable of handling pressures up to 1300 Bar. However, with the advent of superficially porous particle (SPP) column technologies with sub 3-μm particles, the analysis can benefit from UHPLC-like chromatographic resolution without the high operating pressures. For example, a 3.0 mm × 50 mm SPP C18 column with 2.7-μm particles operates at a pressure less than 400 Bar and therefore a much less expensive HPLC system can be employed for cannabinoid analysis.

The HPLC system may employ a variable-wavelength detector (VWD), a multiple-wavelength detector (MWD), or a diode-array detector (DAD) to detect and quantify target cannabinoids. VW detectors monitor a single wavelength at a time and can be programmed to switch to other wavelengths during the analysis, but this switching can be slow to respond and settle, thus interfering with data collection and output. MW detectors can monitor at least 2 wavelengths simultaneously. This is preferable for cannabinoid analyses with simultaneous measurement of 230 and 270 nm wavelengths. Both VW and MW detectors provide two-dimensional data—meaning the resulting chromatogram illustrates the UV absorption (response) of each chemical as it passes through the detector as a function of time (retention time). Diode-array detectors can monitor multiple wavelengths simultaneously and collect a UV spectrum over a defined wavelength range and therefore provides three-dimensional data. UV absorption occurs when valence electrons in molecular orbitals are excited to a higher quantum state corresponding to the wavelength of the impinging radiation [8]. Some structural information can be gleaned through collection of the UV spectrum. For example, use of DAD for cannabinoid analysis can differentiate chemotypes like acids and the corresponding neutrals. According to Beer's Law, absorption is directly proportional to the concentration of the cannabinoid in the original mixture. Over a defined concentration range, the response should be linear, and the concentration is determined through linear regression when plotted in x-y coordinates with $y =$ response of the cannabinoid and $x =$ concentration through Eq. (3). The UV chromatograms for 228 and 270 nm are shown in Fig. 3. In Fig. 4, the UV spectra for the acids and their neutral analogues are shown.

$$Concentration = \left(\frac{Response - Intercept}{Slope} \right) \tag{3}$$

HPLC method parameters for the analysis of cannabinoids using DAD and a quaternary pump [9] are defined in Tables 1 through 3. Using the four channels of the quaternary pump, the HPLC mixes the mobile phases dynamically. In this method, the mobile phase is water/acetonitrile for the first 3.2 min. Then, the acetonitrile percentage is decreased, and the methanol percentage is increased in a linear fashion until 8.2 min. At that point in the method, acetonitrile has been completely exchanged for methanol. This methodology was first described by Kowalski and Laine [10] and leverages their observations that acetonitrile has better resolving power for

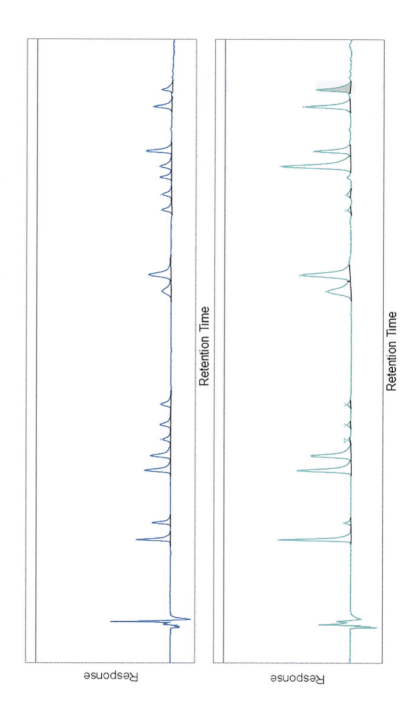

Fig. 3 Top: 228nm DAD chromatogram for 16. Bottom: 270nm DAD chromatogram. From left to right: Acronym (Resolution): CBDVA, CBDV (R = 1.220), CBDA (R = 1.558), CBGA R = 1.101), CBG (R = 1.101), CBD R = 1.089), THCVA (R = 1.103), THCV (R = 1.532), CBN (R = 1.052), Δ[9]-THC (R = 1.193), Δ[8]-THC (R = 1.038), CBL (R = 1.050), CBNA (R = 1.034), CBC (R = 1.096), THCA (R = 1.096), CBCA (R = 1.033).

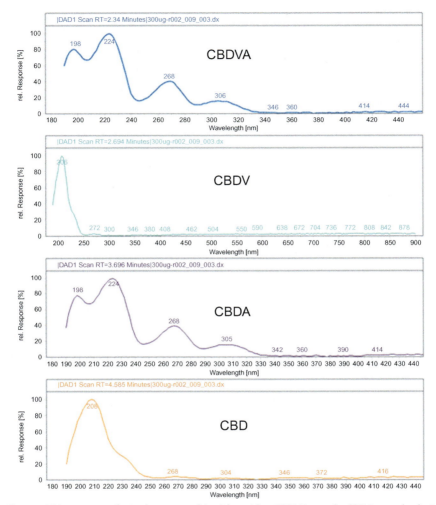

Fig. 4 UV spectra for phytocannabinoid acids CBDVA and CBDA, and their decarboxylated analogues CBDV and CBD, respectively. The COOH moieties add unique spectral signatures allowing them to be identified.

the early eluting cannabinoids and methanol is better for the later eluting cannabinoids on the column phase. Dimensions, and HPLC conditions are given in Tables 1 and 2. Table 3 lists the target cannabinoids.

Data analysis yields identification of each cannabinoid by retention time and determination of each cannabinoid concentration in μg/mL from the linear calibration curves. Total THC and Total CBD are determined as

Table 1 HPLC parameters.

Parameter	Value
Flow	0.5 mL/min
Column	Agilent InfinityLab Poroshell EC-C18 3.0×100, 1.9 μm
Stop time	12.5 min
Post time	3.0 min
Injection volume	0.5 μL
Wavelengths (nm)	228 and 270
UV scan range	225–445 nm
Sampling rate	40 Hz

Table 2 HPLC mobile phase gradient.

Time (min)	A (100% H_2O)	B (100% acetonitrile)	C (100% ethanol)	D (0.1% v/v aqueous formic acid)
0	25	70	0	5
3.2	25	70	0	5
8.2	5	0	90	5

Table 3 Target cannabinoids.

Compound in retention order	Acronym	Resolution
Cannabidivarinic acid	CBDVA	
Cannabidivarin	CBDV	1.22
Cannabidiolic acid	CBDA	1.558
Cannabigerolic acid	CBGA	1.101
Cannabigerol	CBG	1.101
Cannabidiol	CBD	1.089
Δ^9-Tetrahydrocannabivarin	THCV	1.103
Δ^9-Tetrahydrocannabivarinic acid	THCVA	1.532
Cannabinol	CBN	1.052
Δ^9-Tetrahydrocannabinol	THC	1.193
Δ^8-Tetrahydrocannabinol	d8-THC	1.038
Cannabicyclol	CBL	2.924

Table 4 MS acquisition and source parameters.

Acquire mode	Auto
Tune file	atunes.tune
Ion source	ESI
Time filter enabled	On
Target points enabled	On
% SIM	50
Source parameter	Value
Gas temperature	325
Gas flow L/min	13
Nebulizer	55
Capillary voltage	3500

given in Eqs (1) and (2). Conversion from μg/mL to percent by weight is done through Eq. (4).

$$\% \, wt. = \left[\frac{Concentration * V * D}{M * 10,000} \right] \tag{4}$$

where Concentration = concentration of analyte from linear regression analysis (μg/mL), V = volume of extraction solvent (mL), D = dilution factor, M = mass of sample (g), and 10,000 = conversion from μg/g to % wt. The 10,000 factor comes from converting μg to g (10^6) in the denominator and 100 in the numerator to convert to percent thus, 10,000.

3.4 Cannabinoid analyses with liquid chromatography: Mass spectrometry (LC-MS)

The HPLC-UV method defined above employed both DAD and mass spectrometry using electrospray ionization (ESI). The MS method parameters are given in Tables 4 and 5. Like a DAD, mass spectrometry is also 3 dimensional: x = retention time, y = ion abundance (response), and z = mass spectrum for each analyte in the mix. LC-MS using ESI positive (ESI+) mode generally creates a $(M+H)^+$ quasi-molecular ion where M is the molecular mass of the analyte. For example, the exact mass of THC is 314.2246 m/z and its $(M+H)^+$ quasi-molecular ion is 315.2138 m/z. ESI+ mode can also form adducts with metal ions such as Na^+ or K^+ and in these cases the ion is

Table 5 . MS SIM/Scan segments, target ions, and ESI parameters.

Segment	Name	M+H (*m/z*)	Fragmentor	Polarity
SCAN		200–700	100	Positive
SIM	THCA and CBDA	359.2	100	Positive
SIM	CBN	311.2	100	Positive
SIM	CBD, THC, Δ^8-THC, and CBC	315.2	100	Positive
SIM	CBG	317.2	100	Positive
SIM	CBDV and THCV	359.2	110	Positive
SIM	CBGA	361.2	110	Positive

in the form $(M+X)^+$ where X is the metal ion adduct. For example, the $(M+Na)^+$ adduct for THCA is $381.2036\,m/z$. The addition of mass spectrometry to the analysis required no changes to the HPLC method parameters. MS greatly improves the selectivity (specificity) for the accurate identification and quantification of each chemical species vs. using a non-selective UV detector. Fig. 5 illustrates the SIM TIC and DAD chromatograms for 11 cannabinoids.

3.5 Analytical challenges to cannabinoid analyses using HPLC-UV and LC-MS

As noted above, the use of HPLC with UV detection is the most common methodological approach for the analysis of cannabinoids. It is however a non-selective approach relying primarily on retention time for the identification of each target cannabinoid and the chance of mis-identifying an interfering unknown in the sample as a cannabinoid can happen. For example, some processed cannabinoid products contain artificially added concentrations of certain terpenes. If a terpene elutes at the same retention time as a target cannabinoid, the analyst may mis-identify the terpene as a cannabinoid or quantitate an artificially high value for the cannabinoid. This chance of a false-positive result is generally mitigated by the fact that contemporary plant samples or cannabinoid products have 20–30% cannabinoid content (primarily THC or CBD or both) and the natural terpene content is in the 1–3% range. Under these conditions, it's not unusual to dilute the sample 1000-fold to as much as 8000-fold to get the 200–300 mg/g cannabinoid content on scale with the calibration curve. This common sample preparation procedure tends to dilute out the terpenes and minimize their

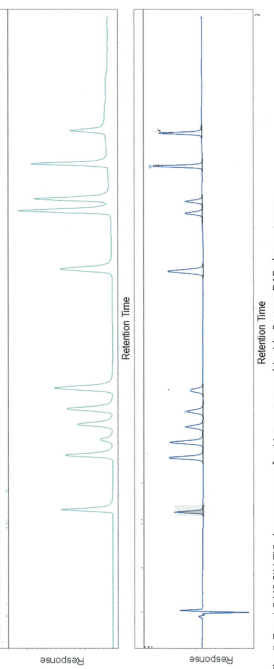

Fig. 5 Top: LC-MS SIM TIC chromatogram for 11 target cannabinoids. Bottom DAD chromatogram.

potential interference with the analysis. Another mitigating factor is the use of 228 and 270 nm wavelengths in the UV detectors. Terpenes tend not to absorb well at these wavelengths.

There is a caveat that must be mentioned about the use of LC-MS for this analysis. As noted above, phytocannabinoid acids spontaneously decarboxylate after harvest and upon exposure the heat and light. This decarboxylation is readily observed when using gas phase systems for cannabinoid analyses. For example, when a liquid sample is injected into the hot inlet of a GC system, decarboxylation occurs and the neutrals are observed. This phenomenon has been leveraged for the semi-quantitative determination of THC content in bulk drug seizures to answer the forensic question: is it marijuana? ESI sources also have heated "zones." For example, heated drying gas to improve desolvation, and heated sheath gas to maintain a columnated ion beam. In recent work [11], the question was posed: do acid phytocannabinoids decarboxylate in the heated zones of an ESI source? The null hypothesis was decarboxylation does not occur when using ESI LC-MS systems. The work used single quadrupole and time-of-flight mass spectrometry and evaluated acid phytocannabinoid stability as a function of drying gas temperature in ESI + and ESI negative (ESI −) modes. The outcome proved that decarboxylation as a function of drying gas temperature does not occur in ESI + mode. A $(M + H)^+$ chemical species was posited, and the lack of a putative decarboxylation mechanism was discussed. Conversely, in ESI − mode decarboxylation of acid phytocannabinoids was observed as a function of drying gas temperature, with complete depletion of THCA through conversion to THC and potentially, degradation products like CBN [12] at temperatures greater than 200–250 °C. A mechanism for the conversion of the acid to the neutral analogue was given. This phenomenon was further confirmed through an independent investigation (unpublished data, Jean-François Roy, Agilent Technologies, Inc.). LC-tandem mass spectrometry (LC-MS/MS) has also been shown for the analysis of cannabinoids and ESI − was used for the detection and quantification of THCA [13]. In that work, flattening of the THCA calibration curve at the higher concentration was observed. This was most likely an indication of THCA decarboxylation in ESI −, but it was not explored.

A third caveat to the analysis of cannabinoids must also be presented and it pertains to solvents used in the sample preparation process. In the presence of acids like HCl, CBD oils and concentrates transform into THC through an acid catalysed stable carbocation intermediate. This phenomenon has also been observed with 35% sulfuric acid, and vinegar [14].

Furthermore, common solvents like dichloromethane (DCM) also become acidic over time through hydrolysis processes as shown in Eq. (5).

$$CH_2Cl_2 + H2O \rightarrow CH_2O + 2HCl \qquad (5)$$

It is therefore suggested that acidic and halogenated reagents like DCM are avoided during sample preparation of processed cannabinoid samples that may contain high CBD concentrations.

3.6 Best practices

Many laboratories perform cannabinoid testing without matrix matching or the use of internal standards to correct for extraction variability and other bias, and surrogates for recovery determinations. Most regions where medicinal or adult use recreational programs exist do not stipulate matrix matching or the use of internal standards and surrogates. However, New York State, which as of April 2020 has only a medicinal cannabis program, defines sample preparation procedures for matrix matching calibrators, quality controls, etc., using norgestrel as the internal standard, and 4-pentylphenyl-4-methylbenzoate as the surrogate [15,16].

4. Analysis of residual pesticides and mycotoxins
4.1 Background

Arguably, the analysis of residual pesticides and mycotoxins is the most challenging of all the regulatory cannabis tests. This reality stems from differences in the requirements of medicinal vs. adult recreation use programs, multiple sample preparation procedures to address the myriad cannabis and cannabinoid products on the market, and the regional differences in the defined target lists and action (not to exceed) levels. Add to this the fact that limits of detection (LOD) and limits of quantitation (LOQ) need to be in the low ng/g (ppb) range in a matrix that typically contains 100s of mg/g of cannabinoids, 10s of mg/g of terpenes, and 100s of other endogenous chemicals. Anecdotally, it is not uncommon to hear a laboratorian exclaim "cannabis, hemp, and cannabinoid product matrices are unlike any other typically encountered for the analysis of residual pesticides in environmental sources or food." From an analytical chemistry perspective, compiling these challenges clearly points to the sensitivity and selectivity requirements only tandem mass spectrometry (MS/MS) can offer. Nonetheless, even MS/MS

Cannabis and hemp testing 19

techniques can struggle in this matrix when it comes to certain pesticides and one must leverage a multi-platform approach in many cases.

4.2 Sample preparation

Like the sample preparation needs to quantify total THC (potency) and other cannabinoids, sample preparation for residual pesticides and myco-toxins requires different procedures for each sample type. However, for residual pesticides and mycotoxins analyses, sample preparation is even more critical to accurately and repeatably quantify ppb levels of the target analytes in a matrix that contains hundreds of mg/g of endogenous chemicals. Again, there are myriad products that require residual pesticides and mycotoxins testing and each may require a unique sample preparation approach. For brevity, this text will focus solely on dry flower (inflorescence) with moisture content less than 13%.

4.2.1 No QuEChERS for dry cannabis inflorescence

QuEChERS (an acronym for quick, easy, cheap, effective, rugged, and safe) is a common extraction procedure develop by U.S.D.A. [17]. Its original intent was for biological analysis, but it was readily adapted for foodstuffs. Its application to cannabis testing ostensibly makes sense because cannabis inflorescence is ingested through inhalation or orally and therefore is food-like. There are two primary approaches to QuEChERS (1) AOAC official method 2007.01 [18], and (2) EN official method 15662 [19]. The AOAC pH is 4.8 [20] and the EN pH is 5–5.5 [21], which affects pH labile pesticides like imazalil, and an exotherm is generated when the sample and salts are mixed which affects thermally labile pesticides like fenthion [22]. Other potential issues with QuEChERS extraction of dry cannabis pertain to dispersive solid-phase extraction (dSPE) not offering enough extraction capacity, primary-secondary amines (PSA) scavenging acidic pesticides, and graphitized carbon black (GCB) scavenging planar pesticides. These issues combined with the need to add water to dry inflo-rescence samples which enhance pH effects, reduces recoveries of many pes-ticides in the various regional target lists. Therefore, QuEChERS is not recommended as a sample preparation procedure for cannabis inflorescence.

4.2.2 Winterization and "dilute and shoot"

Other approaches to sample preparation for residual pesticide and mycotoxin analysis include winterization [23] and "dilute and shoot" [24]. Winterization weighs 200 mg homogenous sample and sonicates in acetonitrile to extract. The samples are then placed in a freezer at $-20\,°C$ overnight. Prior to analysis, the sample is decanted from the frozen matrix components and diluted 1:6 with 75:25 methanol:water (% v/v).

The "dilute and shoot" approach requires 5 g of homogenous sample which is diluted twofold with acidified acetonitrile. The suspension is vortexed to extract and centrifuged. An aliquot of the supernatant is diluted with acidified acetonitrile. An internal standard mix comprised of 30 isotopically labelled compounds is added post-extraction which unfortunately does not correct for extraction bias not to mention the purchase of 30 isotopically labelled compounds may be cost prohibitive.

4.2.3 Caveats to winterization and "dilute and shoot"

As described, winterization is time consuming. Both procedures do not appear to filter particulate matter from the sample extract and may not adequately remove matrix prior to injection into the analytical system. This may result in increased fouling of the ESI or atmospheric pressure chemical ionization (APCI) source. For high-throughput laboratories that do not have multiple analytical instruments for redundancy, increased maintenance equates to lost and unrecoverable revenue.

4.2.4 SPE cleanup and dilution

Another approach to sample preparation for pesticide and mycotoxin analysis is to employ solid-phase extraction followed by dilution. This approach offers the advantages of (1) a single-stream sample preparation procedure amenable to both LC-MS and GC-MS analyses, (2) removing lipids, pigments, and other matrix entities, and (3) high dilution factors to further mitigate chemical noise. From a published method, [25] this procedure consists of:

1. Homogenized 1.0 g cannabis inflorescence in a 50 mL polypropylene (PP) centrifuge tube with two ceramic homogenizers.
2. Add 15 mL of high-purity LC/MS-grade acetonitrile, cap and mechanically shake for 5 min at 3000 rpm.
3. Decant the supernatant into an unconditioned SampliQ C18 EC cartridge (5982-1365, Agilent Technologies, Santa Clara, CA). Gravity-elute into a clean 50 mL tube.
4. Add 5 mL more high-purity LC/MS-grade acetonitrile to the sample in step 2.

5. Decant the supernatant into the SPE cartridge in step 3. Gravity-elute into a clean 50 mL tube.
6. Add 5 mL more high-purity LC/MS-grade acetonitrile to the sample in step 2.
7. Decant the supernatant into the SPE cartridge in step 3. Gravity-elute into a clean 50 mL tube.
8. Bring the final volume of the collected eluents to 25 mL with high-purity LC/MC-grade using the graduations on the side of the PP tube (25-fold dilution factor). Filter 1.0 mL of the eluent for use in the next steps.
9. For LC–MS analysis, transfer 50 µL of extract and mix with 450 µL 25/75% water/methanol (v/v) + 0.1% (v/v) formic acid in a sample vial (250-fold dilution). Cap & vortex for 30 s.
10. For GC–MS analysis, transfer 200 µL of extract and mix with 800 µL high-purity acetonitrile in a sample vial (125-fold dilution). Cap & vortex for 30 s.

The above procedure takes about 45 min for a batch of 20 samples.

4.3 Liquid chromatography-tandem mass spectrometry (LC-MS/MS)

In the current state of pesticide testing in the U.S. and Canada, the largest and most restrictive list has been defined at the federal level by Health Canada [26]. In the U.S., California's list contains 66 pesticides with LOQ as low as 100 ppb [27]. Health Canada has defined three product types: fresh cannabis and plants, dried cannabis, and cannabis oil. The target list includes 96 target pesticides. As of December 2019, the limits of quantitation (LOQ) have been defined for all pesticides in the fresh and dried cannabis, and for 2/3 of the cannabis oil product types. In terms of LOQ, the fresh cannabis and plant, and cannabis oils (where LOQ have been defined) are the most restrictive with about half of the LOQ set to 10 ppb. California has two product types: inhalable cannabis and cannabis products, and other cannabis and cannabis products. However, California has further stratified the 66 pesticides into Category I and Category II where any Category I pesticide detection at any level above the empirically determined limits of detection (LOD) are actionable and the product failed in the certificate of analysis (CoA). There are 21/66 Category I pesticides in the California target list. The remaining 45 Category II pesticides are actionable if quantified above the regulated LOQ.

4.3.1 Method commonalties Canada and the U.S. target lists

With respect to LC-MS/MS analysis, sensitivity, selectivity, accuracy, precision, and robustness are paramount. Commonalities in two methods designed for residual pesticide analysis in Canada and U.S. States with legalized adult use recreational programs, include sample preparation and dilution factors, column phase, dimensions, and particle size, column temperature, injection volume, mobile phases, the ESI-AJS source, and a number of source conditions [28,29]. These methods have been deployed across North America and have demonstrated success for the analysis in cannabis and cannabinoids products. Fig. 6 is a typical chromatogram for 62 LC-MS/MS amenable pesticides in the California list.

4.3.2 Analytical challenges to residual pesticide testing using only LC-MS/MS systems

As noted above, there are pesticides more amenable to analysis with GC-MS systems. For some of these, especially some organochlorine pesticides, LC-MS/MS using APCI has been shown as an alternative. One concern with this approach is throughput. Although multi-mode sources are available that allow intra-analysis switching between ESI and APCI, this approach is not viable for accurate, quantitative analysis within the

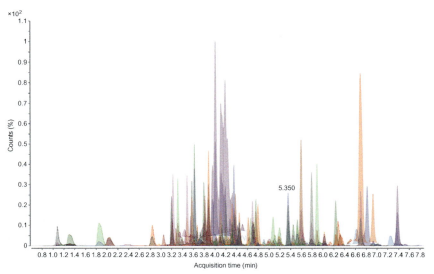

Fig. 6 Overlaid chromatograms of typical pesticides list and mycotoxins in extracted flower matrix. In-vial concentration is 500 ppt after the 250-fold dilution. The actual matrix concentration in the sample is 125 ppb.

chromatographic scale of a typical analytical runtime. The use of ESI and APCI methods therefore requires two independent analyses for each sample with an average throughput of about 1.5 samples per hour. Another concern is selectivity. For example, the APCI negative ionization mechanism for pentacholoronitrobenzene (PCNB) was proposed as the loss of HCl followed by the formation of an ammonium ion adduct [30]. An investigation of this claim was made using LC-quadruple time-of-flight (LC-QTOF) in APCI negative mode for the determination of the true ionization mechanism and this was compared to electron ionization using GC-QTOF and gas chromatography-tandem mass spectrometry (GC-MS/MS) as the quantitative tool [31]. Curtis et al. determined the proper APCI negative mode mechanism where a new chemical species, 2,3,4,5-tetrachloro-6-nitrophenolate is formed *in situ* through a reaction with superoxide anion. This new species is therefore non-selective for PCNB and the resulting collision-induced dissociation (CID) transition products derived from 2,3,4,5-tetrachloro-6-nitrophenolate are not related to PCNB. In another iteration using APCI negative LC-MS/MS to address the Health Canada pesticide list, the EIC-MRM m/z transition is shown to be 275.8/35.10 (2,3,4,5-tetrachloro-6-nitrophenolate $\rightarrow Cl^-$), which is extremely non-specific [32]. Curtis's work did not debate the formation of a quantifiable chemical species when using APCI in negative mode but rather the appropriateness of a targeted analytical approach that uses non-selective precursor-product ion pairs.

4.4 Gas chromatography-tandem mass spectrometry (GC-MS/MS)

As discussed above, most of the pesticides in the various regional lists are amenable to analysis with LC-MS/MS using ESI. However, if one combines the Canadian and California lists and removes the duplicates, the resulting North American list contains 100 pesticides. Of these, there are at least 27 that are more amenable to GC-MS analyses and will be discussed in this section. This determination was made through literature review, empirical experience, and filtering through a proprietary *a priori* algorithm that considers physicochemical properties of each chemical entity and parses each into an appropriate analytical bin: LC/MS or GC/MS.

All samples, calibrators were prepared in matrix as defined above. Based on the resulting target list, a GC-MS/MS method was developed and vetted using a multi-day, multi-replicate model to statistically determine and verify method statistics such as range, linearity, accuracy, precision, LOD, and LOQ [33].

The statistical approach entailed 5 replicate injections at 8 calibration levels over 3 independent days and used these primary equations:

1. Average $= \sum x_i/n$

2. Standard deviation, $(s) = \left[\frac{\Sigma(x - \bar{x})^2}{n-1}\right]^{1/2}$

3. MDL (LOD) $= (s)$ * (Student t-value, $n - 1$ df, 99% Confidence)

4. MDL (LOD) Test = Calculated MDL < Spike Level < 10 * Calculated MDL

5. LOQ $= 10$ * (s)

6. Percent Accuracy $=$ [(spiked concentration $-$ calculated average concentration/spiked concentration)] * 100

7. Precision, (%RSD) $=$ [(s)/Average] * 100

The results of this work demonstrated method statistics far below that required by Health Canada or California.

5. Residual solvent analysis

Residual solvent analysis (RSA) is common in the pharmaceutical industry and well defined in method USP $\langle 467 \rangle$ [34]. However, the types of commercial cannabinoid products are extremely variable and most do not reflect traditional drug-like products. Thus, the application of USP $\langle 467 \rangle$ for cannabinoid products is not appropriate. Like pesticides, California has stratified the 22 residual solvents on their target list into Category I and Category II where any Category I solvent detection at any level above the empirically determined limits of detection (LOD) are actionable and the product failed in the certificate of analysis (CoA). There are 6/22 Category I solvents in the California target list. The remaining 16 Category II solvents are actionable if quantified above the regulated LOQ. Recent work developed a novel analytical approach specifically for RSA of cannabinoid products that defines sample preparation and instrument needs for success [35].

5.1 Reagent and sample preparation
5.1.1 Preparation of α,α,α-trifluorotoluene internal standard (TFT ISTD)
Dilute 27 µL of TFT in 100 mL of N,N-dimethylacetamide (DMA). Aliquot 20 mL of solution and dilute to 1000 mL with DMA for a final ISTD concentration of 6.42 µg/mL.

5.1.2 Preparation of DMA-TFT ISTD sample matrix

Food grade hempseed oil was used as the matrix for all testing. Carefully weigh approximately 2.5 g of matrix, extract or product into a 20 mL scintillation vial and dissolve in 10 mL of TFT ISTD. Transfer the solution to a 50 mL volumetric flask. Dilute to volume with DMA-TFT ISTD and mix thoroughly.

5.1.3 Preparation of saturated saline solution

Add 900 mL of organic free water and 360 g of NaCl to a 1 L volumetric flask. Gently shake and add and organic free water to the 1 L mark. Place clean stir bar in the placed flask and stir plate at ambient temperature. The mixture was stirred for 30 min. Allow the solution to settle and decant the supernatant to a clean 1 L glass bottle.

5.1.4 Sample preparation

Add 1.0 mL of saturated saline, 5 mL DMA-TFT ISTD solution to a headspace vial. Please note an exotherm is generated at this step. Therefore, prior to addition of the sample, the exotherm must be allowed to cool for about 60 min. After the exotherm has cooled, add 0.20 mL of each calibration standard, QC, etc., to the headspace vial and cap prior to analysis.

5.2 Headspace GC-MS analysis (HS GC-MS)

An Agilent 7697A Headspace sampler configured with a 0.5 mL sample loop was plumbed to the Intuvo 9000 gas chromatograph and 5977B mass spectrometer. A 4 mm liner was used, and two DB–Select 624 Ultra Inert columns were used with mid-column backflush enabled to remove unwanted terpenes that elute after the toluenes. A 9 mm extractor lens was used in the EI source of the mass spectrometer.

5.2.1 Results

The work determined that sample preparation is critical and determined the conditions to optimize accuracy and precision. Intra-day and inter-day accuracy, precision, range, linearity, LOD as defined by method detection limits (MDL), and LOQ were determined using the vetting paradigm described in the GC-MS/MS section. The results demonstrated method statistics far below that required California for the analysis of residual solvents in cannabinoid products. Fig. 7 is a typical chromatogram for the analysis of the Category I and II solvents regulated by in California.

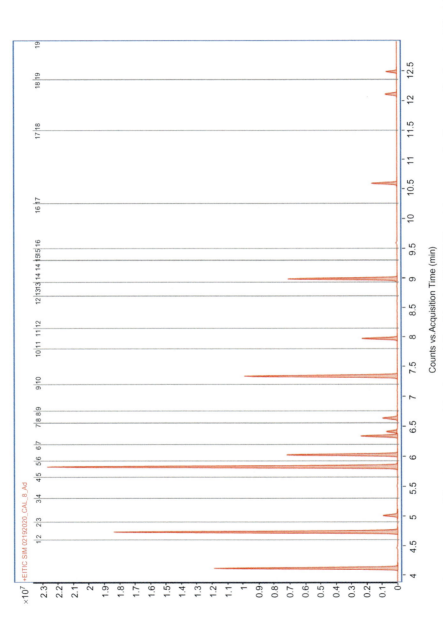

Fig. 7 SIM TIC chromatogram of the Category I and II solvents defined by the CA BCC. The red trace are the Category II solvents at the high calibrator level of 6000ppm. The blue trace is the Category I solvents at the high calibration level of 19ppm.

6. Summary

The information given above should exemplify that sample preparation is critical for the analysis of cannabis and cannabinoid products. Another point of note is that much of the testing methodology has been developed by the various instrument vendors and a lack harmonization must be addressed. The Macherone lab endeavours to show the science of cannabis analyses, align with the most appropriate methodologies, and rigorously vet analytical methods before publication. Laboratorians must leverage the strengths of each analytical platform and not attempt to incorporate disparate chemotypes in one analytical approach or inappropriately pigeonhole analytes into a methodology not well-suited for the analyses.

Acknowledgements

The author would like to thank Anastasia Andrianova, Matthew Curtis, Sue D'Antonio, Eric Fausett, Wendi A. Hale, Terry Harper, Jeffery S. Hollis, Nikolas C. Lau, Bruce Quimby, Peter J.W. Stone, Christy Storm, Jessica Westland, Michael Zumwalt of Agilent Technologies, and Julie Kowalski, of JA Science and Support.

Disclaimers

Agilent products and solutions are intended to be used for cannabis quality control and safety testing in laboratories where such use is permitted under state/country law.

The material presented in this chapter does not necessarily reflect to views and opinions of Johns Hopkins University or Johns Hopkins University School of Medicine.

The material presented in this chapter is for informational purposes only. The author does not advocate the use of marijuana or cannabinoid products.

References

[1] DEA, Drug Scheduling. "Drug Schedules", https://www.dea.gov/drug-scheduling, 2020.
[2] Leafly.com "Map of Cannabis Laws by State". April 2020. https://www.leafly.com/news/cannabis-101/where-is-cannabis-legal.
[3] A. Macherone, Cannabis Industry Journal, Hemp in the United States: An Opinion, https://cannabisindustryjournal.com/column/hemp-in-the-united-states-an-opinion/, 2019.
[4] Federal Register Rules and Regulations, Establishment of a Domestic Hemp Production Program, 84(211)2019, (58522-58564).
[5] B.F. Thomas, M. Elsohly, Biosynthesis and pharmacology of phytocannabinoids and related chemical constituents, in: B.F. Thomas (Ed.), The Analytical Chemistry of Cannabis: Quality Assessment, Assurance, and Regulation of Medicinal Marijuana and Cannabinoid Preparations, Elsevier, Amsterdam, 2016, pp. 27–41.
[6] M.M. Lewis, et al., Chemical profiling of medical Cannabis extracts, ACS Omega 2 (2017) 6091–6103.

[7] C. Storm, M. Zumwalt, A. Macherone, Dedicated Cannabinoid Potency Testing in Cannabis or Hemp Products Using the Agilent 1220 Infinity II LC System, Application note 5991-9285. Agilent Technologies, Inc., 2019.

[8] D.A. Skoog, J.L. Leary, Application of molecular ultraviolet/visible absorption spectroscopy, in: D.A. Skoog, J.L. Leary (Eds.), Principles of Instrumental Analysis, Saunders College Publishing, Orlando, 1992, pp. 150–155.

[9] S. D'Antonio, G. Li, A. Macherone, Quantitation of Phytocannabinoid Oils Using the Agilent Infinity II 1260 Prime/InfinityLab LC/MSD iQ LC/MS System, Application note 5994-1706. Agilent Technologies, Inc., 2020.

[10] J. Kowalski, D. Laine, Improved routine cannabinoids analysis with liquid chromatography-diode array ultraviolet detection for the current Cannabis market, in: Oral presentation, AOAC International Conference, August 26–August 29, Toronto, Ontario, Canada, 2018.

[11] P.J.W. Stone, et al., The stability of acid phytocannabinoids using electrospray ionization LC–MS in positive and negative modes, Cannabis Sci. Technol. 3 (3) (2020) 34–40.

[12] F.E. Dussy, et al., Isolation of Δ^9-THCA-A from hemp and analytical aspects concerning the determination of D9-THC in cannabis products, Forensic Sci. Int. 149 (2005) 3–10.

[13] Shimadzu Scientific Instruments, Application news, in: Quantitative Analysis of Cannabinoids Using the LCMS-8040 Triple Quad MS, Shimadzu Scientific Instruments, 2012. SSI-LCMS-046.

[14] T.D. Kiselak, R. Koerber, G.F. Verbeck, Synthetic route sourcing of illicit at home cannabidiol (CBD) isomerization to psychoactive cannabinoids using ion mobility-coupled-LC–MS/MS, Forensic Sci. Int. 308 (2020) 1–8.

[15] NYS DOH MML-301, "Medical Marijuana Sample Preparation Protocols for Potency Analysis". 2019: https://www.wadsworth.org/sites/default/files/WebDoc/NYS%20DOH%20MML-301-05.pdf.

[16] NYS DOH MML-300, "Measurement of Phytocannabinoids in Medical Marijuana Using HPLC-PDA", 2019: https://www.wadsworth.org/sites/default/files/WebDoc/NYS%20DOH%20MML-300-04.pdf.

[17] M. Anastassiades, S.J. Lehotay, D. Stajnbaher, F.J. Schenck, Fast and easy multiresidue method employing acetonitrile extraction/partitioning and "dispersive solid-phase extraction" for the determination of pesticide residues in produce, J. AOAC Int. 86 (2003) 412–431.

[18] S.J. Lehotay, Determination of pesticide residues in foods by acetonitrile extraction and partitioning with magnesium sulfate: collaborative study, J. AOAC Int. 90 (2) (2007) 485–520.

[19] CEN - EN 15662, Foods of Plant Origin—Multimethod for the Determination of Pesticide Residues Using GC- and LC-Based Analysis Following Acetonitrile Extraction/Partitioning and Clean-up by Dispersive SPE - Modular QuEChERS-Method, https://www.cen.eu/Pages/default.aspx, 2008.

[20] S.J. Lehotay, K. Mastovská, A.R. Lightfield, Use of buffering and other means to improve results of problematic pesticides in a fast and easy method for residue analysis of fruits and vegetables, J. AOAC Int. 88 (2) (2005) 615–629.

[21] M. Anastassiades, E. Scherbaum, B. Tasdelen, D. Stajnbaher, Crop protection, public health, environmental safety, in: H. Ohkawa, H. Miyagawa, P.W. Lee (Eds.), Pesticide Chemistry: Crop Protection, Public Health, Environmental Safety, Wiley-VCH, Weinheim, 2007, pp. 439.

[22] J. Stevens, D. Jones, QuEChERS 101: The Basics and Beyond, https://www.agilent.com/cs/library/eseminars/Public/QuEChERS_101_10_11_01.pdf, 2010.

[23] C.M. Butt, R. Di Lorenzo, D. Tran, A. Romanelli, C. Borton, Analysis of the Massachusetts Cannabis Pesticides List Using the SCIEX QTRAP® 6500+ System, Document number: RUO-MKT-02-8937-A. AB Sciex, 2019.

[24] A. Dalmia, S. Hariri, J. Jalali, E. Cudjoe, T. Astill, C. Schmidt, F. Qin, Novel ESI and APCI LC/MS/MS Analytical Method for Testing Cannabis and Hemp Concentrate Sample Types, 78758 (69954A), PerkinElmer, Inc., 2020.

[25] J.S. Hollis, E. Fausett, J. Westland, A. Macherone, Analysis of Challenging Pesticides Regulated in the Cannabis and Hemp Industry with the Agilent Intuvo 9000–7010 GC/MS/MS System: The Fast-5, Application note 5994-1604, Agilent Technologies, Inc., 2019.

[26] Government of Canada, Publication 190301, Mandatory Cannabis Testing for Pesticide Active Ingredients, Government of Canada, 2019.

[27] Bureau of Cannabis Control Text of Regulations, California Code of Regulations Title 16 Division 42, https://www.bcc.ca.gov/law_regs/cannabis_order_of_adoption.pdf, 2019.

[28] J.-F. Roy, et al., A Sensitive and Robust Workflow to Measure Residual Pesticides and Mycotoxins from the Canadian Target List in Dry Cannabis Flower, Application note 5994-0429. Agilent Technologies, Inc., 2018.

[29] P.J.W. Stone, et al., Determination of Pesticides and Mycotoxins in Cannabis Flower as Defined by Legalized U.S. State Recreational Cannabis Regulations, Application note 5994-1734. Agilent Technologies, Inc., 2020.

[30] D. Tran, et al., DuoSpray™ Ionization: A Novel Approach to Analyzing the California Mandated List of Pesticides in Cannabis, Document number: RUO-MKT-02-7607-A. AB Sciex, 2018.

[31] M. Curtis, et al., Up in smoke: the naked truth for LC–MS/MS and GC–MS/MS technologies for the analysis of certain pesticides in Cannabis flower, Cannabis Sci. Technol. 2 (5) (2019) 6–11.

[32] A. Dalmia, et al., A Novel ESI and APCI LC/MS/MS Analytical Method for Meeting the Canadian Cannabis Pesticide Residues Regulatory Requirements, 21563. PerkinElmer, Inc., 2019.

[33] A.A. Andrianova, et al., Analysis of Twenty-Seven GC-Amenable Pesticides Regulated in the Cannabis Industry in North America with the Agilent 8890/7010B Triple Quadrupole GC/MS System, Application note 5994-1786. Agilent Technologies, Inc., 2020.

[34] Pharmacopeia United States Pharmacopeia and National Formulary (USP 29-NF 24). Rockville, MD: United States Pharmacopeial Forum: 31(5). http://pharmacopeia.cn/v29240/usp29nf24s0_c467.html. Accessed April 2, 2020, 2020.

[35] T. Harper, et al., Novel Residual Solvents Analysis of Cannabinoid Products with the Agilent Headspace-GC/MS System, Agilent Application note 5994-1926EN. Agilent Technologies, Inc., 2020.

CHAPTER TWO

Medicinal cannabis: Pharmaceutical forms and recent analytical methodologies

María Alejandra Fanovich[a], María Sandra Churio[b,c], Cristina Luján Ramirez[b,*]

[a]Instituto de Investigaciones en Ciencia y Tecnología de Materiales (INTEMA) CONICET—Universidad Nacional de Mar del Plata, Mar del Plata, Argentina
[b]Departamento de Química y Bioquímica, FCEyN, Universidad Nacional de Mar del Plata, Mar del Plata, Argentina
[c]Instituto de Investigaciones Físicas de Mar del Plata (IFIMAR) CONICET—Universidad Nacional de Mar del Plata, Mar del Plata, Argentina
*Corresponding author: e-mail address: farmramirez@yahoo.com.ar

Contents

1.	Introduction	31
2.	The endocannabinoid system	32
	2.1 Cannabinoid receptors	32
	2.2 Endocannabidnoids	35
3.	*Cannabis sativa* constituents	37
4.	Pharmaceutical forms of cannabinoids and novel delivery systems	41
5.	Pharmacokinetics of cannabis	45
6.	Analysis of cannabinoids and cannabinoid metabolites in organic matrices	47
7.	Conclusions	52
	Acknowledgements	52
	References	54

1. Introduction

Cannabis sativa has been widely used in ethnomedicine. Despite the benefits of the species, recreational use due to the psychoactive effects has led in the 20th century to classify it as an illegal drug, thus obstructing further survey [1]. With the worldwide extension of legalization of cannabis for medicinal purposes, there is a revived and growing interest in this plant, nowadays used by people with different pathologies. However, the number of fields involved in this subject is so vast (i.e. botany, molecular biology, analytical chemistry, physical chemistry, pharmacology, clinical medicine), that compilation of the information altogether has been difficult. This chapter aims to summarize the bioactivity of phytocannabinoids on their specific receptors, the traditional and novel pharmaceutical formulations, the processes to which cannabinoids are subjected in the organism and the analytical methodologies for analysing the organic matrices exposed to them. These contents should contribute to set the rational bases needed for the design of training programs for health care providers, such as pharmacists and physicians, as well as the general audience concerning Good Manufacturing Practices and the problematic around medicinal products without quality control.

2. The endocannabinoid system

The endocannabinoid (eCB) system is made up of receptors, endogenous ligands, and metabolic enzymes. During the last decades, with the advance of research and the discovery of new components, the knowledge about this system has been significantly expanded. Its complexity and magnificence have inspired McPartland et al. to metaphorically call it "a microcosm of psychoneuroimmunology or mind-body medicine" [2]. This dynamic system plays several homeostatic roles modulating the receptors expression as well as the ligands biosynthesis in order to meet the needs of the body. It was already present in primitive vertebrates with a neuronal network, which is consistent with its ancient phylogenetics [1].

2.1 Cannabinoid receptors

In 1965, Gaoni and Mechoulam began to shed light on the path of science related to *Cannabis sativa* and its properties with the isolation from the plant and characterization of the Δ^9-tetrahydrocannabinol (THC) molecule [3]. More than 20 years later, Devane et al. [4], were able to isolate the first cannabinoid receptor (CBR) that bounds with high affinity to THC in rats. It was called cannabinoid receptor type 1 (CB1). Receptors of this class are found in high density in the central and peripheral nervous system but also in several other places of the body. They are involved in a plethora of biological processes such as movement, reproduction, emotions, memory, reward, pain, food intake, nausea and vomiting [5].

The understanding of the mechanisms of action involved in the activation or modulation of CB1 receptors has been of great importance due to their broad and complex pharmacological profile. CB1 is a G protein-coupled receptor (GPCR) which uses the G(i/o) family as a major mediator, diminishing AMPc through adenylyl cyclases inhibition and regulating calcium and potassium channels. This process ends with blockage of presynaptic calcium channels, reducing the release of neurotransmitters in the synaptic space (Fig. 1) [6]. Shao et al. [7] reported for the first time the crystalline structure of this receptor, bound to an inhibitor (taranabant), achieved by using GPCR engineering and cubic phase lipid crystallization. They found that the extracellular surface is different from other lipid activated GPCRs, forming a highly hydrophobic pocket in which other agonists can be accommodated by modulating their action.

In 1993, a second type of cannabinoid receptors (CB2) was discovered, found mostly in the periphery of the organism with high densities in the immune system [8], the gastrointestinal system, liver, spleen, kidney, bones,

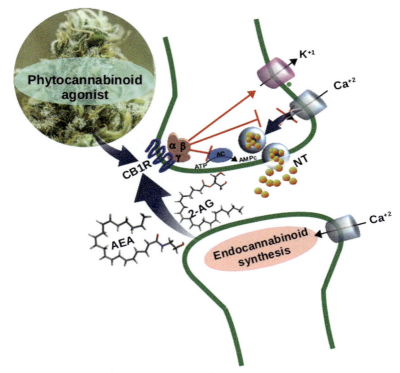

Fig. 1 Representation of the retrograde endocannabinoid signalling. NT, neurotransmitters; ATP, adenosine triphosphate; AMPc, cyclic adenosine monophosphate; AC, adenylyl cyclase.

heart [9] and peripheral nervous system [10]. However, it was recently determined that these CBRs also have biological activities in the central nervous system [11], but unlike CB1, their stimulation do not trigger psychoactive effects [12]. Li et al. obtained the crystal structure of human CB2 receptor in complex with a synthetic antagonist, AM10257, at 2.8 Å resolution. The structure showed a different binding position compared to CB1 (see Fig. 2). Nevertheless, the extracellular portion of the agonist-bound CB2 shares a high degree of conformational similarity with that of CB1 [13]. Both CBRs are phosphorylated by G protein receptor kinases (GRKs) and associated to β-arrestin1 or β-arrestin2. CB1 and CB2 stimulate extracellular signal regulated kinase (ERK) 1 and 2, involving either Gβγ or β-arrestin interactions [14,15].

A possible cross-talk between the endocannabinoid and endovanilloid (vanilloid type 1, TRPV1) systems has been proposed by Perchuk et al. [16]. They inferred that the interaction between these two systems can be exploited in pharmacological applications. Moreover, the TRP vanilloid (TRPV1, TRPV2, TRPV3, TRPV4), TRP ankyrin (TRPA1), and TRP melastatin (TRPM8) subfamilies, showed to have channels that can be modulated by several endogenous cannabinoids as well as phyto- and synthetic cannabinoids. These receptors are considered by some researchers as the "ionotropic cannabinoid receptors."

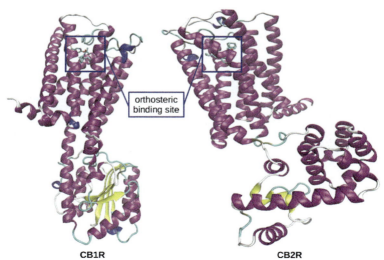

Fig. 2 Overall structures of CB1R-inverse agonist (taranabant) and CB2R-antagonist (AM10257) co-crystallization complexes.

Some orphan G protein-coupled receptors were proposed as putative cannabinoid receptors. In this context, the receptors GPR18 and GPR55 have been named as new cannabinoid receptors [17,18]. N-arachidonyl glycine receptor GPR18 activity, as well as its distribution, suggests that these receptor blockers could be targeted in treatment of endometriosis, cancer, and neurodegenerative disorders. These receptors are also involved in the inhibition of apoptosis, facilitating tumour cell survival in melanoma metastases. Δ^9-THC and the endocannabinoid anandamide are full agonists at GPR18. The activity of the NAGly-GPR18 complex in microglia signalling and its relationship with the endocannabinoid system have been recently reported [19,20].

On the other hand, the G coupled receptor (GPR55) shares numerous cannabinoid ligands with the CB1 and CB2 receptors although it has a low homology with these CBRs. Its pharmacology is complex since it uses different signalling pathways and downstream cascades. Recently the heteromerization of GPR55-CB1 and GPR55-CB2 has been reported throughout the central nervous system, giving rise to complex and even controversial pharmacology [21]. Some phytocannabinoids perform as antagonists, agonists or lysophosphatidylinositol inhibitors [22].

2.2 Endocannabidnoids

Endocannabinoids are lipidic signalling molecules. They are synthesized on demand from phospholipids in membranes and released into the intracellular space. Activation by these compounds takes place on presynaptic receptors, leading to the aforementioned events. The completion of this action takes place through the reuptake of the eCBs and subsequent metabolization by specific hydrolases.

Almost 30 years after the discovering of Δ^9-THC from *Cannabis sativa* in 1964 [3], the first endogenous cannabinoid was isolated. In 1992 this eCB was characterized as N-arachidonoylethanolamine and baptized as anandamide (AEA), from the Sanskrit "ananda," the "amide of the inner bliss" [23]. This multifaceted molecule is a partial agonist of CB1 receptor and showed the ability to impact on many systems of the human body, targeting receptors that include, not only CB1 and CB2, but also TRPV1 channels, GPR55, GPR119, and peroxisome proliferator activated receptors [24–30].

A couple of years after the identification of AEA, a major eCB was discovered, the 2-arachidonoylglycerol (2-AG) [31,32]. This is a full agonist of both CB1 and CB2 receptors and the tissue levels of 2-AG can be a 1000 times higher than those of AEA. Therefore, this lipid is considered the *true* endogenous ligand of CBRs, having more impact on biological activities in vivo than AEA. The 2-AG activity is synergistically increased by "entourage compounds" that can inhibit its metabolization via substrate competition. These compounds are N-palmitoylethanolamide (PEA), N-oleoylethanolamide (SEA), and cis-9-octadecenamide (OEA, oleamide) (Fig. 3).

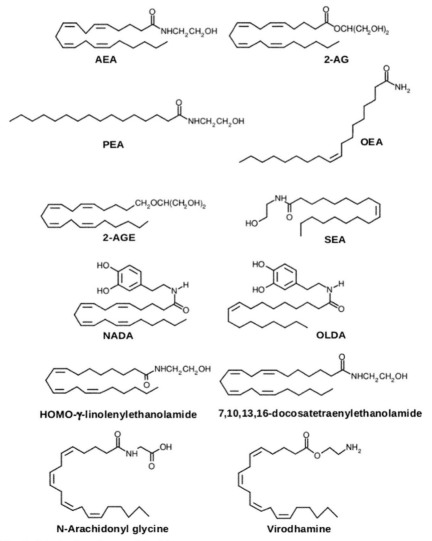

Fig. 3 Principal endocannabinoids.

The 2-arachidonyl glyceryl ether (2–AGE) is another minor lipidic eCB that selectively binds to CB1 but not to CB2. This molecule induces the typical cannabimimetic responses in mouse, namely the "tetrad" of behavioural tests in vivo [33,34].

Minor recently studied lipidic molecules with cannabinoid activity are N-arachidonoyl dopamine (NADA), virodhamine and N-oleoyldopamine (OLDA). NADA and OLDA increase intracellular calcium influx through activation of TRP receptors, especially the nonselective cation channel TRPV1 in mammalian nervous tissues. It activates cannabinoid CB1 receptors, but not dopamine Dl and D2 receptors. NADA and OLDA have been described as agonists of CB1 [35]. On the other hand, virodhamine, consisted of arachidonic acid and ethanolamine units bonded by an ester, is an antagonist for CB1 cannabinoid receptor [36].

Two other polyunsaturated fatty acid-ethanolamides, homo-γ-linolenylethanolamide and 7,10,13,16-docosatetraenylethanolamide, isolated from porcine brain, showed high affinity to CB1 receptor. Besides, PEA was reported to be a CB2 receptor agonist on rat mast cells, but it showed very low affinity and only at high concentrations on human cloned CB2 receptors, so that the classification of this lipid as an eCB still remains controversial [37].

N-arachidonyl glycine is, as discussed before, a high affinity ligand to the G protein-coupled receptors GPR18.

3. *Cannabis sativa* constituents

Cannabis sativa has millenarian therapeutic and recreational uses that have been related to the initial agricultural cultures of Asia. This annual dioecious plant belongs to the Cannabaceae family and can reach 5 m high. It has serrated leaves with a distinctive vein pattern that extends to the tip and depends on the variety [38]. The masculine flowers are histaminated and the feminine ones are pistilated. The inflorescences contain most of the active compounds in the trichomes. Female flowers develop large amounts of resin with a high content of cannabinoids and terpenes [1,39]. Men have learned to use all parts of the plant for various purposes, whether therapeutic or recreational, such as the preparation of medicinal oils from seeds and inflorescences and the consumption of the latter cooked or smoked for their psychoactive effects. It has also been used as a textile fibre source [40,41]. The psychoactive properties of *Cannabis sativa* have led this species to be classified as an illegal drug, thus hindering the research on its components and pharmacological effects.

The secondary metabolism in *Cannabis sativa* gives rise to several bioactive compounds such as cannabinoids, terpenes and phenolic compounds [42]. Phytocannabinoids are terpenophenolic compounds synthesized by the plant in their acidic form that may undergo non-enzymatic decarboxylation upon heating or ageing, in the plant and after harvesting [40,43]. They have 19 or 21 carbons in their non-acidic form, being most abundant within the cannabis resin. Over 70 different cannabinoids are known that can be classified according to the following types: $(-)$-Δ^9-trans-tetrahydrocannabinol $(\Delta^9$-THC), $(-)$-Δ^8-trans-tetrahydrocannabinol $(\Delta^8$-THC), cannabigerol (CBG), cannabichromene (CBC), cannabidiol (CBD), cannabicyclol (CBL), cannabielsoin (CBE), cannabinol (CBN), cannabinodiol (CBND) and cannabitriol (CBT) type [44]. Due to the emergence of a myriad of plant varieties, the relative proportions of cannabinoids became errant; however, these components remain the majority (Fig. 4) [45].

Δ^9-THC is responsible (although not the only one cannabinoid) for the psychoactive effect of cannabis plant. This occurs only for the decarboxylated compound since it must get through the hematoencephalic barrier (HEB) to reach the CB1 receptors in the central nervous system that mediate this effect. Other major cannabinoids are CBN, which is the oxidative degradation product of Δ9-THC, CBD, and CBG. As formerly described, the interaction of agonists with CB1 receptors is also responsible for the analgesic effect of some phytocannabinoids. This cannabinoid participation in the nociceptive transmission has been intensively investigated [40]. On the other hand, CBD does not have psychoactivity since it has low or no affinity for the CBRs orthosteric site. This compound also shows antioxidant and anti-inflammatory activities, and antimicrobial, anxiolytic and anticonvulsant properties [46,47]. The immunomodulatory activity of Δ^9-THC is related to the affinity with CB2 receptors that are highly expressed in the immune system. In addition, CB2 receptors are also considered to be involved in neuro-inflammation, atherosclerosis and bone remodelling [39]. Cannabinoids are generally active in various non-cannabinoid receptors, which explains the variety of pharmacologically recognized actions and those that have not yet been investigated [22].

The biosynthetic pathways for terpenes and cannabinoids are closely related and have common precursors that lead to one or another set of compounds, depending on the selected enzymatic pathway (Fig. 5). As well as cannabinoids, terpenoid molecules have multiple biological and pharmacological activities. Some of these have been revised [39,48] and are summarized in Table 1.

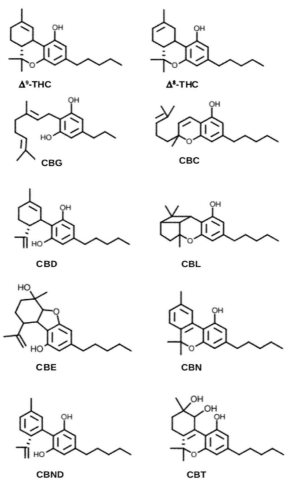

Fig. 4 Main phytocannabinoids according to the classification given by ElSohly and Slade [44].

In addition to the psychoactive and major components of cannabis, there are many other secondary metabolites with pharmacological activity within its components, for example flavonoids, such as cannflavin A and cannflavin B. These prenilated compounds are found only in cannabis and have no psychoactive activity [63–65]. Cannflavins inhibit the production of PGE 2 in rheumatoid synovial cells. Its anti-inflammatory activity is even more potent than aspirin [66,67] and is related to the inhibition in vivo of the production of two pro-inflammatory mediators, prostaglandin E2 and leukotrienes [39].

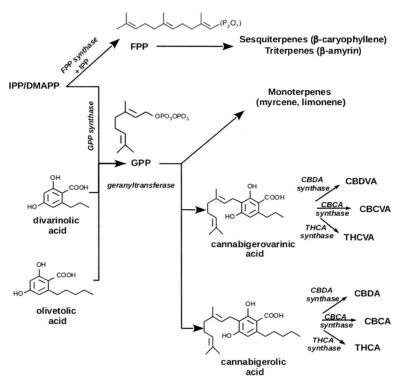

Fig. 5 Summary of biosynthetic pathways leading to the compounds discussed in the text. IPP, isopentenyl diphosphate; DMAPP, dimethylallyl diphosphate; GPP, geranyl diphosphate; FPP, farnesyl diphosphate; CBDA, cannabidiolic acid; CBCA, cannabichromenolic acid; THCA, tetrahydrocannabinolic acid; CBDVA, cannabidivarinic acid; CBCVA, cannabichromevarinic acid; THCVA, tetrahydrocannabivarinic acid.

Table 1 Pharmacological activities for some of the major terpenoids found in cannabis.

Terpenoid	Pharmacological activity	References
Limonene	Anticancer, anxiolytic and immunostimulating, antiacne	[49,50]
α-Pinene	Acetylcholinesteral inhibitor, memory improvement	[51]
β-Myrcene	Anti-inflammatory, analgesic, and anxiolytic, sedative	[52,53]
Linalool	Anxiolytic, anti-leishmanial, analgesic via adenosine	[54–56]
β-Caryophyllene	Anti-inflammatory and gastric cytoprotector, CB2-agonist	[48,57]
Nerolidol	Sedative, antimalarial, anti-leishmanial	[58–60]
β-Amyrin	Anti-bacterial, anti-fungal, anti-inflammatory and anticancer	[61,62]

4. Pharmaceutical forms of cannabinoids and novel delivery systems

Pharmaceutical form (PF) is referred to the individualized provision to which the active ingredients and excipients are adapted to constitute a medicine. PFs have as main objective to protect the active agent from external factors (light, humidity, etc.), facilitate its administration, mask unpleasant tastes and odours and provide physical and chemical stability. These characteristics and functionalities result from the technological process that gives the medicines adequate features as dosage, therapeutic efficacy and stability over time. Given the high technological development and the frequent appearance in the market of new pharmaceutical forms, it becomes increasingly complex to establish a complete classification of these.

For development of phytocannabinoids PFs the physiological effects must be known, however limited information is available. Many factors should be considered. On one side, the content of cannabinoids in plants vary depending on the chemovar (chemical variety) and conditions under which the plant is grown. Extracts from different parts of the cannabis plant yield dissimilar concentration ratios of cannabinoids. Thus, classification arises as THC-predominant (type I cannabis), mixed THC:CBD (type II) and CBD-predominant (type III cannabis) chemovars with broader mechanisms of action and improved therapeutic indexes [68]. Also, cannabinoids and terpenoids concentrations in the extract may change depending on the extraction methods, time and solvents used in the process [69]. Moreover, the acidic forms of cannabinoids possess fascinating pharmacological properties [70,71], despite their concentration will depend on the thermal treatment of the herb.

Conventional methods such as organic solvent extraction may induce chemical modifications related with the thermolability of the extracted components, although they usually allow to obtain a high proportion of the major cannabinoids, THC and CBD. The supercritical fluid extraction (SFE) with pressurized carbon dioxide stands as a favourite innovative and safe methodology producing better results than other techniques. The cannabis extracts obtained by SFE contain a complex mixture of cannabinoids and terpenes with synergistic effect in ratios that are easily controlled by experimental parameters such as the temperature and the pressure applied during the process [69].

The effects on humans of individual components of the plant or in combination may also be determined by the routes of administration. As usual for lipophilic drugs, different formulations will show different mechanisms of absorption, distribution, metabolism and excretion. These data are scarce although is essential in order to undertake clinical studies on individual cannabinoids or any cannabinoid formulation [72].

The adoption of the evidence-based medicine (EBM) approach has been used to claim for or against the efficacy and safety of cannabis as a pharmaceutical, moreover, it is the driving force to advance on the complex regulatory system around prescribing, use and control in the world [73]. There is a gap between the existing evidence of efficacy and its perception in the media. The therapeutic application of cannabis has promoted the development of galenic quality products.

Cannabinoids show very low aqueous solubility (2–10 µg/mL), they are susceptible to degradation in solution, by the action of light and temperature as well as via auto-oxidation [69,74–76].Thus, formulation is very important not only in increasing the solubility but also for the physicochemical stability of the product. Usual medical preparations comprise salt formation, cosolvency, micellization, emulsification, complexation, and encapsulation in lipid-based formulations and nanoparticles [77]. The concept of quality by design (QbD) advocates that quality should be built into the process and product during the development, unlike an earlier concept where quality is tested in the final product [78]. So, new developments show stages controlled from the growth of the cannabis plant to final product using advanced techniques of isolation of active ingredients that meet good manufacturing practices [69].

Koltai et al. [79] reported the steps required to promote the quality of cannabis products. Basically, they propose a first extractive step, where a comprehensive chemical profile of the given variety used should be generated and reported. Second, they mention the need to determine suitable synergistic mixtures for different therapeutic effects and for the generation of new strains of grass [80]. Third, they propose to isolate plant compounds or synthesize them (concept rejected by other researchers [81]) to mimic the biological activity of compositions defined as active. Fourth, the study of characterization of the biological pathways affected in human cells and tissues could contribute to produce drug innovation. This logical sequence to obtain cannabis products would be successful if it were contemplated together with QbD.

The most popular pharmacological form for medicinal cannabis is the cannabis oil [81]. There are oils in the market with very varied qualities and compositions, being available with different flavours. CBD is not a controlled drug and regulations are minimal compared to THC. This has led to the emergence in the international market of numerous extracts rich in CBD. Most of these extracts contain low levels of CBD and high levels of CBD acid, the natural component of the fresh cannabis plant. Charlotte's Web oils (USA), regulated as dietary supplement, contain between 17 and 50 mg/mL CBD. Bedrocan (The Netherland), Aphria and Tilray (Canada) produce quality medical cannabis strains that are known for their consistency and effect, from type I, type II and type III cannabis, thus offering a wide range of compositions of medical cannabis oils. Other forms for cannabis oral administration are reported as capsules or edibles (gums, cookies, brownies); however, the last form may be more difficult to dose [68]. Epidiolex®, an oral solution of CBD (100 mg/mL), is used to treat seizures associated with Lennox-Gastaut syndrome or Dravet syndrome in patients 2 years of age and older, approved by FDA. The oral administration of cannabis follows a path of digestive absorption. This pharmaceutical form is increasingly popular due to convenience and accuracy of dosing. The onset effects of oral cannabis begin between 60 and 180 min and persist for 6–8 h. This form of administration is advantageous for chronic diseases, resulting suitable because it can be flavoured.

Sativex® is a pharmaceutical product produced from the British company GW Pharmaceuticals, it is a patented oral mint-flavoured spray. A dosage (100 μL) contains 2.7 mg THC and 2.5 mg of CBD from *Cannabis sativa* L. and up to 0.04 g of ethanol. The active ingredients of Sativex® are extracted from the plant, subsequently purified, and then applied as an oral spray. The onset effects of cannabis from this oromucosal administration begin between 15 and 45 min and persist for 6–8 h. The qualitative and quantitative determination of THC and CBD in oromucosal sprays requires a good standardized method given the importance of controlling the stability of cannabis products [82].

Also, there is a Medic® vaporizer (Storz and Bickel®), which is currently the only approved medical device in the world for the vaporization of herbal cannabis. Vaporization allows a rapid action (onset 5–10 min) that is advantageous for acute episodic symptoms. Generally, the vaporizers are expensive and frequently not portable. The most important disadvantage is that training is required of the user for an optimal administration. The inhalation of phytocannabinoids produces significantly less harmful byproducts than smoking [68].

Topical administration routes of phytocannabinoids (dermatological creams) are developed for dermatological conditions or arthritis, but with incomplete research evidences. Vaginal ovules containing cannabinoids are a potential route of administration for female conditions (such as endometriosis or menstrual cramps) [83]. Also, suppositories can be indicated for cancer, and gastrointestinal symptoms with variable absorption [68].

The PFs of cannabinoids described above are typical formulations and can be grouped according to the administration pathways (oral, inhalation, oromucosal/mucosal, topical). On the other hand, nanoscience research has opened a very promising field to increase the efficacy of the therapeutic application of cannabinoids. The bioavailability of cannabinoids can be improved, and their undesirable effects limited through advanced nanosized drug delivery systems. For instance, nanoparticle drones have been developed lately to precisely target cancer cells which can be programmed to deliver cannabinoids as therapeutic agents [84].

Liposomes have been frequently chosen as drug delivery systems, in spite of the low encapsulation efficiency for cannabinoids. A formulation of liposomes prepared with di-palmitoyl-phosphatidylcholine and cholesterol and containing 0.3 mg/mL THC showed rapid bioavailability and onset in the pulmonary administration [85].

Micellar and liposomal preparations have also been proposed for delivering terpenes and cannabinoids. The stability ranges from a few days, for micelles, to several months for liposomes [86].

Nanostructured lipid carrier particles are considered suitable transporter systems for THC and CBD since these solid particle matrices are able to host and protect the cannabinoids from degradation. They can also slow the diffusion of THC from inside the particle to the particle surface [75]. Such systems have been developed for nasal delivery of THC with high stabilizing effect [87].

The self-emulsifying drug delivery (SEDD) technology is applied for a THC:CBD (1:1) formulation called PTL401. This is based on an isotropic mixture of the active compounds combined with lipids, surfactants and a co-solvent (the pro-nano liposphere, PNL), which is ingested as a soft gelatine capsule. When the PNL reaches the aqueous phase of the gastrointestinal tract, it spontaneously forms an oil/water micro-emulsion which encapsulates the cannabinoids in nanometric particles. This procedure favours the solubility and oral bioavailability of the liposoluble drugs [88]. Absorption enhancers such as piperine can be incorporated in the PNL, thus increasing the oral bioavailability of CBD, according to in vivo evaluations in a rat model [89].

Polymeric carriers have shown to be effective for encapsulation and controlled release of cannabinoids. The poly(lactic-*co*-glycolic acid) (PLGA), a biocompatible, mechanically strong, and hydrophobic polymer, was used by Martin-Banderas et al. for the preparation of THC-loaded nanoparticles as an anticancer agent [90].

It has been found that poly-ε-caprolactone (PCL) particles loaded with CBD achieve high entrapment efficiency. In addition, in vitro tests indicated that CBD was slowly released over within 10 days when dissolved in these polymeric microspheres [91].

An interesting approach is the design of pro-drugs. This has been used by Adelli et al. who managed to tune the penetration of THC in the anterior segment of the eye following topical application by using more hydrophilic THC pro-drugs [92]. They synthesized mono- and di-valine esters with enhanced solubility and permeability in vitro in comparison to those of THC. In that way, the bioavailability significantly improved the intra-ocular pressure reducing activity.

In the same line, a light activatable multifunctional pro-drug has been developed by Ling et al. It consists of a phthalocyanine photosensitizer (PS), reactive oxygen species (ROS)-sensitive linker, and a cannabinoid. Upon red light irradiation, the PS produces cytotoxic ROS, which simultaneously cleaves the ROS-sensitive linker and releases the cannabinoid drug. The targets are the cannabinoid receptors, which are overexpressed in the cells of various types of cancers, thus improving the therapeutic efficacy [93].

5. Pharmacokinetics of cannabis

Pharmacokinetics is the branch of pharmacology that studies the processes to which a drug is subjected in its passage through the body, from the moment it is administered until its total clearance from the organism.

As mentioned above, Δ^9-THC, its analogues and metabolites are water insoluble and lipophilic [74]. Their absorption depends on the formulation and route of administration. Inhalation and vaping allow reaching a plasma peak only a few seconds after the first puff [94]. Bioavailability is variable since it depends on the smoking technique or depth [95], ranging from 10% to 35%. Pipe smoking might improve the effectiveness up to 45%.

In oral use the absorption is slower reaching the plasma peak after 60–120 min [96]. Other studies showed plasma peaks after 4 h and even more than one plasma peak [46,95]. However, the absorption by oral administration is almost complete, specially using oil vehicles [97]. Δ^9-THC is subject to first-pass liver metabolism, greatly reducing its bioavailability.

In oromucosal administration, cannabinoids are rapidly absorbed through the oral mucosa (and hence, are useful for symptoms requiring rapid relief) producing higher plasmatic concentrations than oral administration, but lower than inhaled cannabis [98]. Nevertheless, part of the dose may be swallowed and orally absorbed [99].

Transdermal administration prevents the hepatic first-pass metabolism. The diffusion is poor due to the hydrophobicity of the cannabinoids to pass through the aqueous layer of the skin and must be improved by pharmacotechnics. However, CBD showed higher permeability than THC isomers [100–102].

Distribution volume (DV) of cannabinoids is large due to their lipophilicity and therefore they are able to deposit within tissues, in spite of their high affinity to plasma lipoproteins [103,104]. Fig. 6 schematizes the distribution for Δ^9-THC in blood and tissues. DV for Δ^9-THC has been reported as 2.55–6.38 L for smoked cannabis and 4.934 for oromucosal CBD/THC combinations after two puffs [105]. In this latter report, DV for CBD was found to be larger (22,169 L). Both are stored with great affinity in fatty and irrigated tissues such as heart, fat, lung, jejunum, kidney, spleen, mammary gland, placenta, pituitary gland, thyroid, adrenal cortex and muscle [106].

Metabolism of cannabinoids has been thoroughly reviewed for Δ^9-THC and CBD [107–109]. Studies in animals, such as rodents and dogs, showed that an important percentage of CBD is excreted unchanged or as its glucuronide [110,111]. Pharmacokinetics of CBD is rather complicated, having high affinity for the CYP450 enzymes. Hydroxylated 7-COOH and 3″ OH-CBD derivatives are the most found, together with CBD-7-oic acid (7-COOH-CBD). However, almost 100 CBD minor metabolites have been identified from various organisms [112,113].

In the case of Δ^9-THC, it also has a hepatic metabolism through the microsomal enzymes, and most of the metabolites found in urine are the

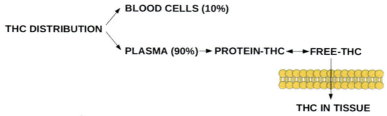

Fig. 6 Scheme of Δ^9-THC distribution in blood and tissues.

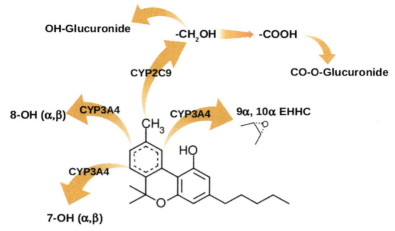

Fig. 7 Summary of the main transformations on cannabinoid structures through cytochrome P450: 7α-OH, 7α-hydroxylation of Δ^8-THC; 7β-OH, 7β-hydroxylation of Δ^8-THC; 8β-OH, 8β-hydroxylation of Δ^9-THC; 8α-OH, 8α-hydroxylation of CBN; 9α,10α-EHHC, 9α,10α-epoxidation of Δ^9-THC.

acidic forms [114,115], being the acidic glucuronide of THC-COOH the most abundant compound. [116]. Nevertheless, free THC-COOH is not excreted in significant concentrations in urine [96,105,117]. Δ^9-THC and 11-OH-THC were found in very small proportions by using an enzymatic hydrolysis step during the extraction [118]. Permanency of 11-OH-THC in brain tissue can be associated to its fast perfusion in and out through the HEB. 11-OH-THC has also psychoactive properties [119,120]. Δ^8-THC and CBN are subjected to analogous metabolic transformations by the same cytochromes, as schematized in Fig. 7.

Because of an important enterohepatic recirculation and high protein binding of cannabinoids, they are greatly excreted with the faeces and the acidic and neutral forms of the metabolites in this matrix, are found in the non-conjugated form [121,122].

6. Analysis of cannabinoids and cannabinoid metabolites in organic matrices

An increasing number of clinical trials are being carried out in recent years for conditions such as neuropathic pain and spasticity, fibromyalgia pain, hand osteoarthritis, psoriatic arthritis and chronic pain among others [123,124].

As discussed before, human liver microsomes metabolizes Δ^9-THC to a short life active metabolite, 11-OH-THC and then, to the inactive THC-COOH (Fig. 7). This latter has been used as a biomarker for the analysis of medicinal cannabis use and forensic fields [125].

Several screening methods are used to investigate cannabis in different organic matrices, either for medicinal or legal purposes. Qualitative determination is usually performed by testing cannabinoids on immunoassay. On the other hand, quantitative and confirmation essays are carried out through chromatographic techniques on blood, plasma, serum, saliva and urine by using HPLC [126], GC–MS [127], and HPLC–MS [128]. A simple but rapid quantification of D9-THC, CBD, 11-OH-THC and THC-COOH in human plasma was reported, using ultra-performance liquid chromatography-mass spectrometry (UPLC-MS). In order to avoid the matrix effects generated by phospholipids, the authors optimized the chromatographic conditions to separate the elution region of phospholipids and cannabinoids. They used a post-column divert valve to send the unwanted portion of chromatographic runs. Also, they studied matrix effect with a post-column infusion method [129]. This technique was applied in clinical research experiments [130]. Recently, a highly sensitive and selective liquid chromatography method, in tandem to high resolution mass spectrometry (HRMS), was performed for the quantification of Δ^9-THC, 11-OH-THC, THC-COOH and CBD [131]. An orbitrap-based instrument allowed for parallel reaction monitoring (PRM) quantification in positive polarity with a negative polarity switching for 11-OH-THC and THC-COOH. This method was fully validated under the guidelines of the Société Française des Sciences et des Techniques Pharmaceutiques. This method resulted fast, sensitive, reproducible and robust. Besides, PRM analyses give access to full MS/MS allowing an improved control over the quantification experiment and orbitrap has a high-resolution power being an alternative over triple quadrupole instruments.

A study aimed at quantifying levels of specific cannabinoids detectable in human milk, on a relatively large sample of breasting women, was performed by Bertrand et al. [132]. Δ^9-THC, 11-OH-THC, and CBD were detected at 1 ng/mL or above in human milk samples. Δ^9-THC was detected in 34 (63%) of the 54 samples analysed. Since 11-OH-THC is less lipophilic than Δ^9-THC, only five of the milk samples had measurable concentrations and it was detectable in a majority of breast milk samples up to ca. 6 days after maternal marijuana use. All samples came from smoked cannabis. In this study, C-18 cartridges were used for solid-phase extraction.

Separation and quantification were performed by HPLC on a reverse-phase C-18 column. The mobile phases consisted of ammonium formiate in formic acid and acetonitrile in formic acid. The oral absorption in breastfeeding infants, metabolism, accumulation patterns, and pharmacological effects on neuro-development in children are unknown and require further insight.

Recently, Pichini et al. used ultra-high-performance liquid chromatography coupled with tandem mass spectrometry (UPLC-MS/MS) for analysing Δ^9-THC, CBD, THCA-A, CBDA, THC-COOH in pharmaceutical preparations, and 11-OH-THC, Δ^9-THC-glucuronide (THC-GLUC) and THC-COOH-glucuronide (THC-COOH-GLUC) biological matrices such blood, urine, saliva and sweat [133]. The detection method was the positive mode electrospray ionization (ESI) operated in multiple reaction monitoring (MRM) mode. This methodology allowed reaching high sensitivity with quantification limits ranging from 0.2 to 0.5 ng/mL or ng per patch, in case of sweat. The extraction procedures from biological matrices required low time and no laborious procedures with good analytical recovery and minimal matrix effects. The procedure was validated according to the standards of analytical toxicology [134,135]. This technique was successfully applied in clinical trials involving cannabis.

Umbilical cord is a very useful non-conventional matrix for determination of in utero drug exposure. It implies abundant and non-invasive collection, since it is a waste product [136,137], and it lacks the limitations of meconium, neonatal hair and urine, such as the difficulty for collection and sample availability. This matrix showed to have a similar detection window as meconium [138–140]. In a recent study, Kim et al. [141] developed a sensitive, specific and validated method for the simultaneous determination of Δ^9-THC, CBD, THC-COOH, 11-OH-THC, 8-β-11-dihydroxy-THC (THC-di-OH), THC-COOH-glucuronide and THC-glucuronide in umbilical cord. This determination was performed by (UPLC-MS/MS) on authentic umbilical cord samples from newborns. The system consisted of a binary LC-30AD pump with an online degassing unit and cooled autosampler C. The triple quadrupole mass spectrometer had a dual ionization source and it was used in the positive ionization mode and each molecule was monitored by two transitions in the MRM mode. A F5 column, 100×2.1 mm, 1.7 µm was used for the separation and a gradient of 0.1% formic acid in water and 0.1% formic acid in acetonitrile were the mobile phases. Meconium samples analysis was performed in order to compare to those on umbilical cord. The extracts were analysed by LC-MS/MS and the chromatographic separation was carried out on a C18 column 50×2.1 mm, 2.6 µm using a gradient of

0.1% formic acid in water and acetonitrile. Two transitions were monitored in MRM positive mode for each compound. The methodology herein proposed, was fully validated under the Scientific Working Group for Forensic Toxicology (SWGTOX) guidelines [142]. The analysis performed on real umbilical cord samples showed that THC-COOH-glucuronide was the only Δ^9-THC metabolite detected.

Drug detection in hair depends on the methodology of analysis, the doses, route of administration, and pH among other parameters. Hair is very simple to obtain, store and process. This matrix has a wide window for detection, ranging from a week to months or even years depending on the length of the hair [143]. One of the variables to consider in the analysis and segments of hair is the growth rate and the collection of samples, since drugs can take about 2 weeks to deposit in newly formed hair [144]. Decontamination from contaminants such as sweat, sebum, powder, cosmetics, etc. by washing to avoid false positives must be taken into account. The method should not remove the Δ^9-THC from the matrix, this can be accomplished with solvents such as dichloromethane or methanol [145,146]. Some authors advise to collect hair samples from all parts of the scalp in order to overcome the variability in Δ^9-THC distribution [147]. They performed digestion and extraction of the hair sample by incubating in a sodium hydroxide solution and extracting with n-hexane: ethyl acetate (9:1). Quantification was performed with GC–MS. Δ^9-THC was detected until the 45th day and 90th day, respectively, at a limit of detection (LOD) of 0.1 ng/mg. The range for the concentration of this cannabinoid in the hair sample was 0.16–2.3 ng/mg (mean, 0.95 ng/mg). Rodrigues et al. [148] investigated THC, CBN and CBD in hair samples in order to determine if there was a correlation between CBD consume and this molecule in hair. They carried out a validated GC–MS/MS method and found no correlation between CBD concentrations in hair and daily doses. On the other hand, no CBN was detected and only one sample was positive for THC. In this study, they used Strata—X 33 µm polymeric reversed phase (100 mg/3 mL tubes) extraction cartridges and a GC–MS/MS system consisting of a 7890A gas chromatography, with an AS 7693 automatic injector and coupled to a 7000C triple quadrupole mass spectrometer. This method was validated, and it was useful to investigate consumers of CBD rich extracts. Kim and Kyo [149] used GC–MS in negative ion chemical ionization mode for the quantification of 11-nor-Δ^9-tetrahydrocannabinol-9-carboxylic acid in hair since the determination of this metabolite is recommended to distinguish passive exposure from active consumption [150]. The concentration level of acidic metabolite into the hair matrix is lower than that of Δ^9-THC [151]. They carried out GC–MS/MS analyses using a micro GC tandem quadrupole mass spectrometer and a capillary column (DB-5 MS, 30 m 0.25 mm i.d., 0.25 mm). The sensitivity was evaluated by determining the LOD and the limit of quantification (LOQ) for the analyte. Pentafluoropropyl was used for derivatizing the analytes after acidic hydrolysis. Liquid-liquid extraction and detection was performed by using the negative ion chemical ionization mode and MS/MS analysis, respectively. This resulted in a higher sensitivity and a very informative fragmentation. Míguez-Framil et al. developed a validated fast HPLC–MS/MS method for the determination of THC, CBD and CBN in hair samples [152]. The advantage of LC-MS over GC-MS lies on the fact that it does not requires derivatization. They achieved short chromatographic times (5.0 min), allowing to set this methodology for routine screening and quantitative confirmation analysis. Δ^9-THC and CBD, could be resolved by a simultaneous determination. This methodology achieved high sensitivity because of the use of tandem MS detector for the determination of targets in hair samples. They used a LC-MS/MS system, with a binary chromatographic pump. Separations were carried out on a XDB-C8 column (2.1 mm × 50 mm, 3.5 µm). Mercolini et al. [153] developed a LC–ESI-(QqQ)MS/MS method for determination of Δ^9-THC and THC-COOH in hair. This is a sensitive method but since the concentration of THC-COOH is very low as compared to Δ^9-THC, it may undergo false-positive results. They used a triple quadrupole mass spectrometer coupled to a chromatographic pump with autosampler and a SB-C18 column (50 mm × 2.1 mm i.d., 3.5 m). Mass spectra were acquired in multiple reaction monitoring (MRM) mode, using an electrospray ionization (ESI) source. Δ^9-THC and THC-d3 were determined using positive ionization mode with significant improvements in terms of sensitivity, extraction yields and lower matrix effect. Dulaurent et al. [154] carried out the simultaneous quantification of cannabinoids based on only a 20 mg hair sample by means of a quadrupole/quadrupole/iontrap (QTRAP) mass spectrometer system, being a highly specific and sensitive method, even more than GC–MS. The field of application of a LC-MS/MS system, such as an AB SCIEX API 5500 QTRAP, resulted wider than that of a GC–MS/MS system.

Other sensitive methods have also been employed such as matrix-assisted laser desorption/ionization mass spectrometry (MALDI–MS) technique, which requires smaller quantities of sample [155] and direct analysis in real time mass spectrometry (DART-MS), demanding shorter times for sample preparation [156].

7. Conclusions

Medicinal cannabis is booming according to the changes in the laws of many countries as well as the technological advances that allowed expanding the knowledge about its multiple pharmacological properties. Given the impact of this millenarian plant on biological systems, it is of great importance to carry out a thorough study on its pharmacological effects and the pharmaceutical forms used for its administration as a medicine, together with its behaviour in relation to organic matrices. Many of the components of cannabis are converted into active metabolites and that is why analytical studies in organisms must be performed with great adjustment. The advances in laboratory technologies allow making precise, reproducible, and shorting time determinations. Many of these techniques have been developed for toxicological analysis purposes; however, they allow their application for the research on the pharmacokinetics of different formulations and combinations of cannabinoids. Regarding pharmacodynamics, the prediction and determination of ligand–receptor interactions, studied by means of molecular modelling and the recently published crystallographic structures, could shed light in the way of rational drug design for these systems. The integration of the different fields of research on cannabis-based drug systems shall be a great advantage in order to achieve the implementation of controlled, safe and effective pharmaceutical devices. This will be the cornerstone for the construction of effective, contemplative and broad legislation on medicinal cannabis.

Acknowledgements

The authors gratefully acknowledge grants from Universidad Nacional de Mar del Plata (EXA 759/16- 15/E710; EXA 1008/20- 15/E966) and Mr. Diego Nutter (F77) for helpful discussions.

Abbreviations

Δ^9-THC	delta-9-tetrahydrocannabinol
11-OH-THC	11-hydroxo-tetrahydrocannabinol
2-AG	2-arachidonoylglycerol
2-AGE	2-arachidonoylglycerol-ether
AC	adenylyl cyclase
AEA	anandamide
AMPc	cyclic adenosine monophospate
ATP	adenosine triphosphate

CB1	cannabinoid receptor 1
CB2	cannabinoid receptor 2
CBC	cannabichromene
CBCA	cannabichromenolic acid
CBCVA	cannabichromevarinic acid
CBD	cannabidiol
CBDA	cannabidiolic acid
CBDVA	cannabidivarinic acid
CBE	cannabielsoin
CBG	cannabigerol
CBL	cannabicyclol
CBN	cannabinol
CBND	cannabinodiol
CBRs	cannabinoid receptors
CBT	cannabitriol
D2	dopamine receptor type 2
DMAPP	dimethylallyl diphosphate
DV	distribution volume
EBM	evidence-based medicine
eCB	endocannabinoid
ESI	electrospray ionization
FDA	Food and Drug Administration
FPP	farnesyl diphosphate
GC–MS	gas chromatography-mass spectrometry
GPCR	G protein-coupled receptor
GPP	geranyl diphosphate
HPLC	high performance liquid chromatography
HRMS	high resolution mass spectrometry
LC	liquid chromatography
LO	limit of quantification
LOD	limit of detection
MRM	multiple reaction monitoring
NADA	N-arachidonoyl dopamine
NAGly	N-arachidonyl glycine
NT	neurotransmitters
OEA	cis-9-octadecenamide
OLDA	N-oleoyldopamine
PCL	poly-ε-caprolactone
PEA	N-palmitoylethanolamide
PF	pharmaceutical form
PGE	prostaglandine E
PLGA	poly(lactic-*co*-glycolic acid)
PNL	pro-nano lipospheres
PRM	parallel reaction monitoring
PS	photosensitizer
QbD	quality by design
QTRAP	quadrupole/quadrupole/iontrap

ROS	reactive oxygenspecie
SEA	N-oleoylethanolamide
SEDD	self-emulsifying drug delivery
SFE	supercritical fluid extraction
SWGTOX	Scientific Working Group for Forensic Toxicology
THCA	tetrahydrocannabinolic acid
THC-COOH	carboxy-tetrahydrocannabinol
THC-COOH-GLUC	carboxy-tetrahydrocannabinol glucuronide
THC-d3	deuterated tetrahydrocannabinol
THC-GLUC	tetrahydrocannabinolglucuronide
THCVA	tetrahydrocannabivarinic acid
TRPV	transient receptor potential cation channels
UPLC-MS	ultra-high-performance-mass spectrometry
Δ^8-THC	delta 8 tetrahydrocannabinol

References

[1] S.A. Bonini, M. Premoli, S. Tambaro, A. Kumar, G. Maccarinelli, M. Memo, A. Mastinu, *Cannabis sativa*: a comprehensive ethnopharmacological review of a medicinal plant with a long history, J. Ethnopharmacol. 227 (2018) 300–315.

[2] L.M. McPartland, G. Guy, V. Di Marzo, Care and feeding of the endocannabinoid system: a systematic review of potential clinical interventions that upregulate the endocannabinoid system, PLoS One 9 (2014) 1–21.

[3] Y. Gaoni, R. Mechoulam, Isolation, structure, and partial synthesis of an active constituent of hashish, J. Am. Chem. Soc. 86 (1964) 1646–1647.

[4] W.A. Devane, F.A. Dysarz, M.R. Johnson, L.S. Melvin, A.C. Howlett, Determination and characterization of a cannabinoid receptor in rat brain, Mol. Pharmacol. 34 (1988) 605–613.

[5] R.G. Pertwee, R.A. Ross, Cannabinoid receptors and their ligands, Prostaglandins Leukot. Essent. Fatty Acids 66 (2002) 101–121.

[6] G. Turu, L.J. Hunyady, Signal transduction of the CB1 cannabinoid receptor, Mol. Endocrinol. 44 (2010) 75–85.

[7] Z. Shao, J. Yin, K. Chapman, M. Grzemska, L. Clark, J. Wang, D.M. Rosenbaum, High-resolution crystal structure of the human CB1 cannabinoid receptor, Nature 540 (2016) 602–606.

[8] S. Munro, K.L. Thomas, M. Abu-Shaar, Molecular characterization of a peripheral receptor for cannabinoids, Nature 365 (1993) 61–65.

[9] F. Weis, A. Beiras-Fernandez, R. Sodian, I. KaczmarekI, B. Reichart, A. Beiras, Substantially altered expression pattern of cannabinoid receptor 2 and activated endocannabinoid system in patients with severe heart failure, J. Mol. Cell. Cardiol. 48 (2010) 1187–1193.

[10] L.K. Vaughn, G. Denning, K.L. Stuhr, H. de Wit, M.N. Hill, C.J. Hillard, Endocannabinoid signalling: has it got rhythm? Br. J. Pharmacol. 160 (2010) 530–543.

[11] D.J. Chen, M. Gao, F.F. Gao, Q.X. Su, J. Wu, Brain cannabinoid receptor 2: expression, function and modulation, Acta Pharmacol. Sin. 38 (2017) 312–316.

[12] T.P. Malan Jr., M.M. Ibrahim, J. Lai, T.W. Vanderah, A. Makriyannis, F. Porreca, CB2 cannabinoid receptor agonists: pain relief without psychoactive effects, Curr. Opin. Pharmacol. 3 (2003) 62–67.

[13] X. Li, T. Hua, K. Vemuri, J.H. Ho, Y. Wu, L. Wu, P. Popov, O. Benchama, N. Zvonok, K. Locke, L. Qu, G.W. Han, M.R. Iyer, R. Cinar, N.J. Coffey, J. Wang, M. Wu, V. Katritch, S. Zhao, G. Kunos, L.M. Bohn, A. Makriyannis, R.C. Stevens, Z.J. Liu, Crystal structure of the human cannabinoid receptor CB2, Cell 176 (2019) 459–467.

[14] X. Chen, C. Zheng, J. Qian, S.W. Sutton, Z. Wang, J. Lv, C. Liu, N. Zhou, Involvement of beta-arrestin-2 and clathrin in agonist-mediated internalization of the human cannabinoid CB2 receptor, Curr. Mol. Pharmacol. 7 (2014) 67–80.

[15] C.S. Breivogel, V. Puri, J.M. Lambert, D.K. Hill, J.W. Huffman, R.K. Razdan, The influence of beta-arrestin2 on cannabinoid CB1 receptor coupling to G-proteins and subcellular localization and relative levels of beta-arrestin1 and 2 in mouse brain, J. Recept. Signal Transduct. Res. 33 (2013) 367–379.

[16] A. Perchuk, S.M. Bierbower, A. Canseco-Alba, Z. Mora, L. Tyrell, N. Joshi, N.N. Schanz, G.G. Gould, E. Onaivi, Developmental and behavioral effects in neonatal and adult mice following prenatal activation of endocannabinoid receptors by capsaicin, Acta Pharmacol. Sin. 40 (2019) 418–424.

[17] K.J. Kuder, T. Karcz, M. Kaleta, K. Kiec-Kononowicz, Molecular modeling of an orphan GPR18 receptor, Lett. Drug Des. Discov. 16 (2019) 1167–1174.

[18] S.S. So, T. Ngo, P. Keov, N.J. Smith, I. Kufareva, Tackling the complexities of orphan GPCR ligand discovery with rationally assisted approaches, in: B. Jastrzebska, P.S.H. Park (Eds.), GPCRs, Academic Press, 2020, pp. 295–334.

[19] D. McHugh, GPR18 in microglia: implications for the CNS and endocannabinoid system signalling, Br. J. Pharmacol. 167 (2012) 1575–1582.

[20] J.C. von Widdern, T. Hohmann, F. Dehghani, Abnormal cannabidiol affects production of proinflammatory mediators and astrocyte wound closure in primary astrocytic-microglial cocultures, Molecules 25 (2020) 496–516.

[21] H. Yang, J. Zhou, C. Lehmann, GPR55—a putative "type 3" cannabinoid receptor in inflammation, J. Basic Clin. Physiol. Pharmacol. 27 (2015) 297–302.

[22] P. Morales, D.P. Hurst, P.H. Reggio, A.D. Kinghorn, H. Falk, S. Gibbons, J. Kobayashi, Molecular targets of the phytocannabinoids: a complex picture, in: Progress in the Chemistry of Organic Natural Products: Phytocannabinoids Unraveling the Complex Chemistry and Pharmacology of *Cannabis sativa*, Springer International Publishing, Switzerland, 2017, pp. 104–131.

[23] W.A. Devane, L. Hanus, A. Breuer, R.G. Pertwee, L.A. Stevenson, G. Griffin, D. Gibson, A. Mandelbaum, A. Etinger, R. Mechoulam, Isolation and structure of a brain constituent that binds to the cannabinoid receptor, Science 258 (1992) 1946–1949.

[24] F. Fezza, M. Bari, R. Florio, E. Talamonti, M. Feole, M. Maccarrone, Endocannabinoids, related compounds and their metabolic routes, Molecules 19 (2014) 17078–17106.

[25] M. Maccarrone, I. Bab, T. Bíró, G.A. Cabral, S.K. Dey, V. Di Marzo, J.C. Konje, G. Kunos, R. Mechoulam, P. Pacher, K.A. Sharkey, A. Zimmer, Endocannabinoid signaling at the periphery: 50 years after THC, Trends Pharmacol. Sci. 36 (2015) 277–296.

[26] V. Di Marzo, F. Piscitelli, The endocannabinoid system and its modulation by phytocannabinoids, Neurotherapeutics 12 (2015) 692–698.

[27] E. Krook-Magnuson, I. Soltesz, Beyond the hammer and the scalpel: selective circuit control for the epilepsies, Nat. Rev. Neurosci. 18 (2015) 331–338.

[28] Z. Benyó, É. Ruisanchez, M. Leszl-Ishiguro, P. Sándor, P. Pacher, Endocannabinoids in cerebrovascular regulation, Am. J. Physiol. Heart Circ. Physiol. 310 (2016) 785–801.

[29] C.J.A. Morgan, T.P. Freeman, J. Powell, H.V. Curran, AKT1 genotype moderates the acute psychotomimetic effects of naturalistically smoked cannabis in young cannabis smokers, Transl. Psychiatry 6 (2016) e738.

[30] L.L. Wang, R. Zhao, J. Li, S.S. Li, M. Liu, M. Wang, M.Z. Zhang, W.W. Dong, S.K. Jiang, M. Zhang, Z.L. Tian, C.S. Liu, D.W. Guan, Pharmacological activation of cannabinoid 2 receptor attenuates inflammation, fibrogenesis, and promotes re-epithelialization during skin wound healing, Eur. J. Pharmacol. 786 (2016) 128–136.

[31] R. Mechoulam, S. Ben-Shabat, L. Hanus, M. Ligumsky, N.E. Kaminski, A.R. Schatz, A. Gophe, S. Almog, R. Martin, D.R. Compton, R.G. Pertwee, G. Griffin, M. Bayewitch, J. Barg, Z. Vogel, Identification of an endogenous 2-monoglyceride, present in canine gut, that binds to cannabinoid receptors, Biochem. Pharmacol. 50 (1995) 83–90.

[32] T. Sugiura, S. Kondo, A. Sukagawa, S. Nakane, A. Shinoda, K. Itoh, A. Yamashita, K. Wakux, 2-Arachidonoylglycerol: a possible endogenous cannabinoid receptor ligand in brain, Biochem. Biophys. Res. Commun. 215 (1995) 89–97.

[33] K. Tsuboi, T. Uyama, Y. Okamoto, N. Ueda, Endocannabinoids and related N-acylethanolamines: biological activities and metabolism, Inflamm. Regen. 38 (2018) 28–48.

[34] T. Sugiura, S. Kishimoto, S. Oka, M. Gokoh, Biochemistry, pharmacology and physiology of 2-arachidonoylglycerol, an endogenous cannabinoid receptor ligand, Prog. Lipid Res. 45 (2006) 405–446.

[35] K.P. Hill, M.D. Palastro, B. Johnson, J.W. Ditre, Cannabis and pain: a clinical review, Cannabis Cannabinoid Res. 2 (2017) 96–104.

[36] P.H. Reggio, Endocannabinoid binding to the cannabinoid receptors: what is known and what remains unknown, Curr. Med. Chem. 17 (2010) 1468–1486.

[37] P.H. Reggio, Endocannabinoid structure–activity relationships for interaction at the cannabinoid receptors, Prostaglandins Leukot. Essent. Fatty Acids 66 (2002) 143–160.

[38] M. Watt, M.G. Breyer-Brandwijk, The Medicinal and Poisonous Plants of Southern and Eastern Africa, second ed., E. & S Livingstone, 1962.

[39] C.M. Andre, J.F. Hausman, G. Guerriero, *Cannabissativa*: the plant of the thousand and one molecules, Front. Plant Sci. 7 (2016) 1–19.

[40] E.B. Russo, Cannabinoids in the management of difficult to treat pain, Ther. Clin. Risk Manag. 4 (2008) 245–259.

[41] G. Skoglund, M. Nockert, B. Holst, Viking and early middle ages northern scandinavian textiles proven to be made with hemp, Sci. Rep. 2686 (2013) 1–6.

[42] I.J. Flores Sanchez, R. Verpoorte, Secondary metabolism in cannabis, Phytochem. Rev. 7 (2008) 615–639.

[43] A. Shani, R. Mechoulam, Cannabielsoic acids: isolation and synthesis by a novel oxidative cyclization, Tetrahedron 30 (1974) 2437–2444.

[44] M.A. ElSohly, D. Slade, Chemical constituents of marijuana: the complex mixture of natural cannabinoids, Life Sci. 78 (2005) 539–548.

[45] A. Hazekamp, K. Tejkalova, S. Papadimitriou, Cannabis: from cultivar to chemovar II—a metabolomics approach to cannabis classification, Cannabis Cannabinoid Res. 1 (1) (2016) 202–215.

[46] E.B. Russo, Taming THC: potential cannabis synergy and phytocannabinoid-terpenoid entourage effects, Br. J. Pharmacol. 163 (2011) 1344–1364.

[47] F. Grotenhermen, K. Müller-Vahl, The therapeutic potential of cannabis and cannabinoids, Dtsch. Arztebl. Int. 109 (2012) 495–501.

[48] B. Singh, R. Sharma, Plant terpenes: defense responses, phylogenetic analysis, regulation and clinical applications, Biotechnology 5 (2015) 129–151.

[49] T. Komori, R. Fujiwara, M. Tanida, J. Nomura, M.M. Yokoyama, Effects of citrus fragrance on immune function and depressive states, Neuroimmunomodulation 2 (1995) 174–180.

[50] S.S. Kim, J.S. Baik, T.H. Oh, W.J. Yoon, N.H. Lee, G.G. Hyun, Biological activities of Korean *Citrus obovoides* and *Citrus natsudaidai* essential oils against acne-inducing bacteria, Biosci. Biotechnol. Biochem. 72 (2008) 2507–2513.

[51] D.O. Kennedy, F.L. Dodd, B.C. Robertson, E.J. Okello, J.L. Reay, A.B. Scholey, Monoterpenoid extract of sage (*Salvia lavandulaefolia*) with cholinesterase inhibiting properties improves cognitive performance and mood in healthy adults, J. Psychopharmacol. 25 (2011) 1088–1100.

[52] M.V. Cleemput, K. Cattoor, K.D. Bosscher, G. Haegeman, D.D. Keukeleire, A. Heyerick, Hop (*Humulus lupulus*)-derived bitter acids as multipotent bioactive compounds, J. Nat. Prod. 72 (2009) 1220–1230.

[53] T.G. do Val, E.C. Furtado, J.G. Santos, G.S. Viana Jr., Central effects of citral, myrcene and limonene, constituents of essential oil chemotypes from *Lippia alba*(Mill.) n.e. Brown, Phytomedicine 9 (2002) 709–714.

[54] E.B. Russo, Handbook of Psychotropic Herbs: A Scientific Analysis of Herbal Remedies for Psychiatric Conditions, Haworth Press, Binghamton, NY, 2001.

[55] S.R. do Socorro, R.R. Mendonca-Filho, H.R. Bizzo, I. de Almeida Rodrigues, R.M. Soares, T. Souto-Padron, C.S. Alviano, A.H. Lopes, Antileishmanial activity of a linalool-rich essential oil from *Croton cajucara*, Antimicrob. Agents Chemother. 47 (2003) 1895–1901.

[56] A.T. Peana, P. Rubattu, G.G. Piga, S. Fumagalli, G. Boatto, P. Pippia, M.G. De Montis, Involvement of adenosine A1 and A2A receptors in (−)-linalool-induced antinociception, Life Sci. 78 (2006) 2471–2474.

[57] J. Gertsch, M. Leonti, S. Raduner, I. Racz, J.Z. Chen, X.Q. Xie, K.H. Altmann, M. Karsak, A. Zimmer, Beta-caryophyllene is a dietary cannabinoid, Proc. Natl. Acad. Sci. U. S. A. 105 (2008) 9099–9104.

[58] L. Binet, P. Binet, M. Miocque, M. Roux, A. Bernier, Recherches sur les proprietes pharmcodynamiques (action sedative et action spasmolytique) de quelques alcools terpeniques aliphatiques, Ann. Pharm. Fr. 30 (1972) 611–616.

[59] H. Rodrigues Goulart, E.A. Kimura, V.J. Peres, A.S. Couto, F.A. Aquino Duarte, A.M. Katzin, Terpenes arrest parasite development and inhibit biosynthesis of isoprenoids in *Plasmodium falciparum*, Antimicrob. Agents Chemother. 48 (2004) 2502–2509.

[60] D.C. Arruda, F.L. D'Alexandri, A.M. Katzin, S.R. Uliana, Antileishmanial activity of the terpene nerolidol, Antimicrob. Agents Chemother. 49 (2005) 1679–1687.

[61] L.H. Vázquez, J. Palazon, A. Navarro-Ocaña, The pentacyclic triterpenes α-, β-amyrins: a review of sources and biological activities, Phytochemistry 23 (2012) 487–502.

[62] T. Moses, J. Pollier, J.M. Thevelein, A. Goossens, Bioengineering of plant (tri)terpenoids: from metabolic engineering of plants to synthetic biology in vivo and in vitro, New Phytol. 200 (2013) 27–43.

[63] G. Vanhoenacker, P. Van Rompaey, D. De Keukeleire, P. Sandra, Chemotaxonomic features associated with flavonoids of cannabinoid-free cannabis (*Cannabissativa* subsp. *sativa* L.) in relation to hops (*Humulus lupulus* L.), Nat. Prod. Lett. 16 (2002) 57–63.

[64] S.A. Ross, M.A. ElSohly, G.N. Sultana, Z. Mehmedic, C.F. Hossain, S. Chandra, Flavonoid glycosides and cannabinoids from the pollen of *Cannabis sativa* L, Phytochem. Anal. 16 (2005) 45–48.

[65] O. Werz, J. Seegers, A.M. Schaible, C. Weinigel, D. Barz, A. Koeberle, G. Allegrone, F. Pollastro, L. Zampieri, G. Grassi, G. Appendino, Cannflavins from hemp sprouts, a novel cannabinoid-free hemp food product, target microsomal prostaglandin E2 synthase-1 and 5-lipoxygenase, Pharm. Nutr. 2 (2014) 53–60.

[66] M.L. Barrett, D. Gordon, F.J. Evans, Isolation from *Cannabissativa* L. of cannflavin-a novel inhibitor of prostaglandin production, Biochem. Pharmacol. 34 (1985) 2019–2024.

[67] M.L. Barrett, A.M. Scutt, F.J. Evans, Cannflavin A and B, prenylated flavones from *Cannabissativa* L, Experientia 42 (1986) 452–453.

[68] C.A. MacCallum, E.B. Russo, Practical considerations in medical cannabis administration and dosing, Eur. J. Int. Med. 49 (2018) 12–19.

[69] C.L. Ramírez, M.A. Fanovich, M.S. Churio, Cannabinoids: extraction methods, analysis and physicochemical characterization, in: Atta-Ur-Rahman (Ed.), Studies in Natural Products Chemistry, Elsevier Science Pub, Amsterdam, 2019, pp. 143–173.

[70] K.C. Verhoeckx, H.A. Korthout, A.P. van Meeteren-Kreikamp, K.A. Ehlert, M. Wang, J. van der Greef, R.J. Rodenburg, R.F. Witkamp, Unheated *Cannabis sativa* extracts and its major compound THC acid have potential immunomodulating properties not mediated by CB1 and CB2 receptor coupled pathways, Int. Immunopharmacol. 6 (2006) 656–665.

[71] X. Nadal, C. del Río, S. Casano, B. Palomares, C. Ferreiro-Vera, C. Navarrete, C. Sánchez-Carnerero, I. Cantarero, M.L. Bellido, S. Meyer, G. Morello, G. Appendino, E. Muñoz, Tetrahydrocannabinolic acid is a potent PPARgamma agonist with neuroprotective activity, Br. J. Pharmacol. 23 (2017) 4263–4276.

[72] J.H. Martin, J. Schneider, C.J. Lucas, P. Galettis, Exogenous cannabinoid efficacy: merely a pharmacokinetic interaction? Clin. Pharmacokinet. 57 (2018) 539–545.

[73] K. O'Brien, Medicinal cannabis: issues of evidence, Eur. J. Int. Med. 28 (2019) 114–120.

[74] E.R. Garrett, C.A. Hunt, Physiochemical properties, solubility, and protein binding of Δ^9-tetrahydrocannabinol, J. Pharm. Sci. 63 (1974) 1056–1064.

[75] F. Grotenhermen, Pharmacokinetics and pharmacodynamics of cannabinoids, Clin. Pharmacokinet. 42 (2003) 327–360.

[76] R. Pacifici, E. Marchei, F. Salvatore, L. Guandalini, F.P. Busardò, S. Pichini, Evaluation of long-term stability of cannabinoids in standardized preparations of cannabis flowering tops and cannabis oil by ultra-high-performance liquid chromatography tandem mass spectrometry, Clin. Chem. Lab. Med. 56 (2018) 94–96.

[77] N. Bruni, C. Della Pepa, S. Oliaro-Bosso, E. Pessione, D. Gastaldi, F. Dosio, Cannabinoid delivery systems for pain and inflammation treatment, Molecules 23 (2018) 2478.

[78] H. Bigares Grangeia, C. Silva, S. Paulo Simões, M.S. Reis, Quality by design in pharmaceutical manufacturing: a systematic review of current status, challenges and future perspectives, Eur. J. Pharm. Biopharm. 147 (2020) 19–37.

[79] H. Koltai, P. Poulin, D. Namdar, Promoting cannabis products to pharmaceutical drugs, Eur. J. Pharm. Sci. 132 (2019) 118–120.

[80] G. Nahler, T.M. Jones, E.B. Russo, Cannabidiol and contributions of major hemp phytocompounds to the "entourage effect"; possible mechanisms, J. Altern. Complement. Integr. Med. 5 (2019) 70.

[81] Y. Landschaft, B. Albo, R. Mechoulam, A. Afek, Medical Grade Cannabis—Clinical Guide, The Israeli Medical Cannabis Agengy (IMCA) Office of the Associated Director General, Ministry of Health, 2017.

[82] W. Saingam, A. Sakunpak, Development and validation of reverse phase high performance liquid chromatography method for the determination of delta-9-tetrahydrocannabinol and cannabidiol in oromucosal spray from cannabis extract, Rev. Bras. Farm. 28 (2018) 669–672.

[83] J. Sinclair, C. Smith, J. Abbott, K.J. Chalmers, D.W. Pate, M. Armour, Cannabis use, a self-management strategy among Australian women with endometriosis, J. Obstet. Gynaecol. Can. 000 (2019) 1–6.

[84] W. Ngwa, R. Kumar, M. Moreau, M.R. Dabney, A. Herman, Nanoparticle drones to target lung cancer with radiosensitizers and cannabinoids, Front. Oncol. 7 (2017) 208.

Medicinal cannabis

[85] O. Hung, J. Zamecnik, P.N. Shek, P. Tikuisis, Pulmonary delivery of liposome-encapsulated cannabinoids. WO2001003668A1, 2001.

[86] M. Donsky, R. Winnicki, Terpene and cannabinoid liposome and micelle formulations. WO2015068052A2, 2015.

[87] G. Hommoss, S.M. Pyo, R.H. Müller, Mucoadhesive tetrahydrocannabinol-loaded NLC—formulation optimization and long-term physicochemical stability, Eur. J. Pharm. Biopharm. 117 (2017) 408–417.

[88] J. Atsmon, I. Cherniakov, D. Izgelov, A. Hoffman, A.J. Domb, L. Deutsch, F. Deutsch, D. Heffetz, H. Sacks, PTL401, a new formulation based on pro-nano dispersion technology, improves oral cannabinoids bioavailability in healthy volunteers, J. Pharm. Sci. 107 (2018) 1423–1429.

[89] I. Cherniakov, D. Izgelov, A.J. Domb, A. Hoffman, The effect of pro nano lipospheres (PNL) formulation containing natural absorption enhancers on the oral bioavailability of delta-9-tetrahydrocannabinol (THC) and cannabidiol (CBD) in a rat model, Eur. J. Pharm. Sci. 109 (2017) 21–30.

[90] L. Martín-Banderas, I. Muñoz-Rubio, J. Prados, J. Álvarez-Fuentes, J.M. Calderón-Montaño, M. López-Lázaro, J.L. Arias, M.C. Leiva, M.A. Holgado, M. Fernández-Arévalo, *In vitro* and *in vivo* evaluation of Δ^9-tetrahidrocannabinol/PLGA nanoparticles for cancer chemotherapy, Int. J. Pharm. 487 (2015) 205–212.

[91] D.H. Perez De La Ossa, A. Ligresti, M.E. Gil-Alegre, M.R. Aberturas, J. Molpeceres, V. Di Marzo, A.I. Torres Suárez, Poly-ε-caprolactone microspheres as a drug delivery system for cannabinoid administration: development, characterization and in vitro evaluation of their antitumoral efficacy, J. Control. Release 161 (2012) 927–932.

[92] G.R. Adelli, P. Bhagav, P. Taskar, T. Hingorani, S. Pettaway, W. Gul, M.A. Elsohly, M.A. Repka, S. Majumdar, Development of a Δ^9-tetrahydrocannabinol amino acid-dicarboxylate prodrug with improved ocular bioavailability, Investig. Ophthalmol. Vis. Sci. 58 (2017) 2167–2179.

[93] X. Ling, S. Zhang, Y. Liu, M. Bai, Light-activatable cannabinoid prodrug for combined and target-specific photodynamic and cannabinoid therapy, J. Biomed. Opt. 23 (2018) 108001.

[94] M.A. Huestis, J.E. Henningfield, E.J. Cone, Blood cannabinoids: I. Absorption of THC and formation of 11-OH-THC and THC-COOH during and after smoking marijuana, J. Anal. Toxicol. 16 (1992) 276–282.

[95] A. Ohlsson, J.E. Lindgren, A. Wahlen, L.E. Hollister, H.K. Gillespie, Plasma delta-9 tetrahydrocannabinol concentrations and clinical effects after oral and intravenous administration and smoking, Clin. Pharmacol. Ther. 28 (1980) 409–416.

[96] B. Law, P.A. Mason, A.C. Moffat, R.I. Gleadle, L.J. King, Forensic aspects of the metabolism and excretion of cannabinoids following oral ingestion of cannabis resin, J. Pharm. Pharmacol. 36 (1984) 289–294.

[97] E. Manno, B.R. Manno, P.M. Kemp, D.D. Alford, I.K. Abukhalaf, M.E. McWilliams, F.N. Hagaman, M.J. Fitzgerald, Temporal indication of marijuana use can be estimated from plasma and urine concentrations of Δ^9-tetrahydrocannabinol, 11-hydroxy-Δ^9-tetrahydrocannabinol, and 11-nor-Δ^9-tetrahydrocannabinol 9-carboxylic acid, J. Anal. Toxicol. 25 (2001) 538–549.

[98] R. Brenneisen, A. Egli, M.A. Elsohly, V. Henn, Y. Spiess, The effect of orally and rectally administered delta 9-tetrahydrocannabinol on spasticity: a pilot study with 2 patients, Int. J. Clin. Pharmacol. Ther. 34 (1996) 446–452.

[99] https://www.tga.gov.au/sites/default/files/auspar-nabiximols-130927.pdf.

[100] A.L. Stinchcomb, S. Valiveti, D.C. Hammell, D.R. Ramsey, Human skin permeation of Delta 8-tetrahydrocannabinol, cannabidiol and cannabinol, J. Pharm. Pharmacol. 56 (2004) 291–297.

[101] P.V. Challapalli, A.L. Stinchcomb, In vitro experiment optimization for measuring tetrahydrocannabinol skin permeation, Int. J. Pharm. 241 (2002) 329–339.

[102] M. Lodzki, B. Godin, L. Rakou, R. Mechoulam, R. Gallily, E. Touitou, Cannabidiol-transdermal delivery and anti-inflammatory effect in a murine model, J. Control. Release 92 (2003) 377–387.

[103] C.A. Hunt, R.T. Jones, Tolerance and disposition of tetrahydrocannabinol in man, J. Pharmacol. Exper. Ther. 215 (1980) 35–44.

[104] P. Kelly, R.T. Jones, Metabolism of tetrahydrocannabinol in frequent and infrequent marijuana users, Anal. Toxicol. 16 (1992) 228–235.

[105] C.G. Stott, L. White, S. Wright, D. Wilbraham, G.W. Guy, A phase I study to assess the single and multiple dose pharmacokinetics of THC/CBD oromucosal spray, Eur. J. Clin. Pharmacol. 69 (2013) 1135–1147.

[106] B.T. Ho, G.E. Fritchie, P.M. Kralikl, Distribution of tritiated-1 delta 9-tetrahydrocannabinol in rat tissues after inhalation, J. Pharm. Pharmacol. 22 (1970) 538–539.

[107] I. Ujváry, L. Hanuš, Human metabolites of cannabidiol: a review on their formation, biological activity, and relevance in therapy, Cann. Cannabinoid Res. 1 (2016) 90–101.

[108] P. Kovacic, S. Ratnasamy, Cannabinoids (CBD, CBDHQ and THC): metabolism, physiological effects, electron transfer, reactive oxygen species and medical use, Nat. Prod. J. 4 (2014) 47–53.

[109] K. Watanabe, S. Yamaori, T. Funahashi, T. Kimura, I. Yamamoto, Cytochrome P450 enzymes involved in the metabolism of tetrahydrocannabinols and cannabinol by human hepatic microsomes, Life Sci. 80 (2007) 1415–1419.

[110] D.J. Harvey, Metabolism and pharmacokinetics of the cannabinoids, in: R.R. Watson (Ed.), Biochemistry and Physiology of Substance Abuse, CRC Press, Boca Raton, 1991, pp. 279–365.

[111] G. Hawksworth, K. McArdle, Metabolism and pharmacokinetics of cannabinoids, in: G.W. Guy, B.A. Whittle, P.J. Robson (Eds.), The Medicinal Uses of Cannabis and Cannabinoids, Pharmaceutical Press, London, 2004, pp. 205–228.

[112] A. Mazur, C.F. Lichti, P.L. Prather, A.K. Zielinska, S.M. Bratton, A. Gallus-Zawada, M. Finel, G.P. Miller, A. Radomińska-Pandya, J.H. Moran, Characterization of human hepatic and extrahepatic UDPglucuronosyltransferase enzymes involved in the metabolism of classic cannabinoids, Drug Metab. Dispos. 37 (2009) 1496.

[113] M.M. Bergamaschi, A. Barnes, R.H.C. Queiroz, Y.L. HurdL, M.A. Huestis, Impact of enzymatic and alkaline hydrolysis on CBD concentration in urine, Anal. Bioanal. Chem. 405 (2013) 4679.

[114] M.M. Halldin, L.K. Andersson, M. Widman, L.E. Hollister, Further urinary metabolites of delta 1- tetrahydrocannabinol in man, Arztl. Forsch. 32 (1982) 1135–1143.

[115] M.M. Halldin, S. Carlsson, S.L. Kanter, M. Widman, S. Agurell, Urinary metabolites of delta 1-tetrahydrocannabinol in man, Arztl. Forsch. 32 (1982) 764–772.

[116] P.L. Williams, A.C. Moffat, Identification in human urine of delta 9-tetrahydrocannabinol-11-oic acid glucuronide: a tetrahydrocannabinol metabolite, J. Pharm. Pharmacol. 32 (1980) 445–453.

[117] M.E. Alburges, M.A. Peat, Profiles of delta 9-tetrahydrocannabinol metabolites in urine of marijuana users: preliminary observations by high performance liquid chromatography-radioimmunoassay, J. Forensic Sci. 31 (1986) 695–706.

[118] C.N. Chiang, R.S. Rapaka, Pharmacokinetics and disposition of cannabinoids, NIDA Res. Monogr. 79 (1987) 173–188.

[119] E.W. Gill, G. Jones, Brain levels of Δ1-tetrahydrocannabinol and its metabolites in mice: correlation with behaviour, and the effect of the metabolic inhibitors SKF 525A and piperonyl butoxide, Biochem. Pharmacol. 21 (1972) 2237–2248.

[120] M. Perez-Reyes, J. Simmons, D. Brine, G.L. Kimmel, K.H. Davis, M.E. Wall, Rate of penetration of Δ^9-tetrahydrocannabinol and 11-hydroxy-Δ^9-tetrahydro-cannabinol to the brain of mice, in: G.G. Nahas (Ed.), Marihuana: Chemistry, Biochemistry, and Cellular Effects, Springer, New York, 1976, pp. 179–187.

[121] M.E. Wall, B.M. Sadler, D. Brine, H. Taylor, M. Perez-Reyes, Metabolism, disposition, and kinetics of delta-9-tetrahydrocannabinol, in men and women, Clin. Pharmacol. Ther. 34 (1983) 352–363.

[122] F. Mikes, A. Hofmann, P.G. Waser, Identification of (−)-delta 9-6a,10a-trans-tetrahydrocannabinol and two of its metabolites in rats by use of combination gas chromatography-mass spectrometry and mass fragmentography, Biochem. Pharmacol. 20 (1971) 2469–2476.

[123] https://www.medicinalcannabis.nsw.gov.au/clinical-trials.

[124] https://www.anzctr.org.au/Trial ID: ACTRN12619001013156; ACTRN1261600 1036404; ACTRN12619001013156; ACTRN12614000622606.

[125] E.L. Karschner, D.M. Schwope, E.W. Schwilke, R.S. Goodwin, D.L. Kelly, D.A. Gorelick, M.A. Huestis, Predictive model accuracy in estimating last Δ^9-tetrahydrocannabinol (THC) intake from plasma and whole blood cannabinoid concentrations in chronic, daily cannabis smokers administered subchronic oral THC, Drug Alcohol Depend. 125 (2012) 313–322.

[126] B. De Backer, B. Debrus, P. Lebrun, L. Theunis, N. Dubois, L. Decock, A. Verstraete, P. Hubert, C. Charlier, Innovative development and validation of an HPLC/DAD method for the qualitative and quantitative determination of major cannabinoids in cannabis plant material, J. Chromatogr. B Analyt. Technol. Biomed. Life Sci. 877 (2009) 4115–4124.

[127] T. Nadulski, F. Sporkert, M. Schnelle, A.M. Stadelmann, P. Roser, T. Schefter, F. Pragst, Simultaneous and sensitive analysis of THC, 11-OH-THC, THC-COOH, CBD, and CBN by GC-MS in plasma after oral application of small doses of THC and cannabis extract, J. Anal. Toxicol. 29 (2005) 782–789.

[128] H. Teixeira, A. Verstraete, P. Proença, F. Corte-Real, P. Monsanto, D.N. Vieira, Validated method for the simultaneous determination of Delta9-THC and Delta9-THC-COOH in oral fluid, urine and whole blood using solid-phase extraction and liquid chromatography-mass spectrometry with electrospray ionization, Forensic Sci. Int. 170 (2007) 148–155.

[129] J.S. Macwan, I.A.I.A. Ionita, M. Dostalek, F. Akhlaghi, Development and validation of a sensitive, simple, and rapid method for simultaneous quantitation of atorvastatin and its acid and lactone metabolites by liquid chromatography-tandem mass spectrometry (LC-MS/MS), Anal. Bioanal Chem. 400 (2011) 423–433.

[130] R. Jamwal, A.R. Topletz, B. Ramratnam, F. Akhlaghia, Ultra-high performance liquid chromatography tandem mass-spectrometry for simple and simultaneous quantification of cannabinoids, J. Chromatogr. B Analyt. Technol. Biomed. Life Sci. 1048 (2017) 10–18.

[131] T. Joye, C. Widmer, B. Favrat, M. Augsburger, A. Thomas, Parallel reaction monitoring-based quantification of cannabinoids in whole blood, J. Anal. Toxicol. (2020) 113–135. https://doi.org/10.1093/jat/bkz113.

[132] K.A. Bertrand, N.J. Hanan, G. Honerkamp-Smith, B.M. Best, C.D. Chambers, Marijuana use by breastfeeding mothers and cannabinoid concentrations in breast milk, Pediatrics 142 (2018) 1076–1086.

[133] S. Pichini, G. Mannocchi, M. Gottardi, A.P. Pérez-Acevedo, L. Poyatos, E. Papaseit, C. Pérez-Mañá, M. Farré, R. Pacifici, F.P. Busardò, Fast and sensitive UHPLC-MS/MS analysis of cannabinoids and their acid precursors in pharmaceutical preparations of medical cannabis and their metabolites in conventional and non-conventional biological matrices of treated individual, Talanta 209 (2019) 120537.

[134] F.T. Peters, D.K. Wissenbach, F.P. Busardò, E. Marchei, S. Pichini, Method development in forensic toxicology, Curr. Pharm. Des. 23 (2017) 5455–5467.
[135] S.M.R. Wille, W. Coucke, T. De Baere, F.T. Peters, Update of standard practices for new method validation in forensic toxicology, Curr. Pharm. Des. 23 (2017) 5442–5454.
[136] T. Gray, M. Huestis, Bioanalytical procedures for monitoring in utero drug exposure, Anal. Bioanal. Chem. 388 (2007) 1455–1465.
[137] A. Concheiro-Guisan, M. Concheiro, Bioanalysis during pregnancy: recent advances and novel sampling strategies, Bioanalysis 6 (2014) 3133–3153.
[138] M. Concheiro, E. Lendoiro, A. de Castro, E. Gónzalez-Colmenero, A. Concheiro-Guisan, P. Peñas-Silva, M. Macias-Cortiña, A. Cruz-Landeira, M. López-Rivadulla, Bioanalysis for cocaine, opiates, methadone, and amphetamines exposure detection during pregnancy, Drug Test. Anal. 9 (2017) 898–904.
[139] D. Montgomery, C. Plate, S.C. Alder, M. Jones, J. Jones, R.D. Christensen, Testing for fetal exposure to illicit drugs using umbilical cord tissue vs meconium, J. Perinatol. 26 (2006) 11–14.
[140] J.M. Colby, Comparison of umbilical cord tissue and meconium for the confirmation of in utero drug exposure, Clin. Biochem. 50 (2017) 784–790.
[141] J. Kim, A. de Castro, E. Lendoiro, A. Cruz-Landeira, M. LópezRivadulla, M. Concheiro, Detection of in utero cannabis exposure by umbilical cord analysis, Drug Test Anal. 10 (2018) 636–643.
[142] Scientific working group for forensic toxicology, Scientific working group for forensic toxicology (SWGTOX) standard practices for method validation in forensic toxicology, J. Anal. Toxicol. 37 (2013) 452–474.
[143] A.M. Bermejo, M.J.S. Tabernero, Determination of drugs of abuse in hair, Bioanalysis 4 (2012) 2091–2094.
[144] M.A. LeBeau, M.A. Montgomery, J.D. Brewer, The role of variations in growth rate and sample collection on interpreting results of segmental analyses of hair, Forensic Sci. Int. 210 (2011) 110–116.
[145] P. Kintz, Segmental hair analysis can demonstrate external contamination in postmortem cases, Forensic Sci. Int. 215 (2012) 73–76.
[146] M. Barroso, E. Gallardo, D.N. Vieira, M. López-Rivadulla, J.A. Queiroz, Hair: a complementary source of bioanalytical information in forensic toxicology, Bioanalysis 3 (2011) 67–79.
[147] H. Khajuria, B.P. Nayak, Detection of Δ^9-tetrahydrocannabinol (THC) in hair using GC–MS, Egypt. J. Forensic Sci. 4 (2014) 17–20.
[148] A. Rodrigues, M. Yegles, N. Van Elsue, S. Schneider, Determination of cannabinoids in hair of CBD rich extracts consumers using gas chromatography with tandem mass spectrometry (GC/MS–MS), Forensic Sci. Int. 292 (2018) 163–166.
[149] J.Y. Kim, M. Kyo, Determination of 11-nor-Δ^9-tetrahydrocannabinol-9-carboxylic acid in hair using gas chromatography/tandem mass spectrometry in negative ion chemical ionization mode, Rapid Commun. Mass Spectrom. 21 (2007) 339–342.
[150] V. Cirimele, H. Sachs, P. Kintz, P.J. Mangin, Testing human hair for cannabis. III. Rapid screening procedure for the simultaneous identification of delta 9-tetrahydrocannabinol, cannabinol, and cannabidiol, Anal. Toxicol. 20 (1996) 13.
[151] K.M. Clauwaert, J.F. Van Bocxlaer, W.E. Lambert, A.P. De Leenheer, Segmental analysis for cocaine and metabolites by HPLC in hair of suspected drug overdose cases, Forensic Sci. Int. 110 (2000) 157–166.
[152] M. Míguez-Framil, J.Á. Cocho, M.J. Tabernero, A.M. Bermejo, A. Moreda-Piñeiro, P. Bermejo-Barrera, An improved method for the determination of Δ^9-tetrahydrocannabinol, cannabinol and cannabidiol in hair by liquid chromatography–tandem mass spectrometry, Microchem. J. 117 (2014) 7–17.

[153] L. Mercolini, R. Mandrioli, M. Protti, M. Conti, G. Serpelloni, M.A. Raggi, Monitoring of chronic cannabis abuse: an LC–MS/MS method for hair analysis, J. Pharm. Biomed. Anal. 76 (2013) 119–125.

[154] S. Dulaurent, J. Gaulier, L. Imbert, A. Morla, G. Lachâtre, Simultaneous determination of Δ^9-tetrahydrocannabinol, cannabidiol, cannabinol and 11-nor-Δ^9-tetrahydrocannabinol-9-carboxylic acid in hair using liquid chromatography–tandem mass spectrometry, Forensic Sci. Int. 236 (2014) 151–156.

[155] E. Beasley, S. Francese, T. Bassindale, Detection and mapping of cannabinoids in single hair samples through rapid derivatization and matrix-assisted laser desorption ionization mass spectrometry, Anal. Chem. 88 (2016) 10328–10334.

[156] W.F. Duvivier, T.A. Beek, E.J.M. Pennings, M.W.F. Nielen, Rapid analysis of Δ^9-tetrahydrocannabinol in hair using direct analysis in real time ambient ionization orbitrap mass spectrometry, Rapid Commun. Mass Spectrom. 28 (2014) 682–690.

CHAPTER THREE

Analysis of cannabinoids in plants, marijuana products and biological tissues

Markel San Nicolas[a,b], Aitor Villate[b], Mara Gallastegi[a], Oier Aizpurua-Olaizola[a], Maitane Olivares[b,c], Nestor Etxebarria[b,c], Aresatz Usobiaga[b,c,*]

[a]Dinafem Seeds (Pot Sistemak S.L.), San Sebastian, Spain
[b]Department of Analytical Chemistry, Faculty of Science and Technology, University of the Basque Country (UPV/EHU), Leioa, Spain
[c]Research Centre for Experimental Marine Biology and Biotechnology (PIE), University of the Basque Country (UPV/EHU), Plentzia, Spain
*Corresponding author: e-mail address: aresatz.usobiaga@ehu.eus

Contents

1. Introduction	65
2. Analysis of cannabinoids in plants	68
3. Analysis of cannabinoids in marijuana products	72
4. Analysis in biological fluids and tissues	78
4.1 Blood and plasma	92
4.2 Oral fluid	93
4.3 Urine	94
4.4 Hair	94
5. Conclusions	96
References	96

1. Introduction

The medical cannabis market is undoubtedly on the rise. It is estimated that by the year 2025 the medical cannabis sector will generate around 50 billion dollars [1]. The cannabis industry is undergoing a remarkable transition as it is going from being an industry that was once stigmatized, or even illegal, to gradually form part of the pharmaceutical industry and be included in governmental research programs. Nowadays it is quite common to find flowers, extracts, and purified cannabis-derived compounds in clinical trials.

However, the pharmaceutical industry demands high quality standards that are not easily found in the medical cannabis industry. More often than not, cannabis-derived products were prepared following homemade recipes, missing the need of standardized doses and overlooking the presence of contaminants. In parallel, the recreational market, which is slowly being legalized in more and more countries, needs to address local regulations and fulfil quality criteria regarding the proper labelling of the products and the safety profiles as required for other recreational drugs like tobacco or alcohol.

These regulatory frameworks are enforcing a strong demand for cannabis testing. These tests can vary from cannabinoid and terpene analysis to pesticides, heavy metals, mycotoxins, synthases, DNA, etc. Because of this lack of adequate and reliable standardized procedures and products, the potential evidences suggesting the benefits of cannabis for several illnesses can be questioned [2].

The need for a proper testing is somehow attributed to the complexity of the plant and the targeted bioactive products. More than 500 chemical compounds have been identified in the cannabis plant, with at least 113 cannabinoids [3]. In the plant, cannabinoids are synthesized in their acidic form, but when heated or in the presence of light, they lose their carboxyl group and break down to a neutral form. They can even experience other types of degradations and turn into other cannabinoids, as is the case of tetrahydrocannabinol (Δ^9-THC), which can degrade into cannabinol (CBN) and Δ^8-THC [4]. Additionally, each cannabinoid can be transformed into several metabolites when administered, each of them showing a specific biological activity and relevance in therapy [5].

In addition to this, the boom of the cannabis industry has gone hand in hand with the increase of private and public laboratories offering analysis services. However, clear systematic differences in the results are common among laboratories, as has been recently pointed in different testing facilities in Washington [6], regardless of the care to control confounding factors. These discrepancies between laboratories are likely caused by systematic differences in their testing methodologies, mainly related to the sample treatment and/or lack of sensitivity to detect minor cannabinoids. In addition, we cannot leave aside the issue of misleading analytical results, either to boost the real cannabinoid content of the product, or to reduce the Δ^9-THC amount to fulfil with local legal requirements. As a result, cannabis industry should implement quality standards to accredit testing facilities and to produce comparable results across labs, no matter which method is being used. In the same direction, development of reference materials for the analysis of cannabinoids in different samples (plants, tinctures, extracts, etc.) would be a great aid in this process.

In the particular case of cannabis, the route of administration and the posology of the different products are as important as the concentration of the active products. As mentioned before, cannabinoids are bio-synthetized in their acidic form, which they have specific effects on their own. However, these effects are different from the effects of their neutral counterparts. When cannabis is consumed via inhalation, either smoking or vaporizing, decarboxylation of acidic cannabinoids occurs and they are transformed into their neutral form automatically due to the applied heat. However, in the case of oils, creams or even edibles, that step must be prop-erly done by the manufacturer if the target compounds are the neutral ones [7]. On the one hand, using too low temperatures or not enough time will end up with a product that is not fully decarboxylated, and therefore it will not have the desired effects. On the other hand, too high temperatures or too long heating will evaporate the cannabinoids. Consequently, proper quality control analysis during the production lines is mandatory in every cannabis company that makes cannabis-derived products.

From a chemical point of view, cannabis plants are classified depending on tetrahydrocannabinol/cannabidiol (THC/CBD) ratio. If that ratio is above 10, then the plant belongs to chemotype I, when it lies between 0.3 and 3, then the plant is chemotype II, and if the ratio is less than 0.1, then the plant is chemotype III. The latter normally has a THC concentra-tion lower than 1%, and they focused the attention of the pharma industry due to their lack of psychoactivity [8]. There is also chemotype IV, with cannabigerol (CBG) as the main cannabinoid, and chemotype V, which is composed of fibre-type plants which contain almost no cannabinoids.

From the biological point of view, there are three species in cannabis: *Cannabis sativa*, *Cannabis indica* and *Cannabis ruderalis*. While *C. ruderalis* is described as a low–THC variety that triggered the "autoflowering" charac-teristics, *sativa* is often described as being uplifting and energetic, whereas *indica* is described as being relaxing and calming. However, this distinction between species effects is meaningless and even potentially dangerous in the context of an unregulated market. Nowadays, it is impossible to guess the biochemical content of a given plant based on its morphology. The degree of hybridization achieved in the last decades is so extended that only a biochemical assay can report what the actual composition of the plant is. As the effect caused by the plants and derived products is closely related to their biochemical content, it is essential that the cannabis industry allows complete and accurate cannabinoid and terpenoid profiles to be available [9]. This work is focused on showing a review of the most relevant works related to the analysis of cannabinoids in various matrices such as plants, manufactured products and biological fluids.

Finally, beyond the standardization of raw and derived products, the medical community requires sound clinical trials to asses all the potential effects that cannabinoids may offer. As a result, accurate analysis methods to assess cannabinoid contents in biological fluids and tissues are also demanded to implement the pharmacokinetic and pharmacodynamic assays and to understand the mechanisms of action. Last but not least, accurate toxicological analyses are as well mandatory, even more with the boom of the "cannabis-light" industry, where consuming cannabis it is not necessarily linked to showing psychoactive effects.

2. Analysis of cannabinoids in plants

Although cannabis is usually known because of its psychoactive effects, it actually hides a high therapeutic potential, which is being introduced slowly into the medical practice. Nowadays, the recreational cultivation and use of this plant is banned in most of the countries around the world, but the debate about the tolerance towards its medical use is slightly settling favourably given the increasing number of studies that are being carried out around this topic. In fact, in order to conduct health-related studies, numerous extraction and analysis procedures are being developed to assure accurate and precise analysis methods [10].

As shown in Table 1, extraction based on sonication is the most used method to extract cannabinoids from plant samples, such as flowers, leaves and seeds. In almost every case, methanol or ethanol is used as extraction solvent, but other co-solvents such as [3,12], acetonitrile [15] or hexane [16] are also used to modify the polarity of the extraction solvent. It is worth pointing that supercritical fluid extraction (SFE) is showing a growing interest as a suitable alternative [13], as it is regularly used by the cannabis industry for large-scale extractions. Specifically, supercritical CO_2 extraction, with ethanol as a modifier, offers some procedural advantages that are worth considering, such as low solvent consumption, avoidance of solvent toxicity and stability of thermo-labile compounds [17].

The analysis of cannabinoids is typically performed by chromatographic systems. Although liquid chromatography (LC) is often used, there is also

Table 1 Cannabinoids analysis in plants.

Sample	Compound	Sample amount	Pre-treatment	Extraction	Analysis	LOD	Apparent recovery (%)	RSD (%)	References
Hemp leaves and flowers (*Cannabis sativa L.*)	CBDA CBGA THCA	100 mg	–	(i) Sonication, EtOH: $CHCl_3$ (9:1, v/v), 1 mL, 15 min. (ii) Filtration (iii) Ethanol, 1:100 dilution	HPLC-DAD	0.1 μg/mL	n.a.	n.a.	[3]
Hemp (*Cannabis sativa L*)	CBD CBDA CBG CBGA	250 mg	Grinding	(i) Dynamic maceration at room temperature, EtOH (10 mL twice and 5 mL once), 45 min. (ii) Filtration (0.45 μm PTFE filter)	HPLC-DAD HPLC-ESI-MS/MS	0.5–0.8 μg/mL	74.17–90.84	0.60–4.46	[10]
Hemp (*Cannabis sativa L* and *Cannabis indica*)	CBC CBD CBDA CBDV CBG CBGA CBN Δ^8-THC Δ^9-THC THCA THCV	500 mg	Pulverization	(i) Sonication, EtOH, 30 min. (ii) Dilution with EtOH (1:20, v/v) (iii) Centrifugation, 10 min., 16,000*g*	HPLC-UV	7–205 ng/mL	88.40–113.30	1.27–9.02	[11]

Continued

Table 1 Cannabinoids analysis in plants.—cont'd

Sample	Compound	Sample amount	Pre-treatment	Extraction	Analysis	LOD	Apparent recovery (%)	RSD (%)	References
Hemp (*Cannabis sativa* L.)	CBC CBD CBDA CBG CBGA CBN Δ^8-THC Δ^9-THC THCA THCV	250 mg	Grinding	(i) Sonication, MeOH: CHCl$_3$ (9:1, v/v), 10 mL, 15 min. (ii) Centrifugation, 5 min., 1620g ($\times$2) (iii) Filtration (0.45 μm PTFE filter)	(i) Derivatization Methylation (300 μL diazomethane, vortex, 30s) or Sylilation (50 μL pyridine + 50 μL MSTFA-TMCS at 60 °C, 15 min) (ii) GC–MS	2.16–58.89 ng/mg	89.33–103.33	1.58–9.52	[12]
Hemp (*Cannabis sativa* L.)	CBD CBG CBN THC THCA THCV	50 mg	Cryo-milling, 660 rpm, 4 min	(i) Addition of 1 g sea sand (ii) SFE with CO$_2$ and EtOH 20%, 35 min., 10 °C, 100 bar, 1 mL/min	HPLC-MS/MS	0.02–0.2 ng/mL	n.a.	2–7	[13]
Hemp (*Cannabis sativa* L.)	CBC CBD CBDA CBDV CBG CBGA CBN Δ^8-THC Δ^9-THC Δ^9-THCA Δ^9-THCV	100 mg	Grinding, 20 min, 6000 rpm	(i) Ultrasonication, MeOH, 1 mL, 15 min. (ii) Centrifugation, 1 min., 14,500 rpm (iii) Filtration, 0.2 μm, filter	UHPLC-ESI-MS/MS	n.a.	n.a.	n.a.	[14]

Hemp flowering buds (*Cannabis sativa L.*)	CBD CBDA CBG CBGA CBN Δ^8-THC Δ^9-THC THCA-A THCV	50 mg	–	Sonication, AcN:MeOH (80:20, v/v), 2.5 mL (×2)	UHPSFC/PDA-MS	1.0–3.0 µg/mL	96.10–107.57	n.a.	[15]
Confiscated Cannabis material (flowers, leaves and stems)	CBD CBDA CBN THC THCA	500 mg	Drying at 40 °C Grinding	(i) Ultrasonication, MeOH:Hex (9:1, v/v), 10 mL, 20 min. (ii) Dilution (1:20)	HPLC-DAD	0.3–0.5 µg/mL	97.4–109.7	0.1–5.4	[16]

AcN, Acetonitrile; CBC, Cannabichromene; CBD, Cannabidiol; CBDA, Cannabidiolic acid; CBDV, Cannabidivarin; CBG, Cannabigerol; CBGA, Cannabigerolic acid; CBN, Cannabinol; CHCl$_3$, Trichloromethane; EtOH, Ethanol; GC–MS, Gas Chromatography-; Mass Spectrometry; Hex, n-Hexane; HPLC-DAD, Liquid Chromatography Coupled to Diode Array Detector; HPLC-MS/MS, Liquid Chromatography tandem Mass Spectrometry; HPLC-UV, Liquid Chromatography coupled to Ultraviolet Detector; MeOH, Methanol; MSTFA-TMCS, N-Methyl-N-trimethylsilyltrifluoroacetamide and Trimethylchlorosilane; SFE, Supercritical Fluid Extraction; THCA, Tetrahydrocannabidiolic acid; THCA-A, Δ^9-Tetrahydrocannabinolic acid A; THCV, Tetrahydrocannabivarin; UHPLC-MS/MS, Ultrahigh Performance Liquid Chromatography tandem Mass Spectrometry; UHPSFC/PDA-MS, Ultrahigh Performance Supercritical Fluid Chromatography coupled to Photodiode Array Detector and Mass Spectrometry; Δ^8-THC, Δ^8-Tetrahydrocannabinol; Δ^9-THC, Δ^9-Tetrahydrocannabinol; n.a., not available.

the possibility to use gas chromatography (GC) after the derivatization of the acidic compounds to avoid the decarboxylation of the carboxyl group of cannabinoids [12]. In both cases, detection via mass spectrometry (MS) offers the lowest limits of detection (see Table 1), being possible the identification of minor cannabinoids. Apart from MS detector, diode array detector (DAD) or ultraviolet detector (UV) are also used in the analysis of cannabinoids. All these detectors allow measures with high precision. Moreover, supercritical fluid chromatography (SFC) offers very suitable performances without the requirement of any derivatization step [15]. The studies about the applicability of SFC are still limited, as it is an analytical technique which has not been fully explored yet.

3. Analysis of cannabinoids in marijuana products

The therapeutic use of cannabinoids in some countries is pushing a wide variety of cannabis products available for consumption. Therapeutic products are mainly in the form of oil preparations, being olive, sesame, hemp and MCT oils (medium-chain triglyceride oils, derived mostly from coconut) the most common ones [18–21], but resins and cannabis extracts are also easy to find on markets [15,22]. A great variety of drinks and foods are commercially available in several countries [23]. In addition to those products, we should consider those produced at home in small scale but often without any control.

Since the legalization in Colorado of both therapeutic and recreational use of marijuana, the lack of control of the products marketed is generating serious problems especially in young people and adolescents [23,24]. Therefore, it is necessary to develop qualitative and quantitative analytical methods that allow the correct characterization of the marketed products.

A proper pretreatment of the sample is required to determine the cannabinoids in cannabis-derived products. Liquid extraction, either solid-liquid extraction (SLE) [15,22,25] or liquid-liquid extraction (LLE) [18,25] are the most widely used technique for that purpose (Table 2). In some cases, if the sample is a non-complex liquid, a simple dilution may be sufficient for sample preparation [21,30]. The extractions are often performed by sonication or agitation without the need for assisted extraction techniques. In order to get higher enrichment factors and cleaner extracts, Calvi and co-workers [19] used solid-phase micro-extraction (SPME) fibres in the headspace for the extraction of the compounds in macerated oils products. In the last year, the use of QuEChERS (Quick, Easy, Cheap, Effective, Rugged & Safe) is gaining interest for the extraction and cleaning of complex samples [32,33].

Table 2 Cannabinoids analysis in marijuana products.

Sample	Compound	Sample amount	Pre-treatment	Extraction	Analysis	LOD	Apparent recovery (%)	RSD (%)	References
Cannabis resin	Δ^9-THC CBD CBN	100 mg (quantification) or 140 mg (profiling)		(i) SLE, 9.9 mL BuAc +100.0 mL toluene, 2 h rotation (ii) Centrifugation, 3000 rpm, 10 min	GC-FID	n.a.	n.a.	n.a.	[22]
Cannabis extracts and hashish	CBD Δ^8-THC THCV Δ^9-THC CBN CBG THCA CBDA CBGA	50 mg		(i) SLE, 2.5 mL AcN: MeOH (80:20, v/v) (ii) Dilution	UHPSFC/PAD A: CO_2 B: isopropanol: AcN (80:20), 1% water	1.0–3.0 μg/mL	96.10–107.57	1.5–7.5	[15]
Cannabis extract	Δ^9-THC CBC, CBD, CBE, CBG, CBL, CBN, CBND, CBCA, CBCV, CBDA, CBDV, CBEA, CBGA, CBLA, CBNA, THCA, THCVA, THV, C^4-CBND, C^4-THCA, C^1-THCRA	30 mg	(i) Decarboxylation, EtOH (2 mL) (ii) MAE, 10 min at 140–170 °C (iii) Concentrated to dryness at 35 °C		UHPLC-MS	n.a.	n.a.	n.a.	[26]

Continued

Table 2 Cannabinoids analysis in marijuana products.—cont'd

Sample	Compound	Sample amount	Pre-treatment	Extraction	Analysis	LOD	Apparent recovery (%)	RSD (%)	References
Olive oil galenic preparation	CBD CBDA CBN THC THCA	50 μL	(i) Addition of cannabis to olive oil (1 g in 10 mL) (ii) Heating at 100 °C for 2 h	LLE, 100 μL isopropanol, vortex, 10 s	UHPLC-MS	5 ng/mg	n.a.	n.a.	[20]
Commercial hemp seed oil	CBDA CBG CBN CBDV THC THCA	100 μL		LLE, 400 μL of 2-propanol, vortex, 1 min	HPLC-UV and HPLC-MS	0.2 μg/mg	n.a.	n.a.	[7]
Olive Oil Preparations	CBD CBDA CBN THC THCA	50 mg	(i) Addition of cannabis to olive oil (5 g in 50 mL) (ii) Stirred for 40 min or 120 min	(i) LLE, 5 mL MeOH, vortex 1 min (×3) (ii) Centrifugation (1789g, 5 min) (iii) To 50 μL of the supernatant 50 μL of the IS solution is added (iv) Evaporated	GC–MS; GC–FID Derivatization: (i) 50 μL of BSTFA-1% TMCS and 50 μL of toluene (ii) vortexed and heated at 70 °C for 30 min	n.a.	n.a.	n.a.	[18]
Macerated oils	CBD CBDA CBG CBGA CBN THC THCA	100 mg	(i) 10 mL of isopropanol + 1 g/mL of IS (ii) 1:9 dilution (mobile phase)	HS-SPME for GC–MS: CAR-PDMS-DVB StableFlex fibre, 37°, 60 min	GC–MS LC-HRMS A: 0.1% HCOOH/H_2O B: ACN	0.01 ng/mL	87–107	4–15	[19]

Consumer products in the US	CBC CBD CBDA CBDV CBG CBGA CBN Δ^8-THC Δ^9-THC THCA THCV	0.03–3.0 g		(i) 0.25–30 mL of EtOH or AcN (ii) Sonicated for 15 min (iii) Filtrated 0.45 μm	GC–MS Derivatization: (i) 200 mL pyridine and 200 mL BSTFA (ii) 30 min and 80 °C	1.0 μg/g	58–111	n.a.	[27]
Consumer products in the US	CBD CBDA CBN Δ^9-THC THCA	0.03–3.0 g		(i) 0.25–30 mL EtOH (ii) Sonicated for 30 min (iii) Filtrated 0.45 μm	HPLC-DAD 66:34 AcN: 0.5% acetic acid	10 μg/g	84–105	0.6–13.0	[28]
Beverages	CBD CBDA CBGA CBN THC THCV THCA	1 mL		(i) 4.0 mL MeOH (ii) Vortexed for 1 min (iii) Centrifuged at 1700*g*, 4 °C for 15 min (iv) 1.0 mL is centrifuged at 9500*g*, 4 °C for 10 min	LC/ESI-QTRAP-MS/MS	0.6 μg/L	84.5–104.5	5.6–17.6	[25]

Continued

Table 2 Cannabinoids analysis in marijuana products.—cont'd

Sample	Compound	Sample amount	Pre-treatment	Extraction	Analysis	LOD	Apparent recovery (%)	RSD (%)	References
Food	THC, CBD, CBN, THCV, THCA, CBGA, CBDA	1 g		(i) SLE: 10 mL of MeOH/chloroform 9/1 (×2)	LC/ESI-QTRAP-MS/MS A: 0.1% HCOOH/H_2O B: 0.1% HCOOH/AcN	6.1 μg/kg	83.4–101.2	5.7–19.5	[25]
Oils, plant materials, and creams/ cosmetics	CBD, CBDA, THC, THCA			(i) extraction solvent + IS to make a 1 mg/mL solution (ii) Sonicated 10 min (iii) Centrifugated 11,000 rpm	HPLC-MS/MS A: 0.1% HCOOH/H_2O B: 0.1% HCOOH/AcN	0.195 ng/mL	94–112	8.1–12.4	[29]
Cannabis olive oil extracts	THC, CBD	40 μL		(i) Dilution, 960 μL of THF (ii) LLE, 50 μL of the mixture +950 μL of MeOH	HPLC-DAD ACN:H_2O 75:25 (5 mM of K2HPO4 pH 3.45)	n.a.	n.a.	10–19	[21]
Resin extracts	CBDA, CBD, CBN, THC, CBC, THCA			(i) Dilution 1:125 MeOH (ii) Sonication, 5 min	UHPLC-DAD A: 0.1% HCOOH/H_2O B: 0.1% HCOOH/ACN	0.02–1 μg/ mL	92–104	0.9–2.8	[30]
Honey	CBDA, CBGA, CBG, CBD, Δ^9-THC, THCA	20 g		(i) 100 mL H_2O, 15 min, 40°C (ii) LLE, 100 mL Hex (x3) (iii) To dryness (iv) 1 mL MeOH	HPLC-ESI-MS/MS A: 2.0 mM aqueous CH3COONH4 B: ACN	0.3 ng/g	n.a.	< 7	[31]

Matrix	Analytes	Amount	Extraction procedure	Technique				Ref.
Honey	CBDA, CBGA, CBG, CBD, Δ^9-THC, THCA	2 g	(i) 8 mL H_2O, agitation (ii) 10 mL ACN + IS (iii) QuEChERS (iv) Centrifuged at 4000 rpm for 5 min (v) To dryness (vi) 150 μL 2.0 mM aqueous CH3COONH4-ACN 50:50	HPLC-ESI-MS/MS A: 2.0 mM aqueous CH3COONH4 B: ACN	0.3 ng/g	n.a.	< 7	[31]
Nyaope street drug	THC, CBD, CBG	10 mg	(i) 3 mL of IS 0.02 mg/mL in t-BuOH	GC–MS	n.a.	n.a.	n.a.	[32]
Cannabis-infused chocolate	THC, CBN, THCA, THCV, CBG, CBD, CBC	1 g	(i) QuEChERS (ii) 15 mL of H_2O, sonicated for 20 min (iii) 15 mL of AcN (iv) Centrifuged 3000g, 5 min	TLC-DESI-MS	20 and 30 μg/mL	n.a.	< 15	[33]

AcN, Acetonitrile; BSTFA, N:O-Bis(Trimethylsilyl)trifluoroacetamide; tBuOH, *tert*-butanol; BuAc, Butyl Acetate; C1-THCRA; C1-tetrahydrocannabiorcolic acid; C4-THCA, C4-tetrahydrocannabinolic acid; CAR-PDMS-DVB, Divinylbenzene/Carboxen/Polydimethylsiloxane fibre; CBC, Cannabichromene; CBCA, Cannabichromenic acid; CBCV, Cannabichromevarin; CBD, Cannabidiol; CBDA, Cannabidiolic acid; CBEA, Cannabielsoic acid; CBDL, Cannabinodiol; CBDV, Cannabidivarin; CBE, Cannabielsoin; CBG, Cannabigerol; CBGA, Cannabigerolic acid; CBL, Cannabicyclol; CBLA, Cannabicyclolic acid; CBN, Cannabinol; CBNA, Cannabinolic acid; CBND, Cannabinodiol; dSPE, Dispersive Solid Phase Extraction; EtOH, Ethanol; GC-FID, Gas Chromatography-Flame Ionization Detector; GC–MS, Gas Chromatography-Mass Spectrometry; Hex, n-Hexane; HPLC-DAD, High Performance Liquid Chromatography-Diode Array Detector; HPLC-ESI-MS/MS, High Performance Liquid Chromatography—Electrospray Ionization tandem Mass Spectrometry; HPLC-UV, High Performance Liquid Chromatography-UltraViolet Detector; HS-SPME, Head Space-Solid Phase MicroExtraction; IS, Internal Standard; LC-Qtrap-MS/MS, Liquid Chromatography coupled to quadrupole-ion trap mass spectrometer; LLE, Liquid-Liquid Extraction; MAE, Microwave Assisted Extraction; MeOH, Methanol; QUECHERS, Quick; Easy; Cheap; Effective; Rugged & Safe; SLE, Solid-Liquid Extraction; THCA, tetrahydrocannabidiolic acid; THCVA, tetrahydrocannabivarinic acid; THF, Tetrahydrofuran; THV, tetrahydrocannabivarin; TLC-DESI-MS, Thin-Layer Chromatography-Desorption Electrospray Ionization- Mass Spectrometry; TMCS, Trimethylsilyl chloride; UHPLC-MS, Ultra High Performance Liquid Chromatography—Mass Spectrometry; UHPSFC/PAD, Ultra High Performance Liquid Chromatography—Pulse Amperometric Detector; Δ^9-THC, Δ^9-tetrahydrocannabinol; n.a., not available.

As pointed before, both LC and GC coupled to MS detectors are widely used for the analysis of the extracts when minor cannabinoids are of interest (see Table 2). HPLC-DAD is used in routine analysis in industry labs because its robustness [21,28,30]. Although GC coupled to flame ionization detector (GC-FID) is still found in the literature to determine high concentration levels of cannabinoids [22], GC–MS is the preferred technique. In any case, derivatization is required to analyse acidic compounds as well [18,27].

Consumption of synthetic cannabinoids is increasing for both therapeutic and recreational purposes [34–37]. These synthetic cannabinoids have very similar chemical structures to natural cannabinoids but their effects are different. Usually, stronger interactions with the endocannabinoid receptors are sought via minor changes in the chemical structure. Although not included in Table 2, the analytical methods for those compounds do not differ from those mentioned in the case of natural cannabinoids.

4. Analysis in biological fluids and tissues

The main psychoactive constituent of cannabis, Δ^9-THC, as well as other major cannabinoids such as CBD and CBN, are distributed into tissues and retained in adipose tissue after consumption [38]. After the administration of cannabinoids in human, Δ^9-THC is metabolized by cytochrome P-450 (CYP) into two major Phase I metabolites known as 11-nor-Δ^9-tetrahydrocannabinol-carboxylic acid (THC-COOH) and 11-hydroxy-Δ^9-tetrahydrocannabinol (OH-THC) [39]. The subsequent conjugation with glucuronic acid originates the Phase II metabolite Δ^9-THC-glucuronide. The cannabinoids Δ^9-THC, CBD, CBN and tetrahydrocannabivarin (THCV) and their metabolites (e.g. Δ^9-THC-glucuronide, OH-THC, THC-COOH, 11-nor-9-carboxy-THCV (THCV-COOH), etc.) are markers of relatively recent cannabis intake and enables the evaluation of occasional or frequent consumption [38,40,41].

Cannabinoids and metabolites can be extracted from several biological matrices (urine, plasma, blood, oral fluid, hair, etc.) together with other compounds (such as proteins, salts and phospholipids), and thus, sample preparation is the key step to get reliable results. Overall, it consists on sample homogenization, analyte extraction, clean–up of interferences and pre-concentration before analysis. Among several sample preparation methods, LLE is the most widely used method for the extraction of cannabinoids from blood, plasma and urine samples (see Table 3). In many cases, solid phase extraction (SPE) is the main alternative to LLE owing to the high

Table 3 Cannabinoids analysis in biological samples.

Sample	Compounds	Sample amount	Pretreatment	Extraction	Measurement	LOD	Apparent recovery (%)	RSD (%)	References
Plasma	THC THC-COOH 11-OH-THC CBC CBD CBN CBDV THCV CBL THCV-COOH CBG CBN-COOH CBDA THCA CBGA	0.5 mL		(i) LLE: 1 mL Hex:EtOAc (80:20: v:v) (ii) acidify the residue with 5 μL acetic acid (iii) LLE: 1 mL diethyl ether (iv) (iv) recombine both organic extracts	LC-Qtrap	n.a.	n.a.	n.a.	[41]
Blood	THC 11-OH-THC THC-COOH CBN CBD	0.5 g	(i) Dilute: with DI water (1:1) (ii) Protein precipitation: 2 mL ACN: vortex (iii) Dilute the final extract with DI water (1:3)	On-line SPE: C18 cartridges (100 mg) load: 5 mL pre-treated sample wash: 1 mL 0.1 M acetic acid: 1 mL ACN:water (2:3:v:v) elute: 1.4 mL ACN	Derivatization: 30 μL MSTFA and 20 μL dry EtOAc: 90°C: 17 min. GC–MS	0.2–2 ng/mL	89.7–102.7	9–20.5	[42]

Continued

Table 3 Cannabinoids analysis in biological samples.—cont'd

Sample	Compounds	Sample amount	Pretreatment	Extraction	Measurement	LOD	Apparent recovery (%)	RSD (%)	References
Plasma	THC 11-OH-THC THC-COOH	0.1 mL	(i) Dilute: with 0.1 M/0.1 M KH2PO4/NaOH buffer solution (pH 6.0) (1:50)	MIP-μ-SPE: 40 °C: 150 rpm: 12 min load: 5 mL pre-treated sample wash: 5 mL 0.1 M/0.1 M KH2PO4/NaOH buffer solution (pH 6.0) elute: 2 mL of MeOHl/HOAc (90:10: v:v): sonicate for 6 min.	UHPLC-QTrap	0.11–0.15 ng/mL	96–99	3–7	[43]
Plasma	THC 11-OH-THC THC-COOH	0.5 mL	(i) Collect the sample with a deviced containing sodium fluoride and potassium oxalate or potassium fluoride and potassium EDTA. (ii) Centrifugate	(i) LLE: 1 mL EtOAc:Hex (1:7: v:v) (ii) vortex (1 min) (iii) collect 0.75 mL supernatant: evaporate till dryness (iv) reconstitute in 40 μL and 110 μL of 5 mM ammonium formiate in water:MeOH (1:1: v/v)	LC-QqQ	0.25–2.5 ng/mL	n.a.	3.3–7	[44]

Plasma	THC 11-OH-THC THC-COOH	1 mL	(i) Dilute sample with 2 mL phosphate buffer (0.15 M: pH 6.0) (ii) Protein precipitation with 0.25 mL AcN	SPE: (strong AXS with a C-8/quaternary amine/silicamaterial: Chromabond® Drug II cartridges (100 mg)) load: 2 mL pre-treated sample wash: 6 mL Milli-Q water elute (THC: 11-OH-THC): (i) 1.6 mL acetone wash: 2 mL AcN/0.1 M acetic acid (30:70: v/v): 1 mL Hex and 1 mL Hex/EtOAc (70:30: v/v): 2 mL acetone elute (THC-COOH): acidic acetone (0.05 M acetic acid in 1.6 mL acetone)	Derivatization: 30 μL MSTFA: 70 °C: 20 min. GC–MS	0.15–2.0 ng/mL	78.7–92.9	0.8–1.2	[45]
Plasma	CBD THC abnCBD Δ^8-THC CBN CBDV CBDA 11-OH-THC THC-COOH	0.2 mL		LLE: 0.75 mL MTBE: vortex 10 min	UHPLC-QqQ	0.5 ng/mL	95.39–109.2	1.14–13.1	[46]

Continued

Table 3 Cannabinoids analysis in biological samples.—cont'd

Sample	Compounds	Sample amount	Pretreatment	Extraction	Measurement	LOD	Apparent recovery (%)	RSD (%)	References
Plasma	CBD THC	0.15 mL	Protein precipitation (0.6 mL cold AcN: 5 min: −20 °C)	(i) Add 0.6 mL Milli-Q water (ii) LLE: 3 mL Hex: vortex 5 min	LC-DAD	10 ng/mL (LOQ)	85–115	< 15	[47]
Plasma	CBD THC	0.05 mL	Protein precipitation (0.1 mL AcN)	LLE: 1 mL Hex: vortex 30 s	LC-QTrap	0.1 ng/mL (LOQ)	93–113	7.00–8.35	[48]
Plasma	THC 11-OH-THC THC-COOH CBD CBG THCV CBN THC-COO-glu THC-glu	1 mL	(i) 2 mL MeOH: centrifugate: evaporate to dryness. (ii) Enzymatic hydrolysis: 1 mL buffer and 0.25 mL of enzyme solution (20,000 U/mL of β-glucuronidase: in 0.1 M potassium phosphate buffer (pH 6.8)): 16 h: 37 °C. Add 80 μL 10 M NaOH: 60 °C: 20 min. Add 50 μL of acetic acid: adjust to pH 7. Add 2 mL of cold ACN: vortex: centrifugate (10 min) Dilute supernatant with 1 mL 2 M sodium acetate buffer (pH 4.0)	SPE: (Strata C18-E: 200 mg) load: whole pre-treated sample wash: 2 mL H_2O: 2 mL of 0.1 M HCl in 95:5 H_2O/AcN (95:5: v/v): 0.2 mL MeOH. elute: 2 mL Hex and 3 mL Hex:EtOAc (1:1: v/v)	LC-APCI-QqQ	0.2–0.5 ng/mL	86–111	1–10	[49]

Sample	Analyte	Amount	Extraction	Cleanup	Detection	LOD/LOQ	Recovery (%)	Precision	Ref.
Plasma	THC CBD 11-OH-THC THC-COOH	0.2 mL	Protein precipitation (1 mL AcN containing 1% formic acid)	(i) Evaporate (ii) Reconstitute in mobile phase	LC-QqQ	1.36–1.78 ng/mL (LOQ)	93.8–97.4	1.52–5.61	[50]
Oral Fluid	THC	0.2 mL oral fluid collected with Intercept Oral Fluid Drug Test kits (0.4 mL saliva/ 0.8 mL buffer)	10 μL trition-X and 70 μL ammonium formate buffer (5 mM: pH 9.3)	(i) SuLE with Isolute96 SLE+: eluted with 700 μL EtOAc/heptane (4:1: v/v) (ii) Repeat extraction (iii) Evaporation of solvents and reconstitution in 100 μL ACN/water (10:90)	LC-ESI-QqQ	0.189 ng/mL	97.3 (0.944 ng/mL)	10.7 (0.944 ng/mL)	[51]
Oral Fluid	THC-COOH CBD CBN	1.2 mL (saliva)		SPME (C18): conditioning: 1.5 mL MeOH/water (50/50: v/v): extraction: immersion: 75 min: 1200 rpm. Rinsed with nanopure water (10s): desorption: immersion in 1.2 mL of ACN/ MeOH/water/ formic acid (40/40/ 19.9/0.1): 60 min: 1200 rpm	LC-QqQ	0.07–0.09 ng/mL (LOQ)	10–13 (10 ng/mL)	12–17 (10 ng/mL)	[52]

Continued

Table 3 Cannabinoids analysis in biological samples.—cont'd

Sample	Compounds	Sample amount	Pretreatment	Extraction	Measurement	LOD	Apparent recovery (%)	RSD (%)	References
Oral Fluid	THC 11-OH-THC THC-COOH THCV CBD CBG	1 mL Oral fluid collected with Quantisal (1 mL Oral fluid/3 mL stabilizing buffer)	(i) 25 µL MeOH (ii) Hydrolysis: 0.3 mL Ammonium acetate buffer (1 M: pH 4) and 40 µL 15:625 U/mL beta glucuronidase solution (625 units): 55 °C: 60 min: vortex (iii) Acidified with 0.5 mL glacial acetic acid: vortex	(i) SPE: Strata X-C cartridges (3 mL/30 mg): conditioning: 3 mL MeOH: water: and 0.1% HCl: load: remaining sample: wash: 2 mL water and 2 mL 0.1% HCl:ACN (70:30: v/v): elution: 2 mL dichloromethane: isopropanol: ammonium hydroxide (78:20:2: v/v/v). (ii) Dry and reconstitution in 150 µL mobile phase A/B (70:30)	LC-APCI-QTrap	0.1–15 µg/L	101.7–105.9 (6 µg/L)	1.4 (6 µg/L)	[53]
Oral Fluid	THC	0.4 mL Oral fluid collected with Quantisal (1 mL Oral fluid/3 mL stabilizing buffer)		SPE	LC-ESI-QqQ	2 ng/mL	n.a.	n.a.	[54]

Oral Fluid	THC THC-COOH	0.2 mL	200 μL ammonium carbonate buffer (2 mM: pH 9.0): vortex	(i) LLE: 1 mL chloroform/ isopropanol (9:1): vortex (5 min: 1500 rpm): centrifugation (2 min: 15000 rpm): discard aqueous layer (ii) 10 μL HCl (0.1 M) in organic layer (iii) Dry and reconstitution in 50 μL MeOH	LC-ESI-QqQ	1 ng/mL (LOQ)	100.9–101.6 (25 ng/mL)	2.5–7.3 (25 ng/mL)	[55]
Oral Fluid	THC THC-COOH CBN CBD	0.5 mL (Oral fluid)		(i) MISPE ((+)-Catechin hydrate—Acrylamide— ethylene glycol dimethacrylate— molecular imprinted polymer): 72 h incubation: desorption: 2 mL MeOH: AcOH (4:1) (pH 2) sonicating: 15 min (ii) Dry and reconstitution in 50 μL MeOH and 50 μL of formic acid 0.1%: centrifugation (14,500 rpm: 10 min): collection of supernatant	LC-ESI-QqQ	0.5 ng/mL	94.1–104 (25 ng/mL)	4.11–6.91 (25 ng/mL)	[56]
Oral Fluid	THC	0.1 mL oral fluid buffer (DDS buffer)	80 μL Ammonia 8% solution (v/v)	(i) SuLE with Isolute SLE200+: eluted with 1000 μL MTBE (ii) Evaporation of solvents and reconstitution in 80 μL MeOH	LC-ESI-QTrap	0.9 ng/mL	101 (12.5 ng/mL)	2 (12.5 ng/mL)	[57]

Continued

Table 3 Cannabinoids analysis in biological samples.—cont'd

Sample	Compounds	Sample amount	Pretreatment	Extraction	Measurement	LOD	Apparent recovery (%)	RSD (%)	References
Urine	THC 11-OH-THC THC-COOH CBN CBD	1.0 g	(i) Enzymatic hydrolysis: 1 mL phosphate buffer (0.1 M: pH 6.8) and 25 μL β-glucuronidase (*E. coli*: 500 units): 16 h: 37 °C (ii) Protein precipitation: 2 mL ACN: vortex (iii) Dilute the final extract with DI water (1:3)	On-line SPE: C18 cartridges (100 mg) load: 5 mL pre-treated sample wash: 1 mL 0.1 M acetic acid: 1 mL ACN:water (2:3:v:v) elute: 1.4 mL ACN	Derivatization: 30 μL MSTFA and 20 μL dry EtOAc: 90 °C: 17 min. GC–MS	0.2–2 ng/L	89.7–102.7	9–20.5	[42]
Urine	THC 11-OH-THC THC-COOH	0.1 mL	(i) dilute: with 0.1 M/0.1 M KH_2PO_4/NaOH buffer solution (pH 6.0) (1:50)	MIP-μ-SPE: 40 °C: 150 rpm: 12 min load: 5 mL pre-treated sample wash: 5 mL 0.1 M/0.1 M KH2PO4/NaOH buffer solution (pH 6.0) elute: 2 mL of MeOH/Acetic Acid (90:10:v:v): sonicate for 6 min	UHPLC-QTrap	0.14–0.17 ng/mL	92–94	4–5	[43]

Urine	THC	1 mL	(i) Enzymatic hydrolysis: 1 mL buffer and 0.25 mL of enzyme solution (20,000 U/mL of β-glucuronidase: in 0.1 M potassium phosphate buffer (pH 6.8)): 16 h: 37 °C. Add 80 µL 10 M NaOH: 60 °C: 20 min. Add 50 µL of acetic acid: adjust to pH 7. Add 2 mL of cold AcN, vortex: (10 min) Dilute supernatant with 1 mL 2 M sodium acetate buffer (pH 4.0)	SPE: (Strata C18-E: 200 mg) load: whole pre-treated sample wash: 2 mL H$_2$O: 2 mL of 0.1 M HCl in 95:5 H$_2$O/AcN (95:5: v/v): 0.2 mL MeOH. elute: 2 mL Hex and 3 mL Hex:EtOAc (1:1: v/v)	LC-APCI-QqQ	0.2–0.5 ng/mL	84–114	1–7	[49]
	11-OH-THC								
	THC-COOH								
	CBD								
	CBG								
	THCV								
	CBN								
	THC-COO-glu								
	THC-glu								
Urine	THC-COOH	5 mL	(i) Alkaline hydrolysis: NaOH (1 N: 300 µL): 56 °C: 30 min. (ii) Acidify solutions with 1 mL HCl 1 M	LLE: 5 mL EtOAc:Hex (1:4:v/v): vortex	Derivatization: 30 µL BSTFA and 50 µL dry EtOAc: 90 °C: 15 min. GC–MS	1 ng/mL	92–113.4	3.9–6.2	[58]
	CBD								
	CBN								

Continued

Table 3 Cannabinoids analysis in biological samples.—cont'd

Sample	Compounds	Sample amount	Pretreatment	Extraction	Measurement	LOD	Apparent recovery (%)	RSD (%)	References
Urine	THC 11-OH-THC THC-COOH CBD CBN CBG THCV THCV-COOH	0.2 mL	(i) Enzymatic hydrolysis: 40 μL β-glucuronidase (2000 IU enzyme): 2 M pH 6.8 sodium phosphate buffer: 37 °C: 16 h. (ii) Protein precipitation: 620 μL AcN: centrifuge 4 °C: 10 min	(i) Add 200 μL 5% aqueous formic acid to supernatant (550 μL) (ii) dSPE (1 mL WAX-S tips)	LC-QTrap	1–2 ng/mL	86.1–109.7	6.9–10	[59]
Urine	THC THC-COOH CBD CBN	0.5 mL		MISPE: 72 h incubation desorption: 2 mL of MeOH:AcOH (4:1: v/v) (pH 2): sonicate: 15 min	LC -QqQ	0.75 ng/mL	94.1–103	3.51–6.69	[56]
Hair	THC CBD CBN 11-OH-THC	50 mg	Washing with dichloromethane	(i) Alkaline Hydrolysis: NaOH (1 M): 90 °C: 20 min: 1 mL (ii) LLE: Hex/EtOAc (9:1: v/v) 4 mL: 10 min (3500 rpm)	Derivatization: MSTFA: GC-EI-QqQ	0.03–1.4 pg/mg (1 ng/mg)	95.40–103.39 (1 ng/mg)	2.85–6.97 (1 ng/mg)	[60]
Hair	THC CBD CBN	20 mg	Washing with water and acetone:	(i) Alkaline Hydrolysis: NaOH (1 N): 60 °C: 90 min: 2 mL (digestion)	Derivatization: MSTFA: GC-EI-QqQ	1 pg/mg	1–2.4	1.1–2.5	[61]

Matrix	Analytes	Amount	Washing	Extraction	Instrument	LOD/LOQ	Recovery	Precision	Ref.
			Pulverization (30 Hz: 2 min)	(ii) SPE: Strata-X (33 µm) cartridges: conditioning: 3 mL water and 3 mL MeOH: wash: 3 mL NaOH (0.1 N) and 70 µL acetone: elution: 2 mL dichloromethane					
Hair	THC	30 mg	Washing with tween 80 (0:1% v/v in water): water and acetone	SLE: MeOH 0.3 mL: overnight sonication: collection of the supernatant 100 µL	LC-HESI-qOrbitrap	30 pg/mg	n.a.	n.a.	[62]
Hair	THC CBD CBN 11-OH-THC THC-COOH	25 mg (hair powder)	(i) Washing with dichloromethane (ii) Ground to powder	(i) Alkaline Hydrolysis: NaOH (1 M): 90°C: 15 min: 1 mL (shaken 6.7 Hz) (ii) SPE HR-XA (100 mg: 45 µm: 1 mL): conditioning: 2 mL water and 2 mL MeOH: load sample: wash: 2.5 mL water and 1.6 mL AcN: elution: 1.2 mL isohexane/ ethyl acetate/acetic acid (80/20/5: v/v/v) (0.8 mL collected)	Derivatization: MSTFA: GC-EI-QqQ	0.2–2 pg/mg	100.1–109.5 (8 pg/mg)	2.3–5.5 (0.8 pg/mg)	[63]
Hair	THC CBD CBN	>10 mg	Washing with MeOH	(i) SLE: MeOH: >4 h in ultrasonic bath (ii) Digestion with NaOH	Derivatization: BSTFA: GC-EI-QqQ	0.01–0.91 pg/mg	n.a.	18–24	[64]

Continued

Table 3 Cannabinoids analysis in biological samples.—cont'd

Sample	Compounds	Sample amount	Pretreatment	Extraction	Measurement	LOD	Apparent recovery (%)	RSD (%)	References
	11-OH-THC THC-COOH			and LLE with chloroform and propan-2-ol (iii) Combination of both extractions (iv) SPE: Isolute HCX cartridges: elution: hexane/ethyl acetate (80/20)					
Hair	11-OH-THC	50 mg	Washing with petroleum benzene	UAE: 3 mL MeOH: 3 h: 55 °C	Derivatization: 2-picolinic acid: LC-ESI-QTrap	0.016 pg/mg	n.a.	n.a.	[65]
Hair	THC CBD CBN THC-COOH	50 mg	(i) Washing with Phosphate buffer (0:1 M: pH 6): isopropanol and dichloromethane: (ii) Homogenization with diatomaceous earth	(i) PLE: water-MeOH (90:10: v/v): 5–6 mL (ii) Centrifugation (6000g: 5 min) (iii) SPE Strata XL (30 mg/1 mL) cartridges: conditioning: 1 mL MeOH 1 mL water/MeOH (90:10 v/v): load 5 mL previous extract: wash: 5 mL water/MeOH (90:10 v/v) and 3 mL water/MeOH (50:50 v/v): elution: 1 mL MeOH	LC-HESI-QTrap	0.03–0.8 pg/mg	90–106 (250 pg/mg)	4–6 (250 pg/mg)	[66]

Hair	THC CBD CBN	20–30 mg	Washing with water and acetone	(i) SLE: MeOH/AcN/ 2 mM of ammonium formate (25:25:50: v/v/v): 0.5 mL: 18 h (shaking: 37° C) (ii) Separate supernatant and repeat SLE (iii) Combination of both extractions	LC-ESI-QqQ	3–10 pg/mg (LOQ)	n.a.	n.a.	[67]
Hair	THC CBD CBN THC-COOH 11-OH-THC THC	< 5 cm	Washing with MeOH		Derivatization: n.a. FMPTS: Coat samples with a spray of CHCA 5 mg/mL in 70:30 ACN/ 0.2% aqueous TFA: MALDI-QTOF		n.a.	n.a.	[68]

abnCBD, Abnormal cannabidiol; AcN, Acetonitrile; APCI, Atmospheric Pressure Chemical Ionization; BSTFA, N:O-Bis(trimethylsilyl)trifluoroacetamide; CBC, cannabichromene; CBD, Cannabidiol; CBDA, Cannabidiolic acid; CBDV, Cannabidivarin; CBG, Cannabigerol; CBGA, Cannbigerolic acid; CBL, Cannabiciclol; CBN, Cannabinol; CBN-COOH, 11-nor-cannabinol-9-COOH; CHCA, α-Cyano-4-hydroxycinnamic acid; dSPE, Dispersive Solid Phase Extraction; ESI, Electrospray Ionization; EtOAc, Ethyl Acetate; FMPTS, 2-fluoro-1-methylpyridinium p-toluenesulfonate; GC–MS, Gas chromatography mass spectrometry; HESI, Heated Electrospray Ionization; Hex, n-Hexane; LC-DAD, Liquid Chromatography coupled to Diode Array Detector; LC-QqQ, Liquid Chromatography triple Quadrupole Mass Spectrometry; LLE, Liquid-Liquid Extraction; MALDI, Matrix-Assisted Laser Desorption/Ionization; MIP-μ-SPE, Molecularly Imprinted Polymer—micro-Solid Phase Extraction; MISPE, Molecularly Imprinted Solid Phase Extraction; MSTFA, N-methyl-N-(trimethylsilyl) trifluoroacetamide; MTBE, Methyl tertbutyl ether; PLE, Pressurized Liquid Extraction; SLE, Solid-Liquid Extraction; SPE, Solid Phase Extraction; SPME, Solid Phase MicroExtraction; SuLE, Supported Liquid Extraction; TFA, Trifluoroacetic acid; THC, Tetrahydrocannabinol; THCA, Tetrahydrocannabinolic acid; THC-COO-glu, 11-nor-carboxy-THC-glucuronide; THC-COOH, 11-nor-carboxy-tetrahydrocannabinol; THC-glu, THC-glucuronide; THCV, Tetrahydrocannabivarin; THCV-COOH, UAE, Ultrasonification Assisted Extraction; UHPLC-qTrap, Ultra High Performance Liquid Chromatography—Quadrupole Ion Trap Mass Spectrometer; 11-OH-THC, 11-hydroxitetrahydrocannabinol; n.a., not available.

extraction efficiency, enrichment factors and clean-up capacities that it offers. Besides those techniques, in order to gain extraction efficiency, reduce sample preparation times and ease miniaturization and automatization of sample preparation, modern micro-extraction techniques such as SPME and liquid-phase micro-extraction (LPME) are also gaining interest in the analysis of cannabinoids in biological tissues [40,69].

Regarding detection, rapid screening tests, frequently based in immuno-assays, are sensitive and inexpensive but they lack high selectivity and may be subject to a false positive result due to cross-reactivity with other non-targeted drugs of similar chemical structure. Thus, recent literature is consistent in the recommendation to use LC and GC coupled to MS as confirmatory techniques to obtain unequivocal analytical results (see Table 3). LC-MS is gradually replacing GC–MS in both screening and targeted procedures for the detection of cannabinoids in biological specimens, because LC does not require the laborious and cumbersome derivatization methods and because it allows the simultaneous determination of the parent drug and most of the metabolites [70]. In fact, LC-MS/MS enables simultaneous quantification of free and conjugated (glucuronides or sulphates) analytes at low concentration levels in a single assay [38,40], whereas LC coupled to high-resolution mass spectrometry (LC-qOrbitrap, LC-qTOF, etc.) can offer high sensitivity and specificity to identify informative cannabinoid markers [71]. The different analytical approaches used in the last 5 years for the screening of cannabinoids and their metabolites in biological specimens are compiled in Table 3 and summarized herein.

4.1 Blood and plasma

Blood or plasma together with urine are the first choices of testing-samples in clinical analysis of cannabinoids. Since Δ^9-THC blood concentrations decrease rapidly as it is distributed into tissues, the study of Δ^9-THC absorption and elimination profiles can help to correlate the concentrations with the observed effects [38]. Cannabinoid markers in blood and plasma should be analysed in the first hours or in 24 h, in the case of occasional and frequent users, respectively [38]. Indeed, some recent works recommend performing the analyses in less than 4 h to detect Δ^9-THC-glucuronide, CBD or CBN in blood of chronic cannabis users [72]. THCV-COOH can be detected after several days after last consumption [73] and Δ^9-THC could be found in blood for as long as 30 days after last use [38], being sampling collection and storage the key steps [44].

Analyses are often performed using plasma, serum or blood (0.01–0.5 mL) after centrifugation and protein precipitation using cold acetonitrile [42,47–50]. Some works use also LLE [46,74] and the dried blood spot approach [75] to treat whole sample. The complexity of plasma/blood and the low levels of cannabinoids and metabolites expected in such specimens makes necessary a pre-concentration and clean-up step by LLE or SPE (see Table 3), involving several commercial reversed phase [49] or strong anion exchanger sorbents [45]. Among new pre-concentration approaches proposed, μ-SPE devices or methods based on molecularly imprinted solid phase extraction (MISPE) showed remarkable selectivity for Δ^9-THC and its major metabolites [43]. In the last years, the detection of target compounds is markedly performed by either LC-QqQ or LC-HRMS approaches (see Table 3), since LC based methods allow for the direct quantification of free and glucuronide-conjugated compounds, improving the interpretation of the results. However, GC–MS after the derivatization of target compounds with silylation reagents such as N-trimethylsilyl-N-methyl trifluoroacetamide (MSTFA) is still used.

4.2 Oral fluid

Oral fluid (OF) is a good testing-sample because it can be easily collected using non-invasive commercial devices (e.g. Salivette® or Drugwipe®). Similar to plasma, OF testing allows for a fast screening analysis of cannabinoids consumption within the first few hours, although there are some commercial sampling devices that allow the determination of cannabinoids after 1–3 months storage [76]. Contrary to urine, OF contains tiny amounts of biotransformation metabolites and, hence, it offers the possibility to measure the free fraction of the cannabinoids [52]. The concentration of cannabinoids detected in OF can be correlated with the ones found in plasma [77]. Indeed, detecting 2 ng/mL Δ^9-THC in OF is considered to be a marker of cannabis intake within the past 24 h, even in chronic frequent smokers [38]. Although OF is almost free from interfering substances and the protein content remains relatively lower in comparison to other physiological fluids, the presence of mucins and food residues [78] makes necessary a pre-concentration and clean-up step prior to cannabinoids' analysis. SPE [54,76], LLE [55], SPME [52], MISPE [56] and automatic supported liquid extraction [51,57] are proposed in the literature as pretreatment methods for cannabinoids' analysis in OF. Regarding detection, although procedures based on GC–MS and LC-MS/MS can be used for the detection of cannabinoids in OF, those based on LC–MS systems are markedly used in recent works (see Table 3). Indeed, most of the efforts to enhance sensitivity of the target compounds present at low concentration are being done for LC–MS applications such as the formation of picolinic acid esters of 11–OH–THC to improve its detection by LC multi stage mass spectrometry (LC–MS^3) [65].

4.3 Urine

As mentioned before, urine is, together with plasma, the first choice of sample for cannabinoids' analysis and is the most commonly sampled biological specimen to screen recent exposure. Δ^9-THC and its metabolites, THC-COOH and THC-COOH glucuronide, are the cannabinoids commonly detected in urine of cannabis consumers [79]. Analytical methods often include enzymatic hydrolysis or tandem alkaline and enzymatic hydrolysis step to liberate free THC-COOH [49]. Sempio and co-workers [59], for example, proposed a fast and automated single-step enzymatic hydrolysis based on an automated supported liquid extraction [43] method for the simultaneous quantification of the cannabinoids (i.e. Δ^9-THC, CBD, CBN, CBG, and THCV) and their metabolites (i.e. 11-OH-THC, THC-COOH, and THCV-COOH). Similarly to what happens with plasma samples, besides LLE using n–hexane/ethyl acetate mixture [58] or SPE with nonpolar sorbent material (e.g. C18) [42,49,80], SPE with anion exchange sorbent received increasing attention for pre-concentration and clean-up purposes. The highly selective strategy based on molecularly imprinted polymers [43,56] or the non-selective "dilute and shoot" after protein precipitation [54,81] strategies are also used in some works. Taking profit of the advantages of LC–MS techniques to determine the entire target compounds simultaneously, most of the works use either LC-QqQ or LC-HRMS both using electrospray (ESI) or atmospheric pressure chemical ionization (APCI) sources to quantify cannabinoids in urine samples (see Table 3).

4.4 Hair

In many works, hair is presented as an alternative specimen for cannabinoids analysis. It is sampled using non-invasive methods and it has a longer time window (weeks to months) for the detection of cannabinoids in comparison to typical biological specimens (blood and urine) [82]. The two main cannabinoids detected in hair are Δ^9-THC and THC-COOH, but other cannabinoids such as CBD, CBN and OH-THC are screened in recent works [60]. The segmental analysis of hair for THC-COOH, a biomarker of cannabinoids consumption that is found at low levels (pg/mg) in hair, could provide information over the dose and period of intake of cannabis [82]; whereas the analysis of Δ^9-tetrahydrocannabinolic acid A (THCA-A), a biogenic non- psychoactive precursor of Δ^9-THC, provides information over the external cannabis exposure [69].

Generally, hair (20–200 mg) for cannabinoids analysis is sampled from the nuchal area of the posterior vertex of the head, cutting as close as possible to the root. In order to exclude any external contamination, the hair is cleaned with one or more washing steps using non-protic (i.e. dichloromethane) and protic (methanol, isopropanol) solvents. Afterwards, the sample is often cut into fragments of 1–3 mm long or ground in a cryogenic mill in order to enhance the efficiency of the extraction. Extraction of the cannabinoids from hair is performed with methanol [62,64,65,67,74], aqueous NaOH [60,61,63] or ionic liquids [83], assisted by ultrasound [62,64,65] or microwave energy and pressurized liquid extraction [66], which degrades the keratinized structure of the hair enhancing the extraction recovery. In order to avoid matrix effect in the detection step, LLE using n-hexane/ethyl acetate mixture [60] or SPE using reverse-phase cartridges [61], strong anion exchanger cartridges [63] or mixed-mode cartridges [64], which produce the cleanest extracts, is performed before cannabinoid detection. Although GC–MS or GC–MS/MS [60,61,63,64] are the techniques of choice for Δ^9-THC, CBD and CBN detection in hair, alternative analytical approaches based on LC–MS/MS instrumentation have been proposed recently [62,65–67]. Remarkably low LODs and LOQs (<2 and 10 pg/mg, respectively, for both analytes) are reported in most of the recent works for the detection of cannabinoids using 20 mg of hair (see Table 3), but the determination of THC-COOH and OH-THC in hair is still challenging due to their low concentration levels. Indeed, the LODs reported are still not satisfactory compared with the cutoff value of 0.2 pg/mg suggested by the Society of Hair Testing for THC-COOH [84]. Cannabinoids are often screened in hair samples together with other drugs and, thus, as pointed out for the rest of biological specimens, the use of chromatographic techniques coupled to HR-MS detectors is gaining importance in recent works [62,74].

5. Conclusions

It is commonly recognized that the general attitude of the laypersons and the regulators against marijuana has become more tolerant and somehow this has triggered not only its extended use as a recreational drug but also a more cautious and scientifically sound analysis of their potential benefits in clinical research or the health risks of the early or excessive consumption. Consequently, the paradigm under which the analysis of cannabinoids is running has evolved in the last years, from the methods to uncover illicit drug trafficking or the screening methods to rapidly identify potential consumers to more elaborate, well-designed and fully validated procedures to support clinical research.

This normalized view of marijuana and cannabinoids is extended from the crops to all the by-products and this has enforced the progressive implementation of technical and scientific standards. Nowadays many crops are being developed from a care selection of seeds or clones to assure a controlled production of certain cannabinoids and terpenes. These compounds can be mixed up according to harmonized procedures to produce a variety of products of pharmaceutical interest that should be tested according to standard procedures to ensure their benefits and therapeutic values. In parallel, the recreational market must fulfil some quality requirements, which can only be achieved via proper testing of the crops and developed products.

In this context, it is worth considering the introduction of the high throughput analytical instrumentation and procedures to bridge many of the existing gaps. Since there are still many constraints to achieve a full knowledge of the effects of the recreational consumption or the benefits of its medical use, we can foresee a steady growth of the analytical tools to satisfy many of these needs. Some of the procedures included in this work can provide not only the state of the art, but also the possibility to be a guide for newer and more reliable methods.

References

[1] Bloomberg Businessweek, Pot of Gold?, Retrieved from Bloomberg database. https://www.bloomberg.com/magazine/businessweek/18_43, 2018. Last access January 2020.

[2] P.F. Whiting, et al., Cannabinoids for medical use: a systematic review and meta-analysis. JAMA 313 (24) (2015) 2456–2473, https://doi.org/10.1001/jama.2015.6358.

[3] O. Aizpurua-Olaizola, et al., Evolution of the cannabinoid and terpene content during the growth of Cannabis sativa plants from different Chemotypes. J. Nat. Prod. 79 (2) (2016) 324–331, https://doi.org/10.1021/acs.jnatprod.5b00949.

[4] R. Mechoulam, Marihuana chemistry. Science 168 (3936) (1970) 1159–1166, https://doi.org/10.1126/science.168.3936.1159.

[5] I. Ujváry, L. Hanuš, Human metabolites of Cannabidiol: a review on their formation, biological activity, and relevance in therapy. Cannabis Cannabinoid Res. 1 (1) (2016) 90–101, https://doi.org/10.1089/can.2015.0012.

[6] N. Jikomes, M. Zoorob, The cannabinoid content of legal cannabis in Washington State varies systematically across testing facilities and popular consumer products. Sci. Rep. 8 (2018) 4519–4534, https://doi.org/10.1038/s41598-018-22755-2.

[7] C. Citti, B. Pacchetti, M.A. Vandelli, F. Forni, G. Cannazza, Analysis of cannabinoids in commercial hemp seed oil and decarboxylation kinetics studies of cannabidiolic acid (CBDA). J. Pharm. Biomed. Anal. 149 (2018) 532–540, https://doi.org/10.1016/j.jpba.2017.11.044.

[8] Y. Cohen, Cannabis Plant Named "Avidekel" Patent and trademark office, 2014, US Patent No. 20140259228A1.

[9] D. Piomelli, E.B. Russo, The Cannabis sativa versus Cannabis indica debate: an interview with Ethan Russo, MD. Cannabis Cannabinoid Res. 1 (1) (2016) 44–46, https://doi.org/10.1089/can.2015.29003.ebr.

[10] V. Brighenti, F. Pellati, M. Steinbach, D. Maran, S. Benvenuti, Development of a new extraction technique and HPLC method for the analysis of non-psychoactive cannabinoids in fibre-type *Cannabis sativa* L. (hemp). J. Pharm. Biomed. Anal. 143 (2017) 228–236, https://doi.org/10.1016/j.jpba.2017.05.049.

[11] M. Križman, A simplified approach for isocratic HPLC analysis of cannabinoids by fine tuning chromatographic selectivity. Eur. Food Res. Technol. 246 (2020) 315–322, https://doi.org/10.1007/s00217-019-03344-7.

[12] V. Cardenia, T.G. Toschi, S. Scappini, R.C. Rubino, M.T. Rodriguez-Estrada, Development and validation of a fast gas chromatography/mass spectrometry method for the determination of cannabinoids in Cannabis sativa L. J. Food Drug Anal. 26 (4) (2018) 1283–1292, https://doi.org/10.1016/j.jfda.2018.06.001.

[13] O. Aizpurua-Olaizola, J. Omar, P. Navarro, M. Olivares, N. Etxebarria, A. Usobiaga, Identification and quantification of cannabinoids in *Cannabis sativa* L. plants by high performance liquid chromatography-mass spectrometry. Anal. Bioanal. Chem. 406 (29) (2014) 7549–7560, https://doi.org/10.1007/s00216-014-8177-x.

[14] S. Fekete, et al., Implementation of a generic liquid chromatographic method development workflow: application to the analysis of phytocannabinoids and Cannabis sativa extracts. J. Pharm. Biomed. Anal. 155 (2018) 116–124, https://doi.org/10.1016/j.jpba.2018.03.059.

[15] M. Wang, et al., Quantitative determination of cannabinoids in Cannabis and Cannabis products using ultra-high-performance supercritical fluid chromatography and diode Array/mass spectrometric detection. J. Forensic Sci. 62 (3) (2017) 602–611, https://doi.org/10.1111/1556-4029.13341.

[16] M. Hädener, S. König, W. Weinmann, Quantitative determination of CBD and THC and their acid precursors in confiscated cannabis samples by HPLC-DAD. Forensic Sci. Int. 299 (2019) 142–150, https://doi.org/10.1016/j.forsciint.2019.03.046.

[17] J. Omar, M. Olivares, M. Alzaga, N. Etxebarria, Optimisation and characterisation of marihuana extracts obtained by supercritical fluid extraction and focused ultrasound extraction and retention time locking GC-MS: gas chromatography. J. Sep. Sci. 36 (8) (2013) 1397–1404, https://doi.org/10.1002/jssc.201201103.

[18] A. Casiraghi, et al., Extraction method and analysis of cannabinoids in Cannabis olive oil preparations. Planta Med. 84 (4) (2018) 242–249, https://doi.org/10.1055/s-0043-123074.

[19] L. Calvi, et al., Comprehensive quality evaluation of medical *Cannabis sativa* L. inflorescence and macerated oils based on HS-SPME coupled to GC–MS and LC-HRMS (q-exactive orbitrap®) approach. J. Pharm. Biomed. Anal. 150 (2018) 208–219, https://doi.org/10.1016/j.jpba.2017.11.073.

[20] C. Carcieri, et al., Cannabinoids concentration variability in cannabis olive oil galenic preparations. J. Pharm. Pharmacol. 70 (1) (2018) 143–149, https://doi.org/10.1111/jphp.12845.

[21] R. Deidda, et al., Analytical quality by design: development and control strategy for a LC method to evaluate the cannabinoids content in cannabis olive oil extracts. J. Pharm. Biomed. Anal. 166 (2019) 326–335, https://doi.org/10.1016/j.jpba.2019.01.032.

[22] K. Grafström, K. Andersson, N. Pettersson, J. Dalgaard, S.J. Dunne, Effects of long term storage on secondary metabolite profiles of cannabis resin. Forensic Sci. Int. 301 (2019) 331–340, https://doi.org/10.1016/j.forsciint.2019.05.035.

[23] J.T. Borodovsky, A.J. Budney, Legal cannabis laws, home cultivation, and use of edible cannabis products: a growing relationship? Int. J. Drug Policy 50 (2017) 102–110, https://doi.org/10.1016/j.drugpo.2017.09.014.

[24] T.S. Ghosh, M. Van Dyke, A. Maffey, E. Whitley, D. Erpelding, L. Wolk, Medical Marijuana's public health lessons—implications for retail marijuana in Colorado. N. Engl. J. Med. 372 (11) (2015) 991–993, https://doi.org/10.1056/NEJMp1500043.

[25] I.D.M. Pisciottano, G. Guadagnuolo, V. Soprano, M.D. Crescenzo, P. Gallo, A rapid method to determine nine natural cannabinoids in beverages and food derived from Cannabis sativa by liquid chromatography coupled to tandem mass spectrometry on a QTRAP 4000. Rapid Commun. Mass Spectrom. 32 (19) (2018) 1728–1736, https://doi.org/10.1002/rcm.8242.

[26] M.M. Lewis, Y. Yang, E. Wasilewski, H.A. Clarke, L.P. Kotra, Chemical profiling of medical Cannabis extracts. ACS Omega 2 (9) (2017) 6091–6103, https://doi.org/10.1021/acsomega.7b00996.

[27] L.A. Ciolino, T.L. Ranieri, A.M. Taylor, Commercial cannabis consumer products part 1: GC–MS qualitative analysis of cannabis cannabinoids. Forensic Sci. Int. 289 (2018) 429–437, https://doi.org/10.1016/j.forsciint.2018.05.032.

[28] L.A. Ciolino, T.L. Ranieri, A.M. Taylor, Commercial cannabis consumer products part 2: HPLC-DAD quantitative analysis of cannabis cannabinoids. Forensic Sci. Int. 289 (2018) 438–447, https://doi.org/10.1016/j.forsciint.2018.05.033.

[29] Q. Meng, B. Buchanan, J. Zuccolo, M.-M. Poulin, J. Gabriele, D.C. Baranowski, A reliable and validated LC-MS/MS method for the simultaneous quantification of 4 cannabinoids in 40 consumer products. PLoS One 13 (5) (2018) e0196396, https://doi.org/10.1371/journal.pone.0196396.

[30] A.C. Elkins, M.A. Deseo, S. Rochfort, V. Ezernieks, G. Spangenberg, Development of a validated method for the qualitative and quantitative analysis of cannabinoids in plant biomass and medicinal cannabis resin extracts obtained by super-critical fluid extraction. J. Chromatogr. B 1109 (2019) 76–83, https://doi.org/10.1016/j.jchromb.2019.01.027.

[31] V. Brighenti, et al., Development of a new method for the analysis of cannabinoids in honey by means of high-performance liquid chromatography coupled with electrospray ionisation-tandem mass spectrometry detection. J. Chromatogr. A 1597 (2019) 179–186, https://doi.org/10.1016/j.chroma.2019.03.034.

[32] P.M. Mthembi, E.M. Mwenesongole, M.D. Cole, Chemical profiling of the street cocktail drug 'nyaope' in South Africa using GC–MS II: stability studies of the

Analysis of cannabinoids 99

cannabinoid, opiate and antiretroviral components during sample storage. Forensic Sci. Int. 300 (2019) 187–192, https://doi.org/10.1016/j.forsciint.2019.04.040.

[33] M. Yousefi-Taemeh, D.R. Ifa, Analysis of tetrahydrocannabinol derivative from cannabis-infused chocolate by QuEChERS-thin layer chromatography-desorption electrospray ionization mass spectrometry. J. Mass Spec. 54 (10) (2019) 1–9, https://doi.org/10.1002/jms.4436.

[34] Z.D. Cooper, J.L. Poklis, F. Liu, Methodology for controlled administration of smoked synthetic cannabinoids JWH-018 and JWH-073. Neuropharmacology 134 (2018) 92–100, https://doi.org/10.1016/j.neuropharm.2017.11.020.

[35] A. Al-Matrouk, M. Alqallaf, A. AlShemmeri, H. BoJbarah, Identification of synthetic cannabinoids that were seized, consumed, or associated with deaths in Kuwait in 2018 using GC–MS and LC–MS-MS analysis. Forensic Sci. Int. 303 (2019) 109960, https://doi.org/10.1016/j.forsciint.2019.109960.

[36] R.F. Somerville, et al., The identification and quantification of synthetic cannabinoids seized in New Zealand in 2017. Forensic Sci. Int. 300 (2019) 19–27, https://doi.org/10.1016/j.forsciint.2019.04.014.

[37] A.V. Oberenko, S.V. Kachin, S.A. Sagalakov, Types of synthetic cannabinoids seized from illicit trafficking in the territory of the Siberian Federal District (Russia) between 2009–2018. Forensic Sci. Int. 302 (2019) 109902, https://doi.org/10.1016/j.forsciint.2019.109902.

[38] M.A. Huestis, Human cannabinoid pharmacokinetics. Chem. Biodivers. 4 (8) (2007) 1770–1804, https://doi.org/10.1002/cbdv.200790152.

[39] M.E. Wall, M. Perez-Reyes, The metabolism of Δ9-tetrahydrocannabinol and related cannabinoids in man. J. Clin. Pharmacol. 21 (S1) (1981) 178S–189S, https://doi.org/10.1002/j.1552-4604.1981.tb02594.x.

[40] W.H. Abd-Elsalam, M.A. Alsherbiny, J.Y. Kung, D.W. Pate, R. Löbenberg, LC-MS/MS quantitation of phytocannabinoids and their metabolites in biological matrices. Talanta 204 (2019) 846–867, https://doi.org/10.1016/j.talanta.2019.06.053.

[41] M. Kraemer, B. Madea, C. Hess, Detectability of various cannabinoids in plasma samples of cannabis users: indicators of recent cannabis use? Drug Test. Anal. 11 (10) (2019) 1498–1506, https://doi.org/10.1002/dta.2682.

[42] U. Meier, F. Dussy, E. Scheurer, K. Mercer-Chalmers-Bender, S. Hangartner, Cannabinoid concentrations in blood and urine after smoking cannabidiol joints. Forensic Sci. Int. 291 (2018) 62–67, https://doi.org/10.1016/j.forsciint.2018.08.009.

[43] J. Sánchez-González, R. Salgueiro-Fernández, P. Cabarcos, A.M. Bermejo, P. Bermejo-Barrera, A. Moreda-Piñeiro, Cannabinoids assessment in plasma and urine by high performance liquid chromatography–tandem mass spectrometry after molecularly imprinted polymer microsolid-phase extraction. Anal. Bioanal. Chem. 409 (5) (2017) 1207–1220, https://doi.org/10.1007/s00216-016-0046-3.

[44] C. Wiedfeld, J. Krueger, G. Skopp, F. Musshoff, Comparison of concentrations of drugs between blood samples with and without fluoride additive—important findings for Δ9-tetrahydrocannabinol and amphetamine. Int. J. Leg. Med. 133 (1) (2019) 109–116, https://doi.org/10.1007/s00414-018-1797-5.

[45] A. Gasse, H. Pfeiffer, H. Köhler, J. Schürenkamp, Development and validation of a solid-phase extraction method using anion exchange sorbent for the analysis of cannabinoids in plasma and serum by gas chromatography-mass spectrometry. Int. J. Leg. Med. 130 (4) (2016) 967–974, https://doi.org/10.1007/s00414-016-1368-6.

[46] A.J. Ocque, C.E. Hagler, R. DiFrancesco, J. Lombardo, G.D. Morse, Development and validation of an assay to measure cannabidiol and Δ9-tetrahydrocannabinol in human EDTA plasma by UHPLC-MS/MS. J. Chromatogr. B 1112 (2019) 56–60, https://doi.org/10.1016/j.jchromb.2019.03.002.

[47] A. Zgair, et al., Development of a simple and sensitive HPLC–UV method for the simultaneous determination of cannabidiol and Δ9-tetrahydrocannabinol in rat plasma. J. Pharm. Biomed. Anal. 114 (2015) 145–151, https://doi.org/10.1016/j.jpba.2015.05.019.

[48] A. Ravula, H. Chandasana, B. Setlow, M. Febo, A.W. Bruijnzeel, H. Derendorf, Simultaneous quantification of cannabinoids tetrahydrocannabinol, cannabidiol and CB1 receptor antagonist in rat plasma: an application to characterize pharmacokinetics after passive cannabis smoke inhalation and co-administration of rimonabant. J. Pharm. Biomed. Anal. 160 (2018) 119–125, https://doi.org/10.1016/j.jpba.2018.07.004.

[49] O. Aizpurua-Olaizola, I. Zarandona, L. Ortiz, P. Navarro, N. Etxebarria, A. Usobiaga, Simultaneous quantification of major cannabinoids and metabolites in human urine and plasma by HPLC-MS/MS and enzyme-alkaline hydrolysis. Drug Test. Anal. 9 (4) (2017) 626–633, https://doi.org/10.1002/dta.1998.

[50] R. Jamwal, A.R. Topletz, B. Ramratnam, F. Akhlaghi, Ultra-high performance liquid chromatography tandem mass-spectrometry for simple and simultaneous quantification of cannabinoids. J. Chromatogr. B 1048 (2017) 10–18, https://doi.org/10.1016/j.jchromb.2017.02.007.

[51] A. Valen, Å.M.L. Øiestad, D.H. Strand, R. Skari, T. Berg, Determination of 21 drugs in oral fluid using fully automated supported liquid extraction and UHPLC-MS/MS. Drug Test. Anal. 9 (5) (2017) 808–823, https://doi.org/10.1002/dta.2045.

[52] V. Bessonneau, E. Boyaci, M. Maciazek-Jurczyk, J. Pawliszyn, In vivo solid phase microextraction sampling of human saliva for non-invasive and on-site monitoring. Anal. Chim. Acta 856 (2015) 35–45, https://doi.org/10.1016/j.aca.2014.11.029.

[53] N.A. Desrosiers, K.B. Scheidweiler, M.A. Huestis, Quantification of six cannabinoids and metabolites in oral fluid by liquid chromatography-tandem mass spectrometry. Drug Test. Anal. 7 (8) (2015) 684–694, https://doi.org/10.1002/dta.1753.

[54] R. West, C. Mikel, D. Hofilena, M. Guevara, Positivity rates of drugs in patients treated for opioid dependence with buprenorphine: a comparison of oral fluid and urine using paired collections and LC-MS/MS. Drug Alcohol Depend. 193 (2018) 183–191, https://doi.org/10.1016/j.drugalcdep.2018.07.023.

[55] M. Di Rago, M. Chu, L.N. Rodda, E. Jenkins, A. Kotsos, D. Gerostamoulos, Ultra-rapid targeted analysis of 40 drugs of abuse in oral fluid by LC-MS/MS using carbon-13 isotopes of methamphetamine and MDMA to reduce detector saturation. Anal. Bioanal. Chem. 408 (14) (2016) 3737–3749, https://doi.org/10.1007/s00216-016-9458-3.

[56] M.C. Cela-Pérez, et al., Water-compatible imprinted pills for sensitive determination of cannabinoids in urine and oral fluid. J. Chromatogr. A 1429 (2016) 53–64, https://doi.org/10.1016/j.chroma.2015.12.011.

[57] J. Rositano, P. Harpas, C. Kostakis, T. Scott, Supported liquid extraction (SLE) for the analysis of methylamphetamine, methylenedioxymethylamphetamine and delta-9-tetrahydrocannabinol in oral fluid and blood of drivers. Forensic Sci. Int. 265 (2016) 125–130, https://doi.org/10.1016/j.forsciint.2016.01.017.

[58] S. Baeck, B. Kim, B. Cho, E. Kim, Analysis of cannabinoids in urine samples of short-term and long-term consumers of hemp seed products. Forensic Sci. Int. 305 (2019) 109997, https://doi.org/10.1016/j.forsciint.2019.109997.

[59] C. Sempio, K.B. Scheidweiler, A.J. Barnes, M.A. Huestis, Optimization of recombinant β-glucuronidase hydrolysis and quantification of eight urinary cannabinoids and metabolites by liquid chromatography tandem mass spectrometry. Drug Test. Anal. 10 (3) (2018) 518–529, https://doi.org/10.1002/dta.2230.

[60] I. Angeli, S. Casati, A. Ravelli, M. Minoli, M. Orioli, A novel single-step GC–MS/MS method for cannabinoids and 11-OH-THC metabolite analysis in hair. J. Pharm. Biomed. Anal. 155 (2018) 1–6, https://doi.org/10.1016/j.jpba.2018.03.031.

[61] A. Rodrigues, M. Yegles, N. Van Elsué, S. Schneider, Determination of cannabinoids in hair of CBD rich extracts consumers using gas chromatography with tandem mass spectrometry (GC/MS–MS). Forensic Sci. Int. 292 (2018) 163–166, https://doi.org/10.1016/j.forsciint.2018.09.015.

[62] S. Odoardi, V. Valentini, N. De Giovanni, V.L. Pascali, S. Strano-Rossi, High-throughput screening for drugs of abuse and pharmaceutical drugs in hair by liquid-chromatography-high resolution mass spectrometry (LC-HRMS). Microchem. J. 133 (2017) 302–310, https://doi.org/10.1016/j.microc.2017.03.050.

[63] T. Kieliba, O. Lerch, H. Andresen-Streichert, M.A. Rothschild, J. Beike, Simultaneous quantification of THC-COOH, OH-THC, and further cannabinoids in human hair by gas chromatography–tandem mass spectrometry with electron ionization applying automated sample preparation. Drug Test. Anal. 11 (2) (2019) 267–278, https://doi.org/10.1002/dta.2490.

[64] R. Paul, R. Williams, V. Hodson, C. Peake, Detection of cannabinoids in hair after cosmetic application of hemp oil. Sci. Rep. 9 (1) (2019) 1–6, https://doi.org/10.1038/s41598-019-39609-0.

[65] D. Thieme, U. Sachs, H. Sachs, C. Moore, Significant enhancement of 11-Hydroxy-THC detection by formation of picolinic acid esters and application of liquid chromatography/multi stage mass spectrometry (LC-MS3): application to hair and oral fluid analysis. Drug Test. Anal. 7 (7) (2015) 577–585, https://doi.org/10.1002/dta.1739.

[66] C. Montesano, et al., Pressurized liquid extraction for the determination of cannabinoids and metabolites in hair: detection of cut-off values by high performance liquid chromatography–high resolution tandem mass spectrometry. J. Chromatogr. A 1406 (2015) 192–200, https://doi.org/10.1016/j.chroma.2015.06.021.

[67] F. Pragst, et al., Hair analysis of more than 140 families with drug consuming parents. Comparison between hair results from adults and their children. Forensic Sci. Int. 297 (2019) 161–170, https://doi.org/10.1016/j.forsciint.2019.01.039.

[68] E. Beasley, S. Francese, T. Bassindale, Detection and mapping of cannabinoids in single hair samples through rapid derivatization and matrix-assisted laser desorption ionization mass spectrometry. Anal. Chem. 88 (20) (2016) 10328–10334, https://doi.org/10.1021/acs.analchem.6b03551.

[69] R. Jain, R. Singh, Microextraction techniques for analysis of cannabinoids. TrAC Trends Anal. Chem. 80 (2016) 156–166, https://doi.org/10.1016/j.trac.2016.03.012.

[70] M. Vincenti, A. Salomone, E. Gerace, V. Pirro, Role of LC–MS/MS in hair testing for the determination of common drugs of abuse and other psychoactive drugs. Bioanalysis 5 (15) (2013) 1919–1938, https://doi.org/10.4155/bio.13.132.

[71] D. Favretto, J.P. Pascali, F. Tagliaro, New challenges and innovation in forensic toxicology: focus on the "new psychoactive substances" J. Chromatogr. A 1287 (2013) 84–95, https://doi.org/10.1016/j.chroma.2012.12.049.

[72] D.M. Schwope, E.L. Karschner, D.A. Gorelick, M.A. Huestis, Identification of recent Cannabis use: whole-blood and plasma free and Glucuronidated cannabinoid pharmacokinetics following controlled smoked Cannabis administration. Clin. Chem. 57 (10) (2011) 1406–1414, https://doi.org/10.1373/clinchem.2011.171777.

[73] M.N. Newmeyer, M.J. Swortwood, A.J. Barnes, O.A. Abulseoud, K.B. Scheidweiler, M.A. Huestis, Free and glucuronide whole blood Cannabinoids' pharmacokinetics after controlled smoked, vaporized, and oral Cannabis administration in frequent and occasional Cannabis users: identification of recent Cannabis intake. Clin. Chem. 62 (12) (2016) 1579–1592, https://doi.org/10.1373/clinchem.2016.263475.

[74] C. Montesano, et al., Multi-class analysis of new psychoactive substances and metabolites in hair by pressurized liquid extraction coupled to HPLC-HRMS. Drug Test. Anal. 9 (5) (2017) 798–807, https://doi.org/10.1002/dta.2043.

[75] L. Mercolini, et al., Dried blood spots: liquid chromatography–mass spectrometry analysis of Δ9-tetrahydrocannabinol and its main metabolites. J. Chromatogr. A 1271 (1) (2013) 33–40, https://doi.org/10.1016/j.chroma.2012.11.030.

[76] K.B. Scheidweiler, M. Andersson, M.J. Swortwood, C. Sempio, M.A. Huestis, Long-term stability of cannabinoids in oral fluid after controlled cannabis administration. Drug Test. Anal. 9 (1) (2017) 143–147, https://doi.org/10.1002/dta.2056.

[77] D. Lee, M.A. Huestis, Current knowledge on cannabinoids in oral fluid. Drug Test. Anal. 6 (1–2) (2014) 88–111, https://doi.org/10.1002/dta.1514.

[78] E. Gallardo, J.A. Queiroz, The role of alternative specimens in toxicological analysis. Biomed. Chromatogr. 22 (8) (2008) 795–821, https://doi.org/10.1002/bmc.1009.

[79] M.A. Huestis, M.L. Smith, Cannabinoid markers in biological fluids and tissues: revealing intake. Trends Mol. Med. 24 (2) (2018) 156–172, https://doi.org/10.1016/j.molmed.2017.12.006.

[80] P.O.M. Gundersen, O. Spigset, M. Josefsson, Screening, quantification, and confirmation of synthetic cannabinoid metabolites in urine by UHPLC–QTOF–MS. Drug Test. Anal. 11 (1) (2019) 51–67, https://doi.org/10.1002/dta.2464.

[81] T.Y. Kong, et al., Rapid analysis of drugs of abuse and their metabolites in human urine using dilute and shoot liquid chromatography–tandem mass spectrometry. Arch. Pharm. Res. 40 (2) (2017) 180–196, https://doi.org/10.1007/s12272-016-0862-1.

[82] S. Vogliardi, M. Tucci, G. Stocchero, S.D. Ferrara, D. Favretto, Sample preparation methods for determination of drugs of abuse in hair samples: a review. Anal. Chim. Acta 857 (2015) 1–27, https://doi.org/10.1016/j.aca.2014.06.053.

[83] J. Restolho, M. Barroso, B. Saramago, M. Dias, C.A.M. Afonso, Contactless decontamination of hair samples: cannabinoids. Drug Test. Anal. 9 (2) (2017) 282–288, https://doi.org/10.1002/dta.1958.

[84] G.A.A. Cooper, R. Kronstrand, P. Kintz, Society of hair testing guidelines for drug testing in hair. Forensic Sci. Int. 218 (1) (2012) 20–24, https://doi.org/10.1016/j.forsciint.2011.10.024.

PART 2

Applications of GC-MS techniques for cannabis analysis

CHAPTER FOUR

State of the art solventless sample preparation alternatives for analytical evaluation of the volatile constituents of different cannabis based products

Gyorgy Vas*

VasAnalytical, Flemington, NJ, United States
*Corresponding author: e-mail address: gyvas70@hotmail.com

Contents

1. Introduction — 105
2. Static head-space based cannabis testing — 107
   2.1 Syringe based head-space sampling — 109
   2.2 Loop based head-space sampling — 110
   2.3 Pressure balanced sampling — 110
   2.4 Analysis of residual solvents for cannabis products using SHS — 111
   2.5 Terpenes and terpenoid analysis by static head space — 113
3. Solid phase microextraction — 116
   3.1 SPME evaluation of cannabis infused gummy bear candies — 122
   3.2 SPME evaluation of CBD oil — 124
   3.3 SPME evaluation of cannabis patches — 125
4. Dynamic head space (DHS) — 127
5. Twister — 129
6. Vacuum assisted sorbent extraction (VASE) — 130
7. Cannabis and electronic delivery devices — 132
8. Summary — 134
References — 134

1. Introduction

Cannabis industry is a continuously shifting dynamic landscape. According to the United States Federal Law, cannabis is categorized as a Schedule I drug, as "substances or chemicals with no currently accepted medical use" [1,2]. The major active substances in the cannabis plant are related to Δ^9 tetrahydrocannabinol (d^9-THC) and more than 100 of them have been identified [3], and products such as hemp with less than 0.3% d^9-THC level are outside the Schedule I DEA classification [4]. The term cannabis is the correct word for describing the plant and its products. The term hemp has historically referred to the use of cannabis as a fibre source which is very low in the psychoactive compound, delta- 9-THC (<0.3%), and has been used for many different products for thousands of years [5]. At state level the cannabis was recognized and legalized either as recreational or substance of medical use. The legalization of medicinal and recreational marijuana (*Cannabis sativa*) in North America resulted in a growing market for different consumer products associated for use of cannabis products in the manufacturing process [6]. Many types of consumer products been developed and are increasing in popularity throughout a wide range of consumer groups from young teens to the elderly population. The regulators, however, need to control the approval of these products to ensure consumer safety and uncompromised and consistent quality of the marketed products. This task requires an urgent need for reliable, accurate and economical analytical methods to control the active and other constituents of the cannabis plant. In addition to the natural constituents of the plants, trace level contaminants also need to be rigorously controlled.

The cannabis plant contains a wide variety of organic and inorganic constituents, many of them are natural constituents of the plants and some of them are impurities related to the farming process of the plants. This category includes different pesticides, and toxic elements which should be controlled and the levels needs to be regulated [7]. Some impurities are related to the manufacturing process of the cannabis products, most impurities are typically from the process related to residual solvents [8,9]. Natural constituents of the plant show a large variety of chemical species due to the vast number of its constituents and their possible interaction with one another. These compounds represent almost all of the chemical classes, e.g., mono- and sesquiterpenes, sugars, hydrocarbons, steroids, flavonoids, nitrogenous compounds and amino acids, among others [10–12]. The cannabinoids are the major active constituents of the plants while others are not considered as active ingredients, they can be synergetic to the effect of the cannabinoids, enhancing its effect to the human body. The volatile terpenes and terpenoids are one of the most abundant class of the cannabis plant and some play an important role to the overall consumer experience, however, some of the terpene constituents are listed as cosmetic allergens (linalool, geraniol) [13]. Since they are volatile, their sensory impact is significant [14], as well as the potential health effect require more studies [15–17]. Similar to cannabinoids more than 100 of the terpenes have been identified and characterized in the cannabis plant and almost all of them are being present in the consumer products [18,19]. More importantly the terpene and terpenoid profile can be characteristic to the farming/growing conditions and can be associated to the geographical origin of the cannabis plant or product [20,21]. Although marijuana odour has long been accepted to be recognizable by dogs, the recognition of the smell by law enforcement officers has recently been challenged in courts by the defence in criminal cases [21].

Mass spectrometry, hyphenated with chromatographic separation has a long successful history of analysing natural products including cannabis. Its versatility, unmatched low detection limit, and capability to detect wide range of organic compounds makes it ideal to characterize cannabis related products from the plant itself to various complex products, such as edible cannabis items [6,22]. Mass spectrometry can be used for high sensitivity targeted analysis, as well as non-targeted screening and it is also an excellent research tool for identifying previously not known chemical entities. Recently the relatively easy access to tandem (MS/MS) and high-resolution accurate mass based systems (HRAM), can lower the detection limit of high risk targeted analytes like pesticides, or can reduce the time and provide higher confidence for identification [23]. To enhance the productivity of the mass spectrometric detection, it is very often combined with automated sample preparation option such as static head-space, SPME or other advanced automated options [24,25].

This chapter is attempting to provide short overview of solventless extraction techniques, that can be used to analyse the cannabis plant itself, as well as different cannabis originated consumer products.

2. Static head-space based cannabis testing

The cannabis plant and the associated cannabis products are representing a very complex matrix and in many cases the target analytes are at trace level concentration, such as low ng/g (pesticide residues), therefore it is necessary to perform a sample preparation technique as part of the analytical process. Sample preparation is one of the most important steps of the analytical process and has the largest impact on the quality and the reliability of the analytical data package. The three major reasons for performing sample preparation are the following.

- Reduce or eliminate sample incompatibility with the analytical systems (analysis of solid samples with gas or liquid chromatography based technique)
- Eliminate heavy matrix components and reduce matrix interferences
- Concentrate the samples

Understanding the complexity of the different analytical matrices and the different analytical techniques, it is not possible to develop an ultimate sample preparation method that is applicable for all of the matrices and compatible with all of the analytical techniques. Because of this, over the years new techniques were developed, which makes the analytical process more effective. A "good" sample preparation method would meet the following requirements:

- Non discriminative for the targets and discriminative for the matrix interferences
- Result of the extraction is predictable
- Quick
- Cost effective
- Automated
- Complies with current regulations and regulatory expectations
- Highly repeatable
- Easy to perform
- Safe for the environment

Depending of the target analyte group, different sample preparation techniques are used for cannabis testing. For the cannabinoid content and potency testing, and residual pesticide testing, solvent based extractions are the most common, while for residual solvents and terpenoids solventless extraction techniques are often used. One of the major issues with the solvent based techniques is that they are very difficult to automate (automated solutions are available) [26,27], and if they are performed manually, the documentation in the regulated laboratories are burdensome, and some of the solvents are not friendly to the environment.

Solventless extraction techniques are getting more and more popular as new methods are being developed in the recent years and being applied for cannabis testing, such as Solid Phase Microextraction (SPME), Dynamic Head-Space (DHS) and Vacuum Assisted Sorbent Extraction (VASE) etc. Those solventless techniques are automated, relatively quick, easy to perform and use minimal or no solvent, which is safe for the environment.

One of the alternatives for solventless sample preparation is Static head-space (SHS). SHS is the most established solventless analytical sample preparation method. It was developed in the late 50s by Leslie Ettre, and the first application was to evaluate the freshness of the potato chips in the packaging bags.

Static Head-Space (SHS) is an equilibrium-based technique; generally, the analyte is not exhaustively removed from the sample matrix. The technique is based on the distribution of a particular analyte between two distinct phases, where one of the phases is a gas phase, while the other phase can be either a condensed liquid or a solid sample. In ideal systems, the target analyte concentration in the head-space is proportional to the concentration in liquid or solid; however, in real systems the head-space concentration is proportional to the analyte partial concentration in the head-space [28]. In practical terms if the analyte partial pressure at a given temperature is high, the chromatographic peak area would be high compared to another analyte that has a lower partial pressure in the head-space at a given temperature. In reality, analytes with a low boiling point are amenable for SHS analysis, while high boilers are outside of the optimum performance of the technique, therefore the "most popular" application for this technique is residual solvent analysis [29].

The static head-space technique is so simple and elegant; therefore, it was one of the first commercially available automated sample preparation system marketed by Perkin Elmer in 1967 [28]. Although the first application of head-space involved a simple gas tight syringe to sample the head-space, later more advanced solutions were developed, such as the loop based sampling and the pressure balanced head-space sampling. The syringe based sampling is also used in an automated fashion, as the option is available for most of the advanced autosamplers, which are available in today's market [30].

2.1 Syringe based head-space sampling

Classically and most simply, headspace may be sampled using a gas tight syringe of appropriate volume, which can be done manually or by an advanced autosampler. The manual sampling approach is the most convenient and inexpensive method for sampling vapours from nontraditional containers such as cans or bags. After equilibrium is reached at a certain temperature, the syringe is inserted into the headspace and an aliquot of appropriate volume is removed. Although seems simple, there are several challenges associated with this method [28,31]:

- If the syringe volume is similar to the headspace volume, equilibrium in the vial may be disrupted. In effect, insertion of the syringe and withdrawal of the syringe plunger cause an increase in the vapour-phase volume, disrupting equilibrium.
- If the syringe is at a lower temperature than the vial, analyte(s) or matrix components may condense in the syringe.
- The syringe may require cleaning or purging with inert gas between each analysis.

The temperature differences between the sample and the vial, and the inert gas flushing is not an issue for automated syringe based systems. Besides the above mentioned obstacles, the syringe based technique is flexible, as different volumes of gas can be sampled and the syringe can be easily varied between different sizes.

2.2 Loop based head-space sampling

In pressure/loop systems the sample head-space is transferred to the analytical system through a transfer line and a pressure loop. The sample vial is pressurized after the equilibrium by the carrier gas of the GC system, using a preset time and pressure to obtain uniform mixture of the volatiles in the head-space. During the sampling, the sample vial is opened temporarily toward the sample loop, and the desired volume (can be as much as the loop size or less) of head-space is transferred to the loop. During injection, the sample collected in the loop is swept into the inlet through the heated transfer line for a preset duration. The major issue of the loop based sampling is associated with the material that a loop is made of. Head-space sampling requires highly inert surfaces for the whole analytical system to minimize the loss of a sensitive or polar analyte. The loop is usually made of stainless steel or silicosteel, which sometimes present active surface spots in the gas path, and as such can have an effect on the analysis of polar analytes, such as chlorinated or sulphur containing species. The other potential performance issue for the loop based system is the dead-volume of the sampling system. It is well known that extra system dead volume has a negative effect on the chromatographic peak shape, and in an ideal system the dead volume should very low, however the loop should not have zero volume, otherwise it cannot be used for sampling.

2.3 Pressure balanced sampling

The pressure balanced system is the most inert and is the most flexible technical solution for sampling the head-space. It offers a similar inert path as the gas chromatography system itself, since it is using a same capillary column as a transfer line as it is used for the chromatographic separation. In a pressure-balanced system the head-space is not drawn by a syringe or suction, instead after equilibrium the vial is pressurized by the carrier gas (equal to the carrier gas pressure of the analytical separation column). The analyte transfer is performed by the carrier gas and the volume is controlled by the column pressure and the sampling time. The dead volume is very minimal since the transfer line uses a small diameter capillary column [28,32] (Fig. 1).

2.4 Analysis of residual solvents for cannabis products using SHS

Cannabis extraction is the process of collecting the desirable compounds from the plant while leaving behind undesirable/unusable materials or solvents that can be used for the manufacturing process of cannabis derived consumer products such as creams, food supplements etc. The cannabis industry uses a wide variety of solvents to extract the active ingredients out of the plant material, and it is desired to eliminate all traces of the solvents from the finished products. However, as we know none of the processes are working with 100% efficiency; therefore, residues remain as process related impurities in the finished product. The level of the residual solvents is controlled by regulations, unfortunately the situation is not that simple, as cannabis derived product can be categorized into different

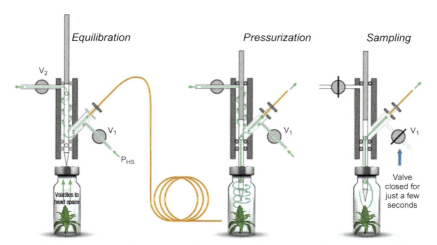

Fig. 1 Head-space extraction steps, using pressure balance sampling. *Redrawn from M. J. Sithersingh, N. H. Snow, Headspace-Gas Chromatography, Chapter 9, Gas Chromatography, Ed. Colin F. Poole Elsevier (2012), fig. 9.3. Courtesy of Perkin Elmer Corporation.*

categories with different applicable regulations. In the pharmaceutical industry USP categorizes residual solvents for finished pharmaceutical products based on their toxicity and possible impact on human health [29]. The allowable limits in finished pharmaceutical products are also listed in the general chapter. It's also important to emphasize that because there are currently no federal regulations in the United States, the allowable concentration limits for each residual solvent are defined by the individual state where the cannabis is grown. The USP residual solvent categories may not directly apply to the cannabis or cannabis derived products; however, the document referenced below is for medical use marijuana, recommended to follow the USP categories and limits [33]. The classifications of residual solvents are listed below:

- Class 1 residual solvents should not be employed in the manufacture of drug substances, excipients, and drug products because of the unacceptable toxicities or deleterious environmental effects of these residual solvents. However, if their use in order to produce a medicinal product is unavoidable, their levels should be restricted. Example of the Class I solvent: benzene, carbon-tetrachloride, 1,2-Dichloroethane, 1,1-Dichloroethane, 1,1,1-Trichloroethane.
- Class 2 residual solvents should be limited in drug substances, excipients, and drug products because of the inherent toxicities of the residual solvents. There are 27 chemicals listed under the class 2 category [33].
- Class 3 residual solvents may be regarded as less toxic and of lower risk to human health than Class 1 and Class 2 residual solvents. Class 3 includes no solvent known as a human health hazard at levels normally accepted in pharmaceuticals. However, there are no long-term toxicity or carcinogenicity studies for many of the residual solvents in Class 3. Available data indicate that they are less toxic in acute or short-term studies and negative in genotoxicity studies. Unless otherwise stated in the individual monograph, Class 3 residual solvents are limited to not more than 50 mg per day (corresponding to 5000 ppm or 0.5% under Option 1). If a Class 3 solvent limit in an individual monograph is greater than 50 mg per day, that residual solvent should be identified and quantified. There are 27 chemicals listed under the class 3 category [33].

The residual solvent testing is mostly associated with cannabis concentrate products, such as extracts, tinctures, edibles, waxes, and oils, which are becoming the most commonly used cannabis products that are legally manufactured for both medicinal and recreational purposes; however, if the cannabis is being marketed for medicinal purposes, it is necessary to provide evidence that the product meets the accepted safety standards. Those samples mostly represent very complex and heavy matrix, therefore a head-space based sample preparation, which is mostly effective for the most volatile constituents and impurities is ideal. Since the residual solvents are highly volatile compared to the other constituents such terpenes and cannabinoids, they are an ideal target for SHS sample preparation.

Many states require taking multiple sampling points from non-homogenous samples (such as waxes and edibles) to ensure a representative sample for analysis. If required, five sampling points from one sample are recommended. For example, 5 replicates of the 1/5th of the required amount (if a state requirement is 500 mg for testing, 5 replicates of a 100 mg sample should be used) should be placed in a vial and brought to a final volume of 10 mL with dimethylacetamide (DMA). Twenty μL of the diluent is then inserted into the HS vial, which is capped and placed onto the HS autosampler for analysis. However, if an average sampling is not required, a 40 mg aliquot of the extract can be directly weighed into the HS vial and analysed [34]. Residual solvent testing usually requires a different length of equilibrium time depending on the target matrix, which could be between 5 and 60 min at 70–105 °C temperature [35].

2.5 Terpenes and terpenoid analysis by static head space

A terpene is a naturally occurring hydrocarbon based on combinations of the isoprene unit. Terpenoids are compounds related to terpenes, which may include some oxygen functionality or rearrangement; however, the two terms are often used interchangeably. The difference between terpenes and terpenoids is that terpenes are hydrocarbons; whereas, terpenoids have been denatured by oxidation (drying and curing the flowers). As it was indicated in the introduction the terpenes and terpenoids are one of the most abundant classes in cannabis, they playing significant role of the sensory properties of the plant and the derived products. Most of them are low molecular weight compounds (below $m/z = 200$), and since they are hydrocarbon types or slightly oxidized their polarity is low, and their logP is high, which means they are good targets for head-space based analysis. The terpene profile is characteristic to the cannabis variety and can support strain identification, as well as can be indicative of the farming conditions.

Multiple samples from different strains from two different varieties (Sativa Dominant and Indica Dominant) have been analysed with static head-space technique hyphenated with enhanced resolution with a triple quadrupole instrument. The samples were legally purchased in a public dispensary in San Diego CA, and the sample analysis was performed in a laboratory with DEA Schedule 1–5 licence. The sample preparation for SHS was very simple and once the sample was prepared, the extraction and the analysis were automated. 50 mg of dried hop was grinded into a fine powder and placed into ultraclean 20 mL size HS vial, spiked with 500 ng of 2-fluorobiphenyl IS (10 ppm concentration) and sealed with a PTFE coated silicone septa. The prepared samples were equilibrated at 85 °C for 10 min and agitated during the equilibrium process at 250 rpm shaking; 1 mL HS injected with 1:10 split injection to a 250 °C PTV injector. Sample analysis was performed on Trace 1310 GC (Thermo scientific, US) coupled to a TSQ Quantum Ultra XLS triple-quadrupole MS (Thermo scientific, US) equipped with a Restek Rxi 5 MS (5% phenylmethyl-silicone) capillary column (40 m × 0.18 mm × 0.18 μm, Restek, US). Hydrogen carrier gas was used for the separation at 40 cm/s linear flow rate (24.1 psi head pressure at 40 °C). The GC oven was programmed from 50 °C (1 min hold), to 300 °C (3 min hold) with a rate of 20 °C/min, and the mass spectrometer was used in a Q3 scan mode at a mass range of 45–450.

The analytical result presented is in Figs. 2 and 3.

The total ion chromatograms (TIC) show significant differences between the two strains. The volatile profiles are abundant and can support

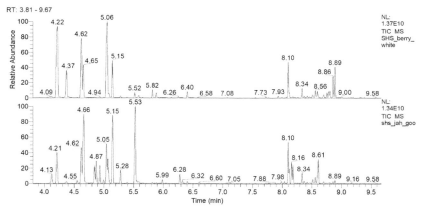

Fig. 2 TIC chromatograms of different cannabis strains, SHS extraction, with a syringe based system. Top trace Berry White (Sativa Dominant variety); bottom trace Jah Goo (Indica Dominant variety).

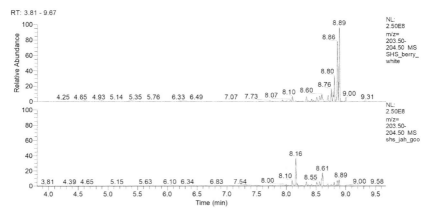

Fig. 3 Extracted ion trace of $m/z = 204$ associated with elemental composition of $C_{15}H_{24}$. Top trace Berry White (Sativa Dominant variety); bottom trace Jah Goo (Indica Dominant variety).

both quantitative and qualitative assessment. The total analysis time with the sample preparation was less than 30 min. The most significant visible difference between the two varieties is the intensity of the chromatographic peak eluting at 5.52 min, and it is the most abundant peak for the Indica variety, and a small peak for the Sativa. The extracted ion chromatogram on Fig. 3 reveal further differences between the two different samples. The $m/z = 204$ ion, which is representative for the $C_{15}H_{24}$ sesquiterpenes was extracted out from the TIC. The Sativa sample shows more abundant peaks (top trace). It is interesting to note that the above referenced paper [19], lists 17 individual terpene species, while the chromatogram on the top presents 26 major individual peaks related to $C_{15}H_{24}$ elemental composition.

The other characteristic feature for the static head-space technique, as it is expected from the nature of the extraction, is that the most volatile peaks show higher relative abundance while the static head-space based technique is not effective for the least volatiles such as $\Delta 9$-THC, compared to the data acquired by different extraction technique such as SPME or SPME-Arrow (see Figs. 4–6).

In summary static head-space is a very efficient technique for analysis of the volatile constituents and the residual solvent content of the cannabis plant and different cannabis related products, and it successfully eliminates the interferences from the heavy sample matrices.

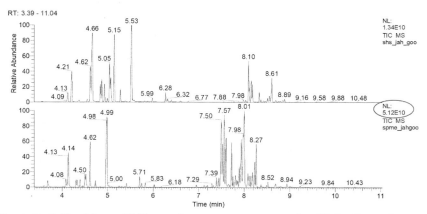

Fig. 4 Static Head Space (SHS) sampling top trace, compared to SPME-HS sampling. TIC of Jah Goo strain. The SPME chromatogram shows about 4 times higher peak abundance.

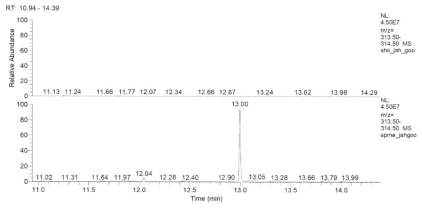

Fig. 5 Static Head Space (SHS) sampling top trace, compared to SPME-HS sampling. Extracted ion trace for m/z = 314 of Jah Goo strain samples. The peak at 13.0 min is Δ9-THC. SHS sampling technique is not sufficient for extraction of high boiling point components, therefore no peak was detected.

3. Solid phase microextraction

Solid Phase Microextraction (SPME) was invented and first published by J. Pawliszyn and Belardi [36], it has become rapidly a commercial product, marketed by Supelco in 1994. SPME integrates sampling, extraction, concentration and sample introduction into a single solvent-free step.

State of the art solventless sample preparation 117

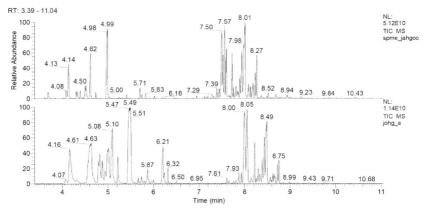

Fig. 6 SPME-HS sampling top trace compared to SPME Arrow sampling, bottom trace. TIC of Jah Goo strain. The SPME Arrow extraction was performed at 40 °C vs. the SPME-HS at 85 °C. The SPME Arrow shows system overload effects for the most volatile species.

Analytes in the sample are directly extracted and concentrated to the extraction fibre. The method saves preparation time and disposal costs and may improve detection limits. It has been routinely used in combination with gas chromatography (GC) and GC/mass spectrometry (GC/MS) and successfully applied to a wide variety of compounds, especially for the extraction of volatile and semi-volatile organic compounds from environmental, biological, food and pharmaceutical samples [37,38]. The concept of SPME may have been derived from the idea of an "inside-out" GC capillary column, where the active polydimethyl-siloxane coating is the outside region, while the supportive quartz fibre is inside. The SPME apparatus is a very simple device. It looks like a modified syringe consisting of a fibre holder and a fibre assembly, the latter containing a 1–2 cm long retractable SPME fibre. The SPME fibre itself is a thin fused-silica optical fibre, coated with a thin polymer film (such as polydimethylsiloxane (PDMS)), conventionally used as a coating material in chromatography. The extraction process is simple and it is available as an option for most of the GC or GC–MS autosamplers [30]. The coated fibre is immersed directly in the sample or the HS of the sample, where the analytes are concentrated. After equilibrium has been reached (from a few minutes to several hours depending on the properties of the analytes measured) or after a defined time, the fibre is withdrawn and transferred either to a GC injection port. The fibre is exposed and the analytes are desorbed thermally in the hot GC injector port and subsequently chromatographed in a conventional manner. The SPME extraction and desorption process is presented in Fig. 7 [39].

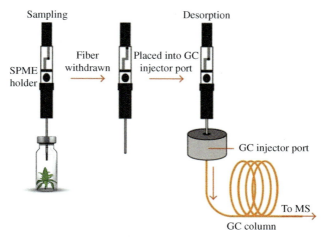

Fig. 7 SPME-HS sampling process. *Figure was modified from fig. 1 in K. Schmidt and I. Podmore, Solid phase microextraction (SPME) method development in analysis of volatile organic compounds (VOCS) as potential biomarkers of cancer, Mol. Biomark Diagn. (2015), 6:6.*

With highly complex matrices such as sludges, biological fluids and food products or using solid samples, the SPME technique is mainly applied for the extraction of analytes from the HS of the sample or an overcoated fibre can be used as immersion sampling technique [40].

An updated and higher capacity version of the classical SPME is the SPME Arrow. The phase carrier is thicker for SPME Arrow and as the phase is distributed to a larger diameter support rod, the phase volume is larger, resulting in higher extraction efficiency. The larger diameter carrier also makes the extraction device more robust and less fragile. The 100 μm PDMS fibre has a phase volume of 0.6 μL and the 250 μm SPME Arrow fibres have a 3.8 μL volume with 9.4 and 44 mm^2 surface area. The differences between the fibres are shown in Fig. 8 [41].

Besides the simplicity of use and the enhanced extraction capability of the SPME extraction the other significant benefit, is that the extraction efficiency can be predicted very accurately for the most common polydimethylsiloxane extraction phase. The extraction efficiency can be plotted vs. the logP of the target analytes, and such a plot can be a significant help for the method development phase. A logP-extraction efficiency plot is shown in Fig. 9, where the extraction efficiency for the 100 μm PDMS, SPME Arrow 1.1 and Twister 0.5 × 10 mm stir bar is being plotted. For practical consideration the extraction efficiency should be at least 15%–20%

State of the art solventless sample preparation 119

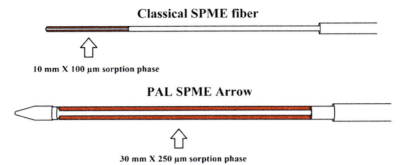

Fig. 8 SPME and SPME-Arrow extraction fibres. *Reproduced from A. Kremser, M. A. Jochmann, T. C. Schmidt, PAL SPME arrow—evaluation of a novel solid-phase, microextraction device for freely dissolved PAHs in water, Anal. Bioanal. Chem. (2016) 408:943–952.*

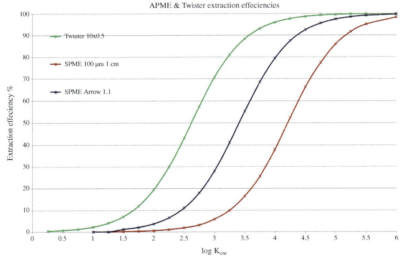

Fig. 9 SPME and twister extraction efficiencies versus logP. Red: SPME 100 μm; blue: SPME Arrow 1.1; green: Twister 10 mm length 500 μm film thickness.

(above 50% is desirable), for a certain analyte. The plot shows that the 100 μm PDMS desired cut-off logP is logP $\geq$ 4.2, for SPME Arrow 1.1 is logP $\geq$ 3.4 and for Twister with a 0.5 mm PDMS film is logP $\geq$ 2.5, it means that the conventional SPME technique could cover all of the terpene compounds, and majority of the terpenoids, while the Twister extraction can be used for wider range of compounds, including analytes with alcohol or other polar groups.

Solid Phase Microextraction is one of the emerging analytical extraction techniques for sample preparation of cannabis, and few papers have been published using SPME as extraction method for analysis of the cannabis plant [24,42,43]. The technique requires minimal sample treatment, and it is easy to automate, and it shows better extraction efficiency for the less volatile terpenoids. Similar to the above explained SHS technique, the matrix interferences are significantly reduced, and it can be used for testing a wide range of cannabis related products. Some examples for the applicability of SPME and SPME Arrow are presented in Figs. 4–6 and 10–12 [44].

On Fig. 4 the SPME extraction was compared to SHS, for the same sample (Indica Dominant, Jah Goo strain). The sample preparation and the instrument conditions for SHS was described in the static head-space section, and for SPME same conditions were used, except that for SPME after the equilibrium time completed, a 5 min extraction was performed for the HS. The chromatograms on Fig. 4 show that SPME has better coverage for the volatiles of the cannabis bud, as well as the peaks are more abundant, and shows about 4-fold intensity enhancement.

The other significant difference between the two techniques, is that the $\Delta 9$-THC is not detected by static head-space, because it has very low volatility, therefore it is not detectable in the static head-space at 85 °C, while SPME is a dynamic technique, and the fibre enriches the head-space concentration, and the THC peak is nicely detectable, as seen in Fig. 5.

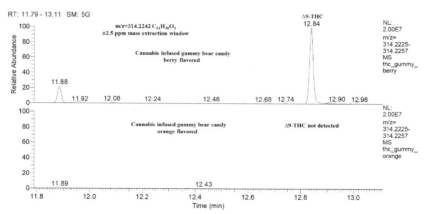

Fig. 10 SPME-HS extracted ion chromatogram ($m/z = 314.2242 \pm 2.5$ ppm), of cannabis infused gummy-bear candies. Top trace: berry flavoured; bottom trace orange flavoured. $\Delta 9$-THC is not present in the orange flavoured product.

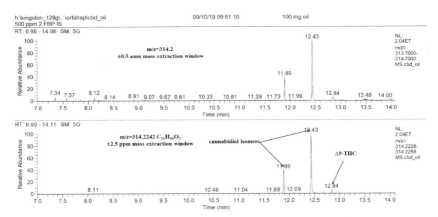

Fig. 11 SPME-HS extracted ion chromatograms of CBD oil sample. Top trace: $m/z = 314.2 \pm 0.5$ amu extraction window; bottom trace: $m/z = 314.2242 \pm 2.5$ ppm extraction window. The HRAM based data processing provides interference free chromatogram.

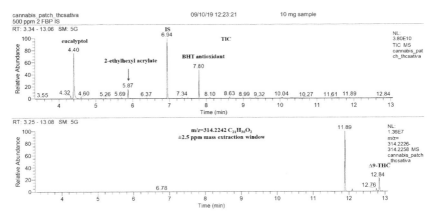

Fig. 12 SPME-HS testing of cannabis patches. Top trace: TIC, eucalyptol, BHT and 2-ethylhexyl-acrylate present as major peaks; bottom trace: extracted ion chromatogram $m/z = 314.2242 \pm 2.5$ ppm, Δ9-THC detected.

This enhanced feature for the SPME makes it more versatile, to check the presence of Δ9-THC in different cannabis and CBD products, even when the THC level is intended to be very low in CBD products. According to the regulations in the US, if the Δ9-THC content is below 0.3% the particular product can be analysed by a laboratory without having a DEA licence, while above this level DEA licence is needed. The SPME based evaluation can provide a quick and reliable limit test for presence or absence of $\triangle 9$-THC peak in the product, therefore a laboratory can make a quick decision if the test object is being accepted for full evaluation, or needs to be rejected.

The SPME Arrow provides similar data package to the SPME, however even at lower extraction temperature at $40\,^{\circ}$C and using 1:50 split, many of the peaks for the early eluting peaks are overloading the analytical system, resulting in compromised peak shapes, showing in Fig. 6. The conventional SPME does not require any modification for the GC-inlet and it can be used with both split/splitless and PTV injectors. However, it is recommended to use low volume inlet liners such with 0.75 or 1.0 mm id, and to minimize septum coring, a pre-drilled septa or Merlin Microseal is recommended. For SPME arrow, some modifications are required for the split/splitless injector, using vendor specific conversion kits [41] and 1.8 mm id inlet liner, due to the larger diameter of the needle, the septum coring is more pronounced, therefore the septa needs to be replaced, or a modified Merlin Microseal can be used. PTV inlet is not compatible with SPME Arrow.

Testing cannabis infused consumer products such as chocolate, coffee, gummies, cosmetics or finished hemp products, requires significant sample preparation work, since cosmetics and edible products represent a heavy matrix with significant interferences, especially for GC–MS testing. Those matrices contain high level of non-volatiles, which can contaminate the analytical system [42]. Many of the applications use some sort of solvent based extraction technique, which may require significant manual labour, or if an automated method is being used, a significant financial investment needs to be made for an autosampler, capable of performing multi-step liquid extractions in automated fashion [45–47]. Solventless sample preparation techniques are offering an economical time saving alternative for testing cannabis and hemp derived products. SPME is one of the solventless methods, which have been introduced early for cannabis testing, and due to the nature of the extraction technique it can be easily adaptable for testing edibles and cosmetics. Head-space base extraction allows direct analysis of those difficult to extract items. This part of the chapter will show some application of SPME for testing cannabis and hemp related edibles and cosmetics.

3.1 SPME evaluation of cannabis infused gummy bear candies

The cannabis market is still relatively new and the laws dictate that edibles be tested prior to packaging. The FDA recently issued an advisory regarding the safety of cannabidiol (CBD), one of the compounds found in cannabis plants, alerting consumers that some companies are marketing products containing the compound in ways that the agency says violates the Federal Food, Drug and Cosmetic Act, but the market for CBD products—and especially edible CBD—is growing rapidly [48]. Potency testing in marijuana-infused edibles is an important problem that analytical labs are facing due to the complexity of the involved matrices, and sometimes the levels are low, since relatively large amount of edible can be consumed as a food item (brownies, chocolate candies). The active ingredient of Δ9-THC can be present as low as parts-per-million level to up to % level, covering a wide concentration range.

Gummy bear candies are very popular edible products and represent a very difficult matrix as it is made from a mixture of sugar, glucose syrup, starch, flavouring, food colouring, citric acid, and gelatin and, in addition to the basic ingredients, cannabis extract is added to the ingredients. Potency testing requires a multiple step extraction prior to the HPLC analysis and an extraction method was developed and published few years ago [49]. The extraction method is a multiple step procedure and requires significant level of manual labour, as it is summarized in Table 1. In contrary the SPME based solventless method requires only two simple manual steps, and the rest of the extraction is fully automated and software controlled.

Two different gummy candies were analysed with the SPME method using high resolution accurate mass (HRAM) instrument for the detection. The SPME sample preparation was described in the previous part. Sample analysis was performed on Trace 1310 GC (Thermo scientific, US) coupled to a Exactive GC which is an Orbitrap based MS system (Thermo scientific, US) equipped with a Restek Rxi 5 MS (5% phenylmethyl-silicone) capillary column (30 m × 0.25 mm × 0.25 μm, Restek, US). Helium carrier gas was used for the separation at 30 cm/s linear flow rate (12.5 psi head pressure at 40 °C). The GC oven was programmed from 50 °C (1 min hold), to 300 °C (3 min hold) with a rate of 20 °C/min, and the mass spectrometer was used in scan mode at a mass range of 45–450 with a resolution set for 60,000 FWHM. The chromatograms of Fig. 10 show significantly different pictures of the two samples when an ion trace of $m/z = 314.2242 \pm 2.5$ ppm is plotted, which is associated to elemental composition of $C_{21}H_{30}O_2$. The top trace acquired from a berry flavoured gummy product shows a nice intense peak at 12.84 min for Δ9-THC and another peak at 11.88 min for one of the cannabidiol isomers. The bottom trace was acquired from an orange flavoured candy product, and although it is from the same manufacturer, and was purchased at the same time, the extracted ion chromatogram shows a flat baseline, and no presence of Δ9-THC and cannabidiol. This example shows why potency testing is so important, and not only from

Table 1 Sample preparation for gummy candies.

Solvent based extraction	Solventless
QuEcHERS based	SPME
Weight the sample	Weight the sample
Dissolve in water	
Spike with IS/RS	Spike with IS/RS
Transfer water for QuEcHERS extraction	Transfer the spiked candy for SPME testing
Add acetonitrile and shake 1 min	
Add salt and shake 5 min	
Centrifuge for 5 min @ 5000 rpm	Equilibrate the sample at the 85 °C extraction temperature for 10 min
Remove top layer for analysis	SPME extraction 85 °C 10 min
Analyse the sample	Analyse the sample
Total manual labour: 26 min	Total manual labour: 5 min
Automated steps: only the final step	Automated steps: all the extraction steps
Total automated steps: 0 min	Total automated steps: 20 min
Non-volatiles injected to the system: significant [50]	Non-volatiles injected to the system: none

a health prospective, it is also important from a product quality prospective, as consumers do not want to spend their money for a very expensive placebo product. The SPME method provides a very quick way of evaluate those type of edible goods, and the laboratory can make a quick decision not to continue the more involved potency testing when no active ingredients are present. This type of head-space based testing can also be used as a QC tool, to check the final product before it is being packaged and distributed, to avoid such defected products to be marketed.

3.2 SPME evaluation of CBD oil

Another emerging area for cannabis use is an excipient of cosmetic products. Cosmetics also represent a wide range of complex matrices from different oil based products, through soaps, creams and lotions. Analysis of those cosmetic products can be even more challenging than edibles, since many of the cosmetics use an oil or petrolatum based matrix. Those non-polar matrices cannot be extracted with water as they form an emulsion, and they are miscible with many organic solvents, therefore the solvent based extraction methods show poor performance for matrix elimination. If the complex matrix is not eliminated it can cause high background, and the laboratory needs to invest more time for system maintenance on a long term. On the other hand, head-space based techniques provide an ideal automated solution for testing cannabis infused cosmetic products. The sample preparation is simple, the product simply needs to be placed in a HS vial (IS spiking can be performed), and after the vial is sealed it is ready for the testing. The chromatograms on Fig. 11 show the extracted ion chromatograms specific to the Δ9-THC and the cannabidiols. The data are acquired from the head-space of 100 mg CBD oil. The major peaks in the chromatograms are associated with the two cannabinoid isomers eluting at 11.89 and 12.43 min, and the peak at 12.84 min is the Δ9-THC peak (4% of the cannabidiol peaks). The product is supposed to be free of THC, however the analysis tells a different story. This analysis is also a good example to show the versatility of the head-space based screening methods, as after a quick screening method is applied and the presence of the THC is confirmed an assay for the potency can confirm the accurate level of the THC, and if the level is above 0.3%, it is considered as a controlled substance and may require further actions.

The top trace shows the extracted ion window with a unit width extraction window, while the bottom trace shows a 5 ppm width extraction window. The two chromatograms seem similar, as three major peaks present on both. Looking closer at the chromatograms the unit resolution chromatogram shows significant noise or peaks in a region of 7–9 min and between 13 and 14 min, while the narrow mass extraction eliminates all the noise. The other significant difference is the signal to noise for the THC peak $S/N = 99$ (RMS) for the unit resolution processing and $S/N = 653$ for the high resolution processing, showing the benefit for using HRAM based technology. Besides it is providing much higher confidence for identification purposes, by reliable elemental composition assessment for each chromatographic peak, it is also reducing the noise, which may be beneficial for quantitation.

3.3 SPME evaluation of cannabis patches

One of the most promising applications of cannabis derived product is pain management. Research data show a promising future for different cannabis based products in pain management [15,17]. To use cannabis based products as medication moves this segment of the industry very close to the pharmaceutical industry, and this approach raises significant technical and regulatory problems. The major roadblocks are:

- Products to be used as medication require approval from the US Food and Drug Administration, and the approval process is lengthy, costly and requires science based evidence of the products efficacy and safety. To generate those data clinical studies are obligatory, as well as more complex chemistry testing needs to be done
- For the manufacturing of the marketed products more control needs to be implemented, and manufacturing sites may need to comply cGMP [51].
- Specification for the cannabis needs to be implemented either when used as active or as excipient. Today more than 200 different strains are available for purchase, and they are not representing the same quality and same potency.

Topical/transdermal pain management is one of the most popular forms for easing pain. Multiple types of delivery systems can be used, such as cream, ointment, spray, and transdermal delivery system (TDS aka patch). While creams are only effective for few hours, the TDS based systems can be effective much longer, even up to few days. The manufacturing process is more complex for TDS than for a cream or spray based systems. It is important to note that up to date no cannabis containing transdermal delivery system have been approved by US FDA for managing pain or other therapeutic areas.

The transdermal systems are multicomponent delivery devices, consisting of multiple layers such as, backing film, drug, adhesive and liner. The adhesive and the polymer part of the delivery system can be an issue for the solvent based sample preparations because the solvents usually remove non-volatiles and injected extracts are trapped in the injector and potentially degrade; to avoid problems head-space based solutions are the best choice for sample preparation. Cannabis containing transdermal systems were tested with SPME head-space. The overall sample preparation was simple: 10 mg of transdermal patch was cut and placed into a 20 mL HS vial and spiked with 5 µg of 2-fluorobiphenyl IS, followed by SPME extraction and GC–MS analysis described above. The result in Fig. 12 shows that eucalyptol is one of the major constituents of the patch and it is often used as a topical analgesic, the patch contains high levels of BHT as antioxidant (the peak at 6.94 min is the IS). The volatiles related to cannabis are barely visible on the TIC (top trace), however the extracted ion chromatograms reveal cannabinoid isomers and the △9-THC peak. Another significant peak present on the TIC, which was identified as 2-ethylhexyl acrylate; this chemical is known and identified as irritant, and can cause irritation on skin [52]. It is important to understand that the tested transdermal system has not been evaluated for safety, and has not been approved by a regulatory agency. In the future if companies try to market them to claim therapeutic benefits, it needs to be evaluated and rigorous analytical testing needs to be performed to avoid using any harmful chemicals.

4. Dynamic head space (DHS)

Dynamic Head Space is a relatively new, highly efficient solventless extraction method, in theory it is a modified purge and trap-based technique. DHS shows a lot of similarities to the more common static head-space technique, with some additional features, which offer significant benefits for the analytical laboratories and offers significantly lower detection limits. Similar to SHS, the samples are heated in a closed head-space vial. After a short equilibrium period, the head space above the sample is purged with a high purity carrier gas and trapped in sorbent filled extraction tube, which is kept at a lower temperature. By purging the head-space above the sample, the analyte is depleted from the head-space and the system can be re-equilibrated. The amount of purge gas is variable between a 20–500 mL at a purging rate of 10–100 mL/min. Compared to the static HS technique, 20–500 times more head space (HS) can be introduced to the analytical system with the DHS approach. The sorbent tube is placed into a heat desorption unit and the extracted components are desorbed by heat. To avoid peak broadening, the GC–MS injector is usually cooled to sub-ambient temperatures for a second trapping cycle and the components are released to the GC column by ultra-fast heating (300–900 °C/min). The technique is ideal for analysis of trace component in solid samples, and it is not recommended to use for high concentration samples, unless a very high split ratio can be used. To perform DHS extraction, a special addition is required to the GC–MS system, which can be an injector addition (Gerstel TDU with DHS), or a stand alone sample preparation station (Markes). The DHS sampling process is explained in Fig. 13 for a Gerstel Thermal Desorption Unit (TDU) based system.

The chromatogram on the Fig. 14 shows the most typical "mistake" that can happen, when DHS is used for sample preparation. A chromatogram was acquired from a 50 mg dried and grinded cannabis bud equilibrated and

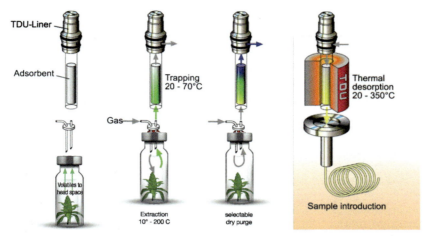

Fig. 13 DHS extraction process. *Redrawn from G. Vas, Evaluating the Volatile Constituents of Different Cannabis Varieties using Solventless Sample Preparation and Orbitrap Based MS Detection, Science of Cannabis Online Symposium. (2019). Courtesy of Gerstel US.*

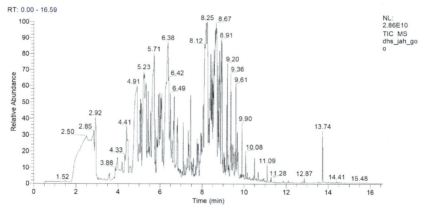

Fig. 14 DHS TIC of Jah Goo cannabis bud. The chromatogram shows typical symptoms of system overloading.

extracted at 40 °C, and a 200 mL of HS was trapped to a Tenax TA sorbent trap. The trap was desorbed at 300 °C, and transferred to the GC–MS system with a 1:400 split. The chromatogram shows massive overload of the terpenes, the peaks are not even separated, the Δ9-THC peak at 13.74 min is eluting as a high intensity peak. To avoid such limited result, a lower sample amount or a much higher split ratio (1:10,000) is recommended.

5. Twister

Stir bar sorptive extraction (SBSE) was introduced in 1999 as a solventless sample preparation method for the extraction and enrichment of organic compounds from aqueous matrices. The method is based on sorptive extraction, whereby the solutes are extracted into a polymer coating on a magnetic stirring rod. The extraction is controlled by the partitioning coefficient of the solutes between the polymer coating, the sample matrix, the phase ratio between the polymer coating and the sample volume. For a polydimethylsiloxane coating and aqueous samples, this partitioning coefficient resembles the octanol–water partitioning coefficient, see Fig. 9. In comparison to solid phase micro-extraction, a larger amount of sorptive extraction phase is used and consequently extremely high sensitivities can be obtained as illustrated by several successful applications in trace analysis in environmental, food and biomedical fields [53]. Initially SBSE was mostly used for the extraction of compounds from aqueous matrices. The technique has also been applied in headspace mode for liquid and solid samples and in passive air sampling mode. Stir bars of 1 or 2 cm long coated with a 0.5 or 1 mm layer have been made commercially available (Twister™, Gerstel GmbH, Mullheim a/d Ruhr, Germany). A magnetic rod is encapsulated in a glass jacket on which a polydimethylsiloxane coating is placed, since the direct contact between metal and PDMS was found to catalyse polymer degradation during thermal desorption. After extraction is completed, either from liquid or from the head-space of the solid test object, the stir bar is removed, dipped on a clean paper tissue to remove water droplets, and introduced in a thermal desorption unit. In some cases, it is recommended to rinse the stir bar slightly with distilled water to remove adsorbed sugars, proteins, or other sample components. This step will avoid the formation of non-volatile material during the thermal desorption step. Rinsing does not cause solute loss, because the sorbed solutes are present inside the polydimethylsiloxane phase. The stir bar then thermally desorbs similar to the DHS process.

A unique application of Twister is presented here, showing the versatility and the extreme low detection limit of the technique. Cleaning is very important for manufacturing cannabis based products, and the manufacturing site should present evidence that cleaning procedures are implemented and are effective to avoid cross contamination between different products [54]. A Twister PDMS coating stir bar can be used as a part of a rolling

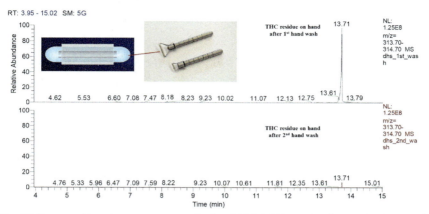

Fig. 15 Extracted ion chromatograms of $m/z = 314.2 \pm 0.5$ amu, to demonstrating residual levels Δ9-THC. Top trace after first hand wash; bottom trace: after second hand wash was performed.

device (Fig. 15), and the stir bar can collect organic species from surfaces. In the present example, the sample was collected from my own hand, after a cannabis plant was used to simulate the joint making progress. After the process was completed the hands were washed with soap and water, and the twister was rolled on the skin, and dipped into water for few seconds remove the unwanted particulates, and consecutively analysed by thermal desorption and GC–MS. After the 1st wash, a cycle was repeated to evaluate the THC residue after the 2nd wash. The data are presented in Fig. 15 and show that even after the 2nd wash a detectable amount of THC is present on the skin. The technique can be used for quick evaluation of different manufacturing surfaces for residues.

6. Vacuum assisted sorbent extraction (VASE)

As an alternative approach to overcome some of the limitations, a technique called vacuum-assisted sorbent extraction (VASE) was introduced to the analytical market very recently [55]. By using commercialized sorbent traps called sorbent pens (SPs) and a headspace extraction environment, VASE combines the advantages of both SBSE and vacuum HS-SPME. The SPs are packed with a large quantity of extraction material (approximately 10 times the volume typically used for SBSE and approximately 500 times the volume typically used for SPME), which favours exhaustive extraction as in SBSE and other HS-extraction techniques. At the same time,

Fig. 16 VASE extraction apparatus and sorbent pens from [56]. Courtesy of Entech Instruments.

VASE operates at near equilibrium conditions, which improves repeatability. To accelerate the extraction kinetics, reduce the sampling time, and extend the range of detectable compounds, in-vial extraction is performed in a reduced-pressure environment during VASE using a commercialized and leak-tight sealing system, see Fig. 16 [56]. The vacuum allows recovery of headspace compounds at lower temperatures (4–40 °C), preventing changes to any heat sensitive sample matrices being studied.

The SPs are thermally desorbed via a unique GC injection port, followed by separation and detection by GC in combination with mass spectrometry (MS). Despite the advantages of this technique, there are very limited number of studies published up to date, that use VASE for analysis of cannabis related products [57]. The extraction with VASE is relatively simple, although it is a multi-step procedure, where many of the steps are automated or multiple samples can be processed at the same time. The extraction steps are:

- Sample preparation and vial extraction; the cannabis product is homogenized, and an aliquot is transferred to a glass vial in the presence of a Headspace Sorbent Pen (HSP). The vial is evacuated using a 2-stage diaphragm pump, helping to promote less-volatile compounds (e.g., THC) into the headspace for capture on the sorbent bed.

- Diffusive Head-Space Extraction: the samples are heated and agitated while under vacuum. Diffusive headspace extraction conditions promote analyte adherence to the front of the sorbent bed, prevents volatile compound breakthrough, enables flash desorption, and promotes tight chromatography.
- Water Management and storage are optional steps, only required for high moisture samples, and if the laboratory schedule requires the loaded sorbent pens they can be stored up to 2 weeks in dedicated isolation racks.
- Sample introduction to the GC–MS by thermal desorption. Sorbent Pen Thermal Desorption Unit (SPDU), Sorbent Pen Thermal Conditioner (SPTC) and Sample Preparation Rail (SPR) positioned on top of a GC–MS. The SPR transfers Sorbent Pens between the SPDU, the SPTC, and air-tight isolation sleeves. The SPDU is connected directly to a capillary column held within the GC-oven, which helps to minimize the flow path and mitigate analyte loss and carryover. Important optimization parameters for desorption include the temperature and duration of the preheat, desorption, and bakeout.

The technique is well fitted for analysis of plants and different consumer products; however, users can face a significant disadvantage similar to DHS. The high efficiency extraction process can overload the analytical system easily, therefore for high concentration samples split flow needs to be used, and injection conditions need to be optimized during method development.

7. Cannabis and electronic delivery devices

Cannabis products are very popular recreational products and the majority of the users experiencing the "recreational benefits" smoke the cannabis plant in a similar fashion to traditional tobacco products. In the past few years, electronic delivery systems became very popular for tobacco products and a whole industry was born. Those electronic delivery systems operate without burning the tobacco and produce none or very low level of the combustion related high safety risk by-products, such as nitrosamines or polynuclear aromatics. Besides smokeless operating conditions, the safety and the long term health effects are not fully understood and are under evaluation. The recent outbreak defined as the "E-cigarette, or Vaping, Product Use Associated Lung Injury (EVALI)" showed evidence of non-sufficient safety data [58]. Even with lack of safety data, electronic delivery systems

State of the art solventless sample preparation 133

were started to be used for delivering the active THC by cannabis users, and unfortunately the EVALI outbreak showed high rates associated with THC containing products. It is very important to note that none of the delivery systems have been fully evaluated and approved by a regulatory agency for use in delivering cannabis products, and the scientific community has insufficient data to support any claim for the safety of those products. In order to generate meaningful data analytical testing should be more extensively used, to understand the chemistry of vaping and the associated toxicological and health risk [59]. By definition those devices are creating fine aerosols of liquid and solid particulates (not vapour, as vaping is used as a common term), therefore they are excellent subjects for head-space based chemical analysis. The generated aerosol is more representative of the consumer exposure than the e-liquid itself. Unfortunately testing the aerosol is more complex than testing the e-juice itself. Testing the leachable chemicals, which may have impact on the safety of the consumers is only appropriate by using the aerosol for the analytical evaluation. To mimic the usage of the finished product requires generation of the aerosol in a repeatable and meaningful way, and the aerosol needs to be collected appropriately for chemical testing (sorbent, liquid, foam based trapping, or cryo-trapping). The experimental set-up on Fig. 17 presents one possible alternative for analytical testing of electronic

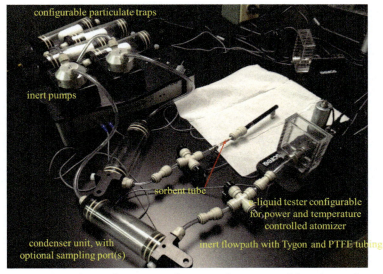

Fig. 17 Sampling device for testing the aerosols generated by electronic smoking devices and from e-liquids.

delivery systems and the associated liquids. For aerosol collection, different sorbent tubes (DHS, VASE) can be used, as well as passive sampling solutions (Twister and diffusion sampling tubes) can be used. If the electronic delivery devices or the associated liquids are evaluated with such a testing set-up, it significantly reduces the interference from polyethylene–glycol or glycerin. Besides the condenser sampling units, the connection tubes also can be used for testing the condensates, directly or rinsed with solvents and the rinse solution can be used for the testing [60].

8. Summary

Head-space based sample preparation techniques hyphenated with mass spectral detection provide comprehensive analytical coverage for testing cannabis and related consumer products. They can simplify sample preparation and reduce the testing time. For some of the testing, like residual solvents, head-space based techniques are the only practical alternatives. Furthermore, for other tests (terpenes and cannabinoid), head space techniques are offering an economical complimentary solution.

References

[1] Diversion Control Division, Title 21 United States Code (USC) Controlled Substances Act, Chapter 13 Drug Abuse Prevention and Control, Part B, Section 811. Authority and criteria for classification of substances, 2016.

[2] https://www.dea.gov/drug-scheduling, n.d.

[3] S.A. Ahmed, S.A. Ross, D. Slade, M.M. Radwan, I.A. Khan, M.A. Elsohly, Minor oxygenated cannabinoids from high potency Cannabis sativa L, Phytochemistry 117 (2015) 194–199.

[4] National Media Affairs Office, https://www.dea.gov/press-releases/2019/08/26/dea-announces-steps-necessary-improve-access-marijuana-research, 2019.

[5] B. Nie, J. Henion, I. Ryona, The role of mass spectrometry in the cannabis industry. J. Am. Soc. Mass Spectrom. 30 (2019) 719Y730, https://doi.org/10.1007/s13361-019-02164-z.

[6] L.A. Ciolino, T.L. Ranieri, A.M. Taylor, Commercial cannabis consumer products part 1: GC–MS qualitative analysis of cannabis cannabinoids, Forensic Sci. Int. 289 (2018) 429–437.

[7] C.B. Craven, N. Wawryk, P. Jiang, Z. Liu, X.F. Li, Pesticides and trace elements in cannabis: analytical and environmental challenges and opportunities, J. Environ. Sci. 85 (2019) 82–93. https://doi.org/10.1016/j.jes.2019.04.028.

[8] Residual Solvents, Residual Solvents in Cannabis, Application Note by Ellutia Chromatography Solutions, n.d.

[9] Shimadzu, Analysis of Residual Solvents in Drug Products using Nexis GC-2030 Combined With HS-20 Head Space Sampler—USP <467> Residual Solvents Procedure A, Application Note G290, Shimadzu, 2017.

[10] M.A. ElSohly, D. Slade, Chemical constituents of marijuana: the complex mixture of natural cannabinoids, Life Sci. 78 (2005) 539–548.

[11] R. Brenneisen, Chemistry and Analysis of Phytocannabinoids and Other Cannabis Constituents, Forensic Sci Med: Marijuana and the Cannabinoids, Chapter 2, Edited by: M. A. ElSohly, Humana Press Inc., Totowa, NJ, n.d.

[12] M.M. Lewis, Y. Yang, E. Wasilewski, H.A. Clarke, L.P. Kotra, Chemical profiling of medical cannabis extracts. ACS Omega 2 (2017) 6091–6103, https://doi.org/10.1021/acsomega.7b00996.

[13] European Commission Scientific Committee on Consumer Safety (SCCS), Scientific Committee on Health and Environmental Risks (SCHER), Scientific Committee on Emerging and Newly Identified Health Risks (SCENIHR), Opinion on Use of the Threshold of Toxicological Concern (TTC) Approach for Human Safety Assessment of Chemical Substances with focus on Cosmetics and Consumer Products, 2012.

[14] B.E. Erickson, Cannabis industry gets crafty with terpenes, C&EN News vol. 97, (29) (2019).

[15] A. Small Howard, A natural product approach to Cannabis based therapies, PharmaEd, Canna Pharma Meeting, November 13–14, 2019, 2019, Marriott Mission Valley, San Diego CA

[16] J. Meehan-Atrash, W. Luo, R.M. Strongin, Toxicant formation in dabbing: the terpene story. ACS Omega 2 (2017) 6112–6117, https://doi.org/10.1021/acsomega.7b01130.

[17] E. Russo, Hot Topics in Therapeutic Cannabis Research, PharmaEd, Canna Pharma Meeting, November 13–14, 2019, 2019, Marriott Mission Valley, San Diego CA

[18] O. Aizpurua-Olaizola, U. Soydaner, E. Öztürk, D. Schibano, Y. Simsir, P. Navarro, N. Etxebarria, A. Usobiaga, Evolution of the cannabinoid and terpene content during the growth of *Cannabis sativa* plants from different chemotypes. J. Nat. Prod. 79 (2016) 324–331, https://doi.org/10.1021/acs.jnatprod.5b00949.

[19] A. Shapira, P. Berman, K. Futoran, O. Guberman, D. Meiri, Tandem mass spectrometric quantification of 93 terpenoids in Cannabis using static headspace injections, Anal. Chem. 91 (2019) 11425–11432.

[20] R. Brenneisen, M.A. ElSohly, Chromatographic and spectroscopic profiles of Cannabis of different origins: part I, J. Forensic Sci. 33 (1988) 1–385.

[21] S.A. Rossi, M.A. ElSohly, The volatile oil composition of fresh and air-dried buds of *Cannabis sativa*, J. Nat. Prod. 59 (1996) 49–51.

[22] G. Vas, Evaluating the Volatile Constituents of Different Cannabis Varieties using Solventless Sample Preparation and Variety of MS Detection, PharmaEd, Canna Pharma Meeting, November 13–14, 2019, 2019. Marriott Mission Valley, San Diego CA.

[23] G. Vas, Evaluating the volatile constituents of different Cannabis varieties using Solventless sample preparation and Orbitrap based MS detection, in: Science of Cannabis Online Symposium, 2019.

[24] K.K. Stenerson, M.R. Halpenny, Analysis of terpenes in Cannabis using headspace solid-phase microextraction and GC–MS, Cannabis Sci. Technol. (2018)

[25] N. Wiebelhaus, D. Hamblin, N.M. Kreitals, J.R. Almirall, Differentiation of marijuana headspace volatiles from other plants and hemp products using capillary microextraction of volatiles (CMV) coupled to gas-chromatography–mass spectrometry (GC–MS), Forensic Chem. 2 (2016) 1–8.

[26] CEM Corporation, EDGE (Energized Dispersive Guided Extraction) System Product Brochure, 2018, CEM Corporation, B140.2 0518.

[27] CEM Corporation, Application Note Extraction and UPLC Analysis of THC, THCA, CBN, CBD, and CBDA from Hemp Flower, 2019 ap0193v1.

[28] B. Kolb, L.S. Ettre, Static Head Space Gas Chromatography, Wiley-VCH Inc, 1997.

[29] United States Pharmacopeia, <467> Residual Solvents, Official. 2019.

[30] Thermo Scientific TriPlus RSH Robotic Sample Handling User Guide P/N 31709620, Ninth Edition, December 2015.

[31] M.J. Sithersingh, N.H. Snow, C.F. Poole (Ed.), Headspace-Gas Chromatography, Chapter 9, Gas Chromatography, Elsevier, 2012.

[32] TurboMatrix Headspace Sampler and HS 40/110 Trap User's Guide, M0413401 F February 2008, Perkin Elmer Corporation.

[33] Association of Public Health Laboratories, Guidance for State Medical Cannabis Testing Programs, Published by Association of Public Health Laboratories, 2016.

[34] L. Marotta, T. Kwoka, D. Scott, M. Snow, T. Astill, Fast, Quantitative Analysis of Residual Solvents in Cannabis Concentrates, PerkinElmer application note. 2018.

[35] L. Vernarelli, J. Whitecavage, J. Stuff, F. Foster, Analysis of Residual Solvents in Hemp Oil by Full Evaporation Head-Space Gas Chromatography and Mass Spectrometry, Pittcon, Chicago, 2020. Poster No. 211.

[36] R.P. Belardi, J.B. Pawliszyn, The application of chemically modified fused silica fibers in the extraction of organics from water matrix samples and their rapid transfer to capillary columns, Water Pollut. Res. J. Can. 24 (1989) 179.

[37] G. Vas, K. Vekey, Solid-phase microextraction: a powerful sample preparation tool prior to mass spectrometric analysis, J. Mass Spectrom. 39 (2004) 233–254.

[38] G. Vas, L. Alquier, C.A. Maryanoff, J. Cohen, G. Reed, Investigation of mass-balance issue in e-beam sterilized paclitaxel eluting coronary stents by SPME/GC–MS, J. Pharm. Biomed. Anal. 48 (2008) 568–572.

[39] K. Schmidt, I. Podmore, Solid phase microextraction (SPME) method development in analysis of volatile organic compounds (VOCS) as potential biomarkers of cancer, Mol. Biomark Diagn. 6 (2015) 6.

[40] K. K. Stenerson, T. Young, R. Shirey, Y. Chen, L. Sidisky, Application of SPME using an overcoated PDMS–DVB fiber to the extraction of pesticides from spaghetti sauce: method evaluation and comparison to QuEChERS, LCGC N. Am. 34, 7, pg 500–509, n.d.

[41] A. Kremser, M.A. Jochmann, T.C. Schmidt, PAL SPME arrow—evaluation of a novel solid-phase, microextraction device for freely dissolved PAHs in water, Anal. Bioanal. Chem. 408 (2016) 943–952.

[42] K.K. Stenerson, Analysis of terpenes in Cannabis using headspace solid-phase micro-extraction and GC–MS, LCGC (2019) 18–30.

[43] J. Westland, SPME Arrow Sampling of Terpenes in Cannabis Plant Material, Poster 1570–1. Pittcon, Philadelphia, 2019.

[44] G. Vas, Evaluating the Volatile Constituents of Different Cannabis Varieties using Solventless Sample Preparation and Orbitrap Based MS Detection, Poster 1570-11. Pittcon, Philadelphia, 2019.

[45] SPE, Automated Solid Phase Extraction SPE with Gerstel MPS, Product Brochure, n.d., Gerstel Global Analytical Solutions.

[46] Teledyne Tekmar, Pesticide Analysis Using the AutoMate-Q40: An Automated Solution to QuEChERS Extractions, Teledyne Tekmar Application Note, n.d.

[47] V. Settle, F. Foster, P. Roberts, P. Stone, J. Stevens, J. Wong, K. Zhang, Automated QuEChERS Extraction for Determination of Pesticide Residues in Foods Using Gas Chromatography/Mass Spectrometry, Gerstel Application Note 4/2010, n.d., Available at https://www.gerstel.com/pdf/p-lc-an-2010-04.pdf.

[48] K. Loria, The Trouble with Edible Testing, Food Qual Safety, 2020.

[49] O. Shimelis; K. K. Stenerson; M. Wesley, Analysis of Active Cannabis Compounds (Cannabinoids) in Edible Food Products: Gummy Bears and Brownies, Analytix Reporter, Issue 4, 2019, Available at https://learning.sepscience.com/hubfs/Companies/Merck%20(Millipore%20Sigma)/Merck_Analytix_Reporter/Issue%204/Analytix_Reporter_4_MSIG_Web.pdf.

[50] K.K. Stenerson, M.R. Halpenny, Analysis of Terpenes in Cannabis Using Headspace Solid-Phase Microextraction and GC–MS, Advancing the Analysis of Medical Cannabis, 2017.

[51] CFR Title 21 Part 211, Current Good Manufacturing Practice for Finished Pharmaceuticals, 2019.

[52] S. Murphy, R.E. Hutchings, L. Finch, S. Welz, K. Wiench, Critical evaluation of 2-ethylhexyl acrylate dermal carcinogenicity studies using contemporary criteria, Toxicol. Lett. 294 (15) (2018) 205–211.

[53] B. Tienpont, F. David, T. Benijts, P. Sandra, Stir bar sorptive extraction-thermal desorption-capillary GC–MS for profiling and target component analysis of pharmaceutical drugs in urine, J. Pharm. Biomed. Anal. 32 (4–58) (2003) 569–579.

[54] M. Moussourakis, Critical Cleaning—The Key to Quality & Safety, Canna Pharma Meeting, Marriott Mission Valley, San Diego CA, 2019.

[55] M.J.T. Rodríguez, J.L. Anderson, S.J.B. Dunham, V.L. Noad, D.B. Cardin, Vacuum-assisted sorbent extraction: an analytical methodology for the determination of ultraviolet filters in environmental samples, Talanta 208 (1) (2020) 120390.

[56] Sorbent Pens™, Cannabis Testing Applications Solvent-Free Solutions for Research & Regulatory Compliance, Entech Instruments e-book, 2019.

[57] V.L. Noad, S.J.B. Dunham, B. Arakelian, D.B. Cardin, Beyond THC and CBD: Solvent Free Headspace Extraction and GC-MS Profiling of Terpenes and Lesser-Known Cannabinoids in Cannabis, Pittcon, Chicago, 2020.

[58] P.Melstrom, E-cig or Vaping Product Use-Associated Lung Injury: Analysis From Epidemiological, Clinical, Forensic and Mechanistic Perspective, Society of Toxicology Webinar, 2020.

[59] C. Dasenbrock, Forensic Approach to Evaluation of Vaping Liquids Associated with Lung Injury of Unknown Origin 2019, Society of toxicology webinar, 2020.

[60] G. Vas, Analytical Challenges and Solutions for Testing e-Cigarettes. How to Deal With Special Gas Based Extraction Methods for Product Evaluation Regulations for E-Cigarettes, 2016. Westin Alexandria, VA, US.

CHAPTER FIVE

Quantitating terpenes/terpenoids and nicotine in plant materials and vaping products using high-temperature headspace gas chromatography–mass spectrometry

Trinh-Don Nguyen[a,b], Seamus Riordan-Short[b], Thu-Thuy T. Dang[a], Rob O'Brien[b,c], Matthew Noestheden[a,b,*]

[a]Department of Chemistry, I.K. Barber School of Arts and Sciences, University of British Columbia, Kelowna, BC, Canada
[b]Supra Research and Development, Kelowna, BC, Canada
[c]Department of Biology, I.K. Barber School of Arts and Sciences, University of British Columbia, Kelowna, BC, Canada
[*]Corresponding author: e-mail address: matt@suprarnd.ca

Contents

1.	Introduction	140
2.	Materials and methods	143
	2.1 Analysis samples, reagents, and chemicals	143
	2.2 Headspace GC–MS analysis	145
	2.3 Liquid-injection GC–MS/MS cross-validation	147
	2.4 Method validation	149
3.	Results and discussion	150
	3.1 Method development and optimisation	150
	3.2 Method validation for plant tissue analysis	154
	3.3 Method validation for e-juices and related vaping products	158
	3.4 Possibility of concurrent cannabis potency test	164
4.	Conclusion	165
	Acknowledgement	166
	References	166

Comprehensive Analytical Chemistry, Volume 90
ISSN 0166-526X
https://doi.org/10.1016/bs.coac.2020.04.006

© 2020 Elsevier B.V.
All rights reserved.

139

1. Introduction

Medicinal and psychoactive plants have been used throughout human history, with bioactive small molecules, also known as natural products or specialised metabolites, being the key components that elicit biological effects. Although significant progresses in biotechnology and chemical synthesis of these valuable molecules have been made, plants remain the main source for most of them [1,2]. Reliable and robust authentication procedures are, therefore, needed to help commercial suppliers of natural products-containing plants and their derived products validate the integrity of their supply chain, while at the same time providing regulatory agencies with tools to evaluate product authenticity and safety [3,4]. This has become a more pressing issue as the market for bioactive plants and their extracts are fast expanding, particularly following the legalisation of recreational cannabis (*Cannabis sativa* L.) consumption in jurisdictions around the world.

Of all the natural products found in plants, terpenoids constitute the largest, most functionally and structurally diverse class with more than 50,000 compounds identified to date [5]. Some terpenes and terpenoids have been reported to possess anticancer, pesticidal, antimicrobial, anti-inflammatory and immuno-modulatory properties [6–9]. While these bioactive properties are of immense academic and industrial interests, terpenes and terpenoids are still best known for their flavour and aroma profiles that contribute to consumer preferences (e.g. essential oils, food flavouring, perfume, etc.). Such preferences are becoming apparent in the way consumers select the cannabis they are purchasing for recreational purposes as well [10,11]. In addition to their distinct sensory attributes, terpenes and terpenoids may exert some "entourage" neurological effects on the cannabis user experience [12,13], all of which makes their accurate quantitation a critical analytical objective.

In cannabis, terpene and terpenoid profiles include many mono- and sesquiterpenoids (Fig. 1) at different abundance levels, in addition to the well-known psychoactive Δ^9-tetrahydrocannabinol (THC) and the bioactive cannabidiol (CBD) [14,15]. Regardless of different opinions on the medicinal value of cannabis, existing evidence suggests that it should be subjected to rigorous quality and safety tests before entering the market, as is required for other bioactive plant and natural health products [16,17]. Given their potential bioactivity, such characterisations should include terpenoid quantitation.

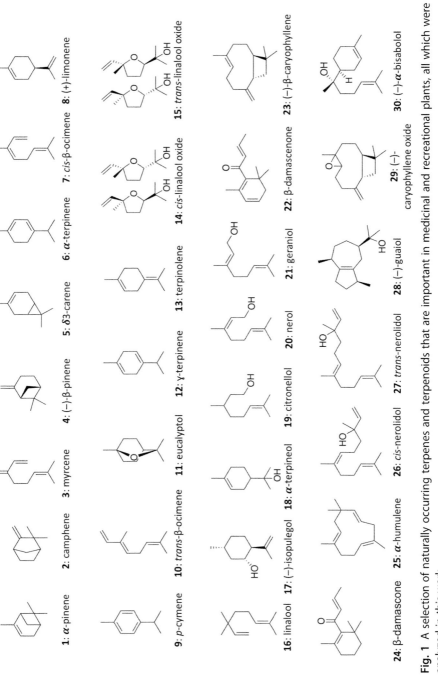

Fig. 1 A selection of naturally occurring terpenes and terpenoids that are important in medicinal and recreational plants, all which were analyzed in this work.

In addition to their natural occurrence in plant materials, terpenoids are found in e-cigarette vaping liquids (e-juices) and supplements, where they impart a wide range of flavours (i.e., piney, sweet, succulent, bitter, citrusy, etc.). Although adding flavours is prohibited in conventional cigarette products in Canada, the United States, and the European Union, at the time of manuscript preparation, this is not the case for e-juices. Such flavourants appeal to users despite the possible health implications of vaping [18]. This is of acute public concern given that vaping has long been advertised as a safer alternative to smoking and it has particularly attracted younger consumers. With recent vaping-related deaths and subsequent legislative proposals to ban or at least discouraging flavourants-added e-juices [19,20], it is important that flavourants including terpenoids in vaping liquids be analysed quantitatively for quality control purposes. Such efforts will support an understanding of the potential safety impacts that they may have.

Analysis of terpenes and terpenoids in plants and other products generally involves organic solvent extraction of plant materials followed by direct injection and analysis by gas chromatography (GC) coupled with flame ionisation or mass spectrometry (MS) detectors [21–23]. While effective at yielding an exhaustive chemical profile from the plant materials' complex organic matrix, several extraction steps using organic solvents and sample concentration may be required [24]. In addition, complex sample matrices could be problematic for direct injection due to co-extracted matrix interferences—particularly for gas chromatography—mass spectrometry (GC–MS) where the co-extracted low-volatility macromolecules can interfere with chromatography and introduce contamination in injection ports.

Headspace sampling techniques are frequently employed, as only compounds with suitable vapour pressures are analysed [25–27]. Furthermore, the resulting gas phase phytochemicals may better reflect those compounds that users of vaporised products are exposed to. In addition, much of sensory attributes to these products are related to terpenes and terpenoids—such users do not employ vigorous, exhaustive extraction with organic solvents prior to product use. This is especially true of materials that are highly processed or intended for inhalation after combustion or high-temperature vaporisation. Headspace solid-phase microextraction methods have been developed for analysing terpenoids in liquid and solid plant products [23,28,29], but variation in chemical selectivity depending on the chemistry of the fibre can make comprehensive terpenoid quantitation difficult. Full evaporative transfer (FET) is an alternate approach that has been recently reported for the quantitation of 93 terpenes and terpenoids in cannabis

[27]. While FET does not suffer from the challenges of SPME fibre chemistry, care must be taken when using FET, as it could be problematic if nontrivial sample amounts (e.g. >5 mg) are used; higher sample quantities are likely requisite to ensure representative results, even after thorough processing [23,30].

Here we present the use of GC–MS with high-temperature headspace sampling for the quantitation of select terpenes and terpenoids in plant samples and e-juices. The reported method is simple, sensitive, and specific, and should be broadly amenable to the quantitation of all mono- and sesquiterpenoids due to the high-temperature incubation utilised. To demonstrate high-temperature headspace GC–MS as an effective tool to analyse various matrices and analyte classes, method validation data are presented for cannabis terpenoid mixtures and e-juice matrices in addition to plant material. We also demonstrated the possibility of including cannabinoids and other bioactive nitrogen-containing small molecules such as caffeine in such analyses.

2. Materials and methods
2.1 Analysis samples, reagents, and chemicals

Dry stinging nettle (*Urtica dioica* L.) was purchased from Westpoint Naturals (British Columbia, Canada). Cannabis terpenoid mixtures were purchased from VapeurTerp (California, USA), and e-juices were products of Canada Vape Lab (Ontario, Canada), Illusions Vape (Ontario, Canada), and Premium Labs (British Columbia, Canada).

Anhydrous, ACS-grade sodium chloride, HPLC-grade solvents (water, glycerol, DMSO, DMF, hexanes, and tetradecane), and nicotine were purchased from Sigma-Aldrich (Ontario, Canada) and used as received.

Terpenoid standards include:
- Cannabis Terpenes Standard 1 from Restek (California, USA) containing 2500 μg/mL of: (−)-α-bisabolol, camphene, δ-3-carene, (−)-β-caryophyllene, geraniol, (−)-guaiol, α-humulene, *p*-cymene, (−)-isopulegol, (+)-limonene, linalool, β-myrcene, nerolidol (*cis* and *trans*), β-ocimene (*cis* and *trans*), α-pinene, (−)-β-pinene, α-terpinene, γ-terpinene, and terpinolene;
- Cannabis Terpenes Standard 2 from Restek (California, USA) containing 2500 μg/mL of: (−)-caryophyllene oxide, and 1,8-cineole (eucalyptol).
- Select analytical standards from Sigma-Aldrich used as received: linalool oxides, citronellol, nerol, β-damascone, β-damascenone, and α-terpineol.

- Mixed terpenoid and nicotine calibration samples (5–800 µg/mL) were prepared in tetradecane and stored at −20 °C for up to 3 months (Tables 1 and 2). We used (±)-linalool-d_3 (CDN Isotopes, Quebec,

Table 1 Calibration scheme for validating the terpenoid quantitation method for plant tissue using HS GC–MS.

Calibration level	Concentration (ng/mL)	Source stock concentration (µg/mL)
Cal 9	4000	800
Cal 8	3000	600
Cal 7	2000	400
Cal 6	1000	200
Cal 5	500	100
Cal 4	200	40
Cal 3	100	20
Cal 2	50	10
Cal 1	25	5

Each calibration curve sample includes 25 µL source stock, 25 µL 400 µg/mL (±)-linalool-d_3 (internal standard), and 5000 µL glycerol.

Table 2 Calibration scheme for validating the terpenoid and nicotine quantitation method for vaping liquids using HS GC–MS.

Calibration level	Concentration		Source stock concentration	
	Terpenoid (ng/mL)	Nicotine (µg/mL)	Terpenoid (ng/mL)	Nicotine (mg/mL)
Cal 9	3000	300	600	60
Cal 8	2000	200	400	40
Cal 7	1000	100	200	20
Cal 6	500	50	100	10
Cal 5	200	20	40	4
Cal 4	100	10	20	2
Cal 3	50	5	10	1
Cal 2	25	2.5	5	0.5
Cal 1	10	1	2	0.2

Each calibration curve sample includes 25 µL source stock, 25 µL 400 µg/mL (±)-linalool-d_3 (internal standard), and 5000 µL glycerol.

Canada) as an internal standard. A $400 \mu g/mL$ linalool-d_3 stock solution was prepared in tetradecane and stored at $-20\,°C$ for up to 12 months. Standard cannabinoids were acquired from Cerilliant Corporation (Texas, USA): cannabidivarin (CBDV), tetrahydrocannabivarin (THCV), cannabichromene (CBC), cannabidiol (CBD), Δ^8-tetrahydrocannabinol (Δ^8-THC), Δ^9-tetrahydrocannabinol (Δ^9-THC), cannabigerol (CBG), and cannabinol (CBN).

2.2 Headspace GC–MS analysis

For dry plant tissue, sample was homogenised using a commercial coffee grinder. Samples and standards were weighed on a VWR-164AC analytical balance (VWR) and analysed using a TriPlus™ 500 GC Headspace Autosampler and a TRACE™ 1300 GC coupled with an ISQ™ 7000 Single Quadrupole MS System (ThermoFisher). The temperatures of the MS transfer line, electron impact source, and ISQ transfer line were 250, 300, and $300\,°C$, respectively.

Water, brine (saturated NaCl solution [stored under ambient conditions]), DMSO, DMF, or glycerol were used as headspace carrier solvents for GC–MS analysis.

Helium (99.999%) was used as carrier gas at $1.5\,mL/min$. The inlet temperature was set at $125\,°C$ with a septum purge flow of $20\,mL/min$ and a split ratio of 100:1 (unless otherwise noted). Data were acquired in either scan or selected ion monitoring (SIM) modes (Table 3), and analysed using the Chromeleon™ Chromatography Data System (ThermoFisher; version 7).

The following protocol was used for headspace analysis:

1. Add $5\,mL$ headspace carrier solvent to 20-mL headspace sampling vial (Chromatographic Specialties, Ontario, Canada). Glycerol can be preheated to $50\,°C$ to facilitate pipetting due to its high viscosity at room temperature.
2. Add sample (50 mg ground plant material or $50\,\mu L$ e-juice), cannabis terpenoid mixtures ($25\,\mu L$), and cannabinoid standards ($25\,\mu L$) to carrier solvent-containing vials, and seal with 18 mm (diameter) $\times$ 1.524 mm (thickness) PTFE/silicone gas tight septa (Supelco, Pennsylvania, USA). Note that samples and standards (terpenoids or cannabinoids) can be added separately or together depending on the purpose of the analysis (i.e. sample analysis or spike-recovery).
3. Set the headspace incubation conditions to 15 min at $85–220\,°C$, and vial pressure to $100\,kPa$ with an equilibration time of 12 s.

Table 3 Molecular ions used for selected ion monitoring (SIM) analysis of terpenes/terpenoids and nicotine in headspace GC–MS.

ID	Compound	Retention time (min)	Quantitation (*m/z*)	Qualifiers (*m/z*)
1	α-Pinene	4.45	93	77, 91
2	Camphene	4.76	79	93, 121
3	β-Myrcene	5.07	93	69, 91
4	(−)-β-Pinene	5.20	93	69, 92
5	δ-3-Carene	5.60	91	79, 93
6	α-Terpinene	5.76	136	93, 121
7	*Cis*-β-ocimene	5.84	79	91, 93
8	(+)-Limonene	5.95	67	69, 107
9	*p*-Cymene	6.02	117	119, 134
10	*Trans*-β-ocimene	6.10	93	79, 91
11	8-Cineole (eucalyptol)	6.22	81	139, 154
12	γ-Terpinene	6.46	91	93, 136
13	Terpinolene	7.06	121	93, 136
14	*Cis*-linalool oxide	7.26	111	59, 94
15	*Trans*-linalool oxide	7.55	111	59, 94
–	(±)-Linalool-d_3	7.63	96	55, 69, 74
16	Linalool	7.64	121	71, 93
17	(−)-Isopulegol	8.61	67	55, 121
18	α-Terpineol	9.20	121	59, 93
19	Citronellol	9.45	95	55, 81
20	Nerol	9.50	93	67, 69
21	Geraniol	9.74	93	68, 69
22	β-Damascenone	11.09	121	69, 105
23	β-Caryophyllene	11.29	161	133, 189
24	β-Damascone	11.34	177	69, 192
25	α-Humulene	11.56	80	93, 147

Table 3 Molecular ions used for selected ion monitoring (SIM) analysis of terpenes/terpenoids and nicotine in headspace GC–MS.—cont'd

ID	Compound	Retention time (min)	Quantitation (*m/z*)	Qualifiers (*m/z*)
26	*Cis*-nerolidol	11.97	69	93, 107
27	*Trans*-nerolidol	12.18	107	69, 136
28	(−)-Guaiol	12.62	161	59, 204
29	(−)-Caryophyllene oxide	12.71	93	79, 121
30	(−)-α-Bisabolol	13.04	119	69, 105
31	Nicotine	10.86	162	84, 133

4. Inject the headspace with a 1-mL loop with an equilibration time of 30 s and an injection time of 30 s, with sample loop temperatures ranging from 85 to 220 °C.

5. For analyses of terpenoids only or terpenoids and nicotine, set the GC oven temperature gradient as follows: 60 °C for 30 s, followed by a ramp of 50 °C/min to 130 °C and 3-min hold, another increase to 140 °C at 5 °C/min, and finally to 280 °C at 22 °C/min and hold for 3 min. The total gradient run time was 16.5 min. Samples were analysed using a 30 m × 0.25 mm × 1.4 μm TG-624SilMS capillary GC column (ThermoFisher).

For analyses of terpenoids and cannabinoids, set the GC oven temperature gradient as follows: 40 °C for 30 s, followed by a ramp of 10 °C/min to 130 °C, another increase at 30 °C/min to 290 °C and hold for 5 min. Samples were analysed using a 30 m × 0.25 mm × 0.25 μm TG-35MS capillary GC column (ThermoFisher).

2.3 Liquid-injection GC–MS/MS cross-validation

Where cross-validation was required, samples were analysed using a TriPlus RSH™ Autosampler, TRACE™ 1310 GC, and TSQ™ 9000 Triple Quadrupole GC–MS/MS system equipped with an advanced electron ionisation (AEI) source (ThermoFisher). Helium (99.999%) was used as the carrier gas (1.5 mL/min) on a TG-624SilMS capillary column (30 m × 0.25 mm ID × 1.4 μm). MS transfer line and ion source temperatures were 250 and 300 °C, respectively. Data were acquired in selected reaction monitoring (SRM) mode (Table 4).

Table 4 Precursor and product ions used for selected reaction monitoring (SRM) analysis of terpenes/terpenoids in GC–MS/MS cross-validation.

ID	Compound	Retention time (min)	Quantitation (m/z)	Qualifier (m/z)
1	α-Pinene	4.74	93 ➜ 77 (12)	93 ➜ 91 (6)
2	Camphene	5.07	93 ➜ 77 (10)	121 ➜ 93 (5)
3	β-Myrcene	5.43	93 ➜ 77 (10)	93 ➜ 91 (6)
4	(−)-β-Pinene	5.53	93 ➜ 77 (10)	93 ➜ 91 (6)
5	δ-3-Carene	5.96	93 ➜ 77 (10)	105 ➜ 79 (5)
6	α-Terpinene	6.13	93 ➜ 77 (12)	105 ➜ 79 (5)
7	*Cis*-β-ocimene	6.22	93 ➜ 77 (10)	121 ➜ 93 (5)
8	(+)-Limonene	6.33	107 ➜ 91 (10)	93 ➜ 97 (12)
9	*p*-Cymene	6.39	134 ➜ 119 (6)	119 ➜ 117 (8)
10	*Trans*-β-ocimene	6.51	93 ➜ 77 (10)	121 ➜ 93 (5)
11	8-Cineole (eucalyptol)	6.60	108 ➜ 93 (6)	108 ➜ 77 (20)
12	γ-Terpinene	6.85	136 ➜ 93 (8)	136 ➜ 121 (6)
13	Terpinolene	7.42	136 ➜ 121 (8)	121 ➜ 93 (6)
14	*Cis*-linalool oxide	7.62	94 ➜ 79 (6)	94 ➜ 77 (20
15	*Trans*-linalool oxide	7.90	94 ➜ 79 (6)	94 ➜ 77 (20
–	(±)-Linalool-d_3	7.95	74 ➜ 43 (8)	96 ➜ 79 (10)
16	Linalool	7.99	93 ➜ 77 (12)	93 ➜ 91 (6)
17	(−)-Isopulegol	8.89	121 ➜ 93 (10)	111 ➜ 55 (10)
18	α-Terpineol	9.45	59 ➜ 31 (8)	59 ➜ 43 (22)
19	Citronellol	9.70	138 ➜ 95 (6)	95 ➜ 67 (8)
20	Nerol	9.75	93 ➜ 77 (12)	93 ➜ 91 (6)
21	Geraniol	9.98	69 ➜ 41 (6)	69 ➜ 39 (14)
22	β-Damascenone	11.32	121 ➜ 105 (8)	121 ➜ 79 (14)
23	β-Caryophyllene	11.50	93 ➜ 77 (12)	133 ➜ 105 (8)
24	β-Damascone	11.57	177 ➜ 121 (12)	177 ➜ 149 (8)
25	α-Humulene	11.78	93 ➜ 77 (12)	93 ➜ 91 (6)

Quantitating terpenes/terpenoids and nicotine | 149

Table 4 Precursor and product ions used for selected reaction monitoring (SRM) analysis of terpenes/terpenoids in GC–MS/MS cross-validation.—cont'd

ID	Compound	Retention time (min)	Quantitation (m/z)	Qualifier (m/z)
26	*Cis*-nerolidol	12.19	93 ➔ 77 (12)	136 ➔ 121 (5)
27	*Trans*-nerolidol	12.36	93 ➔ 77 (12)	136 ➔ 121 (5)
28	(−)-Guaiol	12.71	161 ➔ 105 (8)	161 ➔ 119 (8)
29	(−)-Caryophyllene oxide	12.78	121 ➔ 93 (5)	161 ➔ 105 (5)
30	(−)-α-Bisabolol	13.02	109 ➔ 67 (5)	93 ➔ 77 (12)

Collision energies (V) are included in parentheses following each SRM transition.

The following protocol was used for liquid-injection cross-validation analysis:

1. Extract 200 mg ground plant materials with 20 mL hexane in a 50-mL conical tube by shaking for 1 min, followed by incubation in an ultrasonic bath for 10 min and subsequent 10-s vortex.
2. Centrifuge the sample for 5 min at $3000 \times g$, and collect and dilute the clear supernatant 1000-fold in hexane for liquid-injection GC–MS/MS analysis.
3. Inject 1 µL of sample at an inlet temperature of 220 °C, The same oven temperature gradient as described above was used, and the split flow was 6 mL/min.

2.4 Method validation

Method validation in plant tissue was performed using plant material fortified with select terpenes and terpenoids (*vide infra*). Each headspace vial included 5 mL glycerol, 25 µL internal standard (400 µg/mL linalool-d_3), 25 µL of the fortification (or solvent as method blank), and 50 mg ground, dry stinging nettle where applicable.

Nine calibration levels were prepared at final concentrations of 25–4000 ng/mL in vial (Table 1). Calibration functions for all analytes were fit to linear or quadratic curves, with inverse concentration weighting applied to all analytes. Dynamic range was calculated as the logarithm of the high calibration concentration divided by the low calibration concentration for each analyte. Method detection limits (MDL) were evaluated by determining the minimum measured concentration of an analyte at

25–200 ng/mL in vial (2.5–20 ppm in plant sample; $n=7$ over 3 days) that could be reported as distinct from the method blank with 99% confidence. The method reporting limit (MRL) for each analyte was verified using precision and accuracy specifications as outlined by the United States Environmental Protection Agency [31]. MRLs were established at a fortification concentration of 25–200 ng/mL in vial ($n=7$). Carry-over was assessed by comparing the raw area response of each analyte in the high calibration sample (4000 ng/mL in vial) to that of a method blank analysed immediately after—an area response ratio <0.1% in the method blank was accepted as suitable carry-over. Repeatability ($n=5$; 1250 ng/mL in vial or 125 ppm in plant sample), intermediate precision ($n=17$ over 3 days; 1250 ng/mL in vial or 125 ppm in plant sample) and method recoveries ($n=5$; 3500 ng/mL in vial or 350 ppm in plant sample) were also evaluated.

For vaping liquid (e-juice) method validation, each headspace vial included 5 mL glycerol, 25 µL internal standard (400 µg/mL linalool-d_3), 25 µL of the fortification (or solvent as method blank), and 50 µL glycerol:propylene glycol (1:1 v/v) as a surrogate vaping liquid where applicable. The calibration range included nine levels of 10–3000 ng/mL and 1–300 mg/mL for terpenoids and nicotine, respectively (Table 2). MDLs and MRLs were evaluated by identifying the minimum concentration of an analyte at 10–100 ng/mL (terpenoid) or 1–10 µg/mL (nicotine) in vial (1–10 ppm terpenoids or 0.1–1 mg/mL nicotine in e-juice). Repeatability and intermediate precision were established at fortification concentrations of 750 ng/mL terpenoids and 75 µg/mL nicotine in vial (or 75 ppm terpenoids and 7.5 mg/mL nicotine in surrogate e-juice). All other parameters and settings were identical to those of the plant tissue method validation (*vide supra*).

3. Results and discussion
3.1 Method development and optimisation

The chromatographic gradient was optimised to resolve all 30 mono- and sesquiterpenoids of interest on using SIM mode (Table 3; Figs 1 and 2). A TG-624SilMS column was employed for its low- to mid-polarity stationary phase containing 6% cyanopropylphenyl methylpolysiloxane. Using this thermal program, the stable working conditions for the system were optimised by establishing analyte recoveries and response repeatability. This was evaluated using a range of stock solvents, headspace carrier liquids,

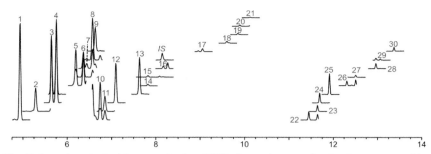

Fig. 2 Sample selected ion monitoring (SIM) chromatogram for the terpenes and terpenoids under investigation. Numbers represent structures depicted in Fig. 1. *IS*: internal standard, (±)-linalool-d_3.

and vial incubation and loop/sample path temperatures (with 10–12 replicate injections under all conditions evaluated).

The first headspace liquid examined was brine, as this is an inexpensive and easily prepared solution that should increase the equilibrium vapour pressure of terpenes. Terpenoid standards were prepared in hexanes at 1 μg/mL and fortified to 5 ng/mL in 5 mL of brine. These samples were analysed with a 5:1 split ratio and vial incubation and sample loop temperatures of 85 °C. The intention with this approach was to leverage the ability of hexane to quantitatively extract terpenes from plant materials with the reduction of co-extraction interferences that headspace analysis would provide. Analyte responses showed low calculated relative standard deviations (RSD; ≤ 5%) for the low-boiling analytes (mostly non-oxygenated monoterpenes, compounds **1–13**, Fig. 1), and larger variations (up to 20% RSD) for the higher boiling components (oxygenated monoterpenoids and sesquiterpenoids, compounds **14–30**, Fig. 1). To improve method precision, different sample path temperatures (95–125 °C) were evaluated, as the higher sample path temperatures might prevent condensation of analytes. In addition, higher septum purge flows (10 and 20 mL/min) were also evaluated for possible flow dynamic effects within the high-temperature headspace sampler. Results showed that an offset of 10–20 °C between vial incubation and sample path temperatures and a purge flow of 20 mL/min improved the precision of instrumental replicates ($n = 10$). At 5 mL/min purge flow, a 100 °C sample path temperature produced all RSD values below 20%, with all but three below 10%. When higher septum purge flows were applied, all RSDs were less than 10% at 100 and 125 °C sample path temperatures (Table 5). Neither increasing septum purge flow beyond 20 mL/min nor changing needle purge flow levels improved the RSDs

Table 5 Calculated absolute response precision (relative standard deviation; RSD) of analytes in brine incubated at 85 °C and analysed with different sample path temperatures and septum purge flows.

		RSD (%)								
	Septum purge flow (mL/min)		**5**			**10**			**20**	
ID	**Sample path temperature (°C)**	**85***	**95***	**100#**	**125***	**100#**	**125#**	**100#**	**125#**	**190#**
1	α-Pinene	2.6	4.1	4.9	2.0	4.3	3.9	2.4	1.6	6.3
2	Camphene	3.5	4.5	5.1	3.8	4.9	4.5	2.4	1.6	6.7
3	β-Myrcene	5.5	5.9	5.6	7.6	5.9	5.9	3.5	2.5	8.1
4	(−)-β-Pinene	5.4	5.5	5.6	7.0	5.7	5.5	3.2	2.3	7.6
5	δ-3-Carene	3.7	4.0	4.9	4.5	4.7	4.6	2.7	1.5	6.4
6	α-Terpinene	9.4	6.3	6.2	9.6	6.0	6.0	3.3	2.8	8.2
7	*Cis*-β-ocimene	3.4	3.8	5.3	3.7	4.7	4.2	2.5	1.9	7.7
8	(+)-Limonene	3.0	4.0	5.0	2.6	4.1	3.9	2.5	1.6	6.5
9	*p*-Cymene	7.4	5.1	5.7	6.8	5.4	5.1	2.8	2.2	6.8
10	*Tans*-β-ocimene	5.5	4.8	5.3	5.9	5.2	4.9	3.1	2.1	7.6
11	8-Cineole (eucalyptol)	4.6	4.6	5.0	4.8	4.8	4.1	2.7	1.3	7.6
12	γ-Terpinene	3.7	3.5	4.8	3.5	4.2	4.1	2.8	1.6	6.4
13	Terpinolene	8.2	4.9	5.8	8.7	6.0	5.3	2.9	2.3	7.6
14	*Cis*-linalool oxide	15	9.1	7.4	15	8.6	7.1	3.7	3.8	9.3
15	*Trans*-linalool oxide	12	7.5	6.5	9.3	6.6	5.2	2.9	2.9	8.0
16	Linalool	11	5.6	6.6	7.0	5.6	4.9	2.9	2.0	10
17	(−)-Isopulegol	10	6.1	7.4	7.2	6.1	5.1	3.2	2.3	9.3
18	α-Terpineol	14	8.7	7.2	13	7.4	6.6	3.4	3.1	11
19	Citronellol	19	9.6	16	36	12	7.5	6.2	5.1	9.0
20	Nerol	7.1	9.3	13	24	10	7.0	4.5	4.6	11
21	Geraniol	20	11	17	41	14	8.0	7.4	7.0	9.6
22	β-Damascenone	4.3	5.4	5.9	3.5	4.6	4.8	2.6	2.2	11
23	β-Caryophyllene	6.1	4.7	5.6	6.1	4.9	5.1	2.5	2.4	7.9
24	β-Damascone	7.8	6.9	6.6	9.1	6.0	6.5	3.6	2.8	11

Table 5 Calculated absolute response precision (relative standard deviation; RSD) of analytes in brine incubated at 85 °C and analysed with different sample path temperatures and septum purge flows.—cont'd

		RSD (%)								
	Septum purge flow (mL/min)	**5**				**10**		**20**		
ID	**Sample path temperature (°C)**	**85***	**95***	**100#**	**125***	**100#**	**125#**	**100#**	**125#**	**190#**
25	α-Humulene	6.3	5.3	5.8	7.6	5.5	5.3	2.8	2.5	7.8
26	*Cis*-nerolidol	11	8.7	8.7	10	6.5	6.3	4.6	6.0	14
27	*Trans*-nerolidol	13	11	9.9	13	8.1	7.0	6.0	7.1	15
28	(−)-Guaiol	14	9.3	8.6	15	7.6	7.0	4.4	5.4	15
29	(−)-Caryophyllene oxide	14	37	9.0	19	7.8	27	5.5	6.2	4.9
30	(−)-α-Bisabolol	14	13	10	14	7.6	6.8	8.2	8.9	15

Values were calculated from 10 to 12 replicates (*: n = 10; #: n = 12).

further. A substantial gap between vial incubation and sample path temperatures (85 and 190 °C) did not improve repeatability (Table 5).

Additional headspace liquids were explored, including dimethyl sulfoxide (DMSO), dimethylformamide (DMF), and glycerol. It was hypothesised that their higher boiling points compared to water might improve the equilibrium vapour pressure of terpenes with lower inherent volatility (i.e. higher boiling terpenes). As well, the ability to use high temperatures with glycerol would facilitate investigations into profiling terpenoids across a wider range of incubation temperatures (*vide infra*). Analysis of terpenes and terpenoids fortified in DMSO and DMF at various incubation temperatures up to 175 °C and 140 °C, respectively, revealed low signal-to-noise ratios for most analytes, and non–detects for many others (data not shown). For this reason, further evaluations with these solvents were not explored. Conversely, when glycerol was used, the most abundant raw responses for higher boiling terpenes were observed (relative to brine), likely due to the utilised vial incubation temperature of 180 °C and a sample path temperature of 200 °C. Of the headspace liquids evaluated, glycerol was unique in permitting the exploration of very high incubation temperatures due to its boiling point of 290 °C, versus 189 °C and 153 °C for DMSO and DMF, respectively. This facilitated better transfer of analytes to the gas phase, and improved precision.

It was interesting to note that analyte recoveries for calibrators and spike-recovery samples using glycerol as the headspace solvent were dependent on the solvent used to prepare the terpenoid stock solutions. The low-boiling solvent hexane (68 °C) and the high-boiling solvent tetradecane (254 °C) were used to investigate this effect. Analyses of samples fortified with terpenoids in the absence of plant material (i.e., a calibration sample) in hexane-based stocks showed that glycerol samples displayed higher absolute responses for low-boiling terpenoids (compounds **1–13**, Fig. 1), but lower absolute responses (as low as 10–20%) for high-boiling terpenoids (compounds **14–30**, Fig. 1), compared to brine samples. When the analogous tetradecane-based terpenoid stocks were used, most analytes displayed higher responses with glycerol compared to brine. This suggested that the solvents used to prepare stock solutions increased the partial pressure in the headspace such that the amount of analyte in the headspace decreased when the low-boiling hexane was used. It is postulated that the higher boiling point of tetradecane mitigated this effect. When plant material was also present in the headspace vials, higher absolute responses relative to the data obtained in brine were observed for the majority of terpenoids when glycerol was used regardless of terpenoid stock solvents (hexane or tetradecane). The reason for this effect is currently being explored.

Based on these investigations, terpenoid stock solutions for calibration samples and spike-recovery tests were prepared in tetradecane and glycerol was selected as the optimal headspace carrier liquid.

3.2 Method validation for plant tissue analysis

The developed method was evaluated for dynamic range, repeatability ($n = 5$), intermediate precision ($n = 17$, over 3 days), carry-over, MDL, MRL, and spike-recovery. Calibration functions for all 30 terpenes and terpenoids were determined from 25 to 4000 ng/mL (equivalent to 2.5–400 ppm in plant material; dynamic range $= 2.2$). The split ratio can be adjusted to accommodate different expected terpenoid concentrations, or to permit for the evaluation of different quantities of plant material (i.e. not 50 mg). However, the issue of suitable sample quantity to overcome potential homogeneity issues needs to be considered; erring on the side of more material sampled for a given analysis is generally the prudent approach, with the authors suggesting 50 mg as a minimum quantity. Across a range of 25–4000 ng/mL, all calibration functions had correlation coefficients > 0.99 and the accuracy of all calibration concentrations were $\pm 20\%$ (data not shown). The calibration functions were

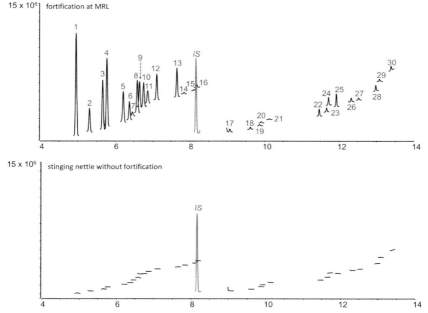

Fig. 3 Sample SIM chromatogram for the 30 terpenes/terpenoids under investigation in the surrogate matrix, dry stinging nettle tissue (bottom), compared to that of standards at method reporting limit concentrations (top). Names and structures of the terpenoids are depicted in Fig. 1. *IS* is internal standard, (±)-linalool-d_3.

consistent across 3 days of method validation and subsequent sample analyses. Additionally, all analytes presented with <0.1% carry-over when a method blank was analysed immediately after a high concentration calibration sample.

Stinging nettle was selected as the matrix for method development and validation as the dry tissues of this plant contain nominal incurred terpenoids, which made it an ideal blank matrix with which spike-recovery studies could be conducted (Fig. 3). This was necessary for the method development reported herein, as the intended matrices for this method (i.e., cannabis and industrial hemp) have high concentrations of many terpenoids, such that performing spike-recovery studies would be problematic due to limitations in dynamic range. Using dry, ground stinging nettle tissue, terpenoid spike-recoveries of at least 70% were achieved, with 27 out of 30 terpenes and terpenoids reaching 80–108% at 10 ppm (Table 6). At the evaluated mid and high spike-recovery concentrations (125 and 400 ppm, respectively), the spike-recovery data ranged from 82% to 104% for all analytes except (−)-caryophyllene oxide, which has a high

Table 6 Method validation summary for quantitating terpenoids in plant materials by headspace GC–MS.

ID	Analyte	Limit (ppm)		Recovery (%)			RSD (%)	
		MDL	MRL	Low (10 ppm)	Mid (125 ppm)	High (350 ppm)	Intraday	Interday
1	α-Pinene	0.95	10.0	84.2	86.4	83.3	1.34	8.70
2	Camphene	0.87	10.0	83.3	86.2	85.1	0.85	7.33
3	β-Myrcene	1.13	10.0	82.8	89.1	87.9	1.19	4.77
4	(−)-β-Pinene	1.03	10.0	80.6	88.5	86.4	1.07	4.12
5	δ-3-Carene	1.28	10.0	83.4	89.0	85.5	0.90	3.41
6	α-Terpinene	1.72	10.0	99.5	93.6	84.8	1.90	11.2
7	*Cis*-β-ocimene	1.57	10.0	74.2	97.4	88.5	1.95	7.60
8	(+)-Limonene	1.37	10.0	80.4	91.5	85.2	1.25	6.38
9	*p*-Cymene	1.19	10.0	70.2	81.8	82.8	1.30	6.92
10	*Trans*-β-ocimene	1.36	10.0	90.5	95.6	89.3	1.07	6.36
11	8-Cineole (eucalyptol)	0.93	10.0	81.8	95.1	92.7	1.39	8.71
12	γ-Terpinene	0.94	10.0	84.0	101	89.1	0.90	6.02
13	Terpinolene	1.39	10.0	93.4	87.5	85.9	2.11	5.77
14	*Cis*-linalool oxide	2.10	10.0	93.3	96.9	91.6	3.65	6.76
15	*Trans*-linalool oxide	1.32	10.0	95.3	102	102	2.32	5.76

16	Linalool	1.09	10.0	97.4	103	99.0	1.24	2.29
17	(−)-Isopulegol	1.19	10.0	94.9	95.0	101	1.81	4.79
18	α-Terpineol	1.41	10.0	91.5	96.1	102	2.58	6.21
19	Citronellol	1.30	10.0	79.5	90.4	101	1.76	4.91
20	Nerol	3.15	10.0	104	98.0	96.6	2.13	9.41
21	Geraniol	2.01	10.0	79.3	93.5	90.8	3.74	5.67
22	β-Damascenone	1.15	10.0	96.5	95.9	91.5	2.14	3.15
23	β-Caryophyllene	2.43	10.0	110	99.9	85.3	1.87	5.10
24	β-Damascone	0.98	10.0	88.5	90.1	91.5	2.02	6.41
25	α-Humulene	2.49	10.0	87.6	98.2	84.8	2.38	5.61
26	*Cis*-nerolidol	2.02	10.0	86.9	86.8	94.0	2.38	8.10
27	*Trans*-nerolidol	2.54	10.0	90.1	87.8	95.3	1.50	4.29
28	(−)-Guaiol	1.77	10.0	95.0	92.5	99.4	2.20	4.75
29	(−)-Caryophyllene oxide	3.10	10.0	108	104	129	3.46	6.05
30	(−)-α-Bisabolol	2.42	10.0	87.5	87.1	93.0	2.83	5.80

Concentrations are presented based on plant weight. MDL: method detection limit. MRL: method reporting limit. The calibration ranges were from 2.5 to 400 ppm (25–4000 ppb in vial), except nerol and (−)-caryophyllene oxide (5–400 ppm in samples or 50–4000 ppb in vial).

concentration spike-recovery of 129% (Table 6). MDLs for all analytes were less than 4 ppm in plant (40 ppm in vial), and MRLs were verified at 10 ppm in plant (100 pm in vial) for 50 mg of plant material (Table 6). Lower MRLs could be achieved for many analytes, but given the expected concentrations in cannabis and industrial hemp, it was not necessary to fully optimise this statistically verified reporting limit. All of the method validation data reported herein demonstrates the suitability of the method for the intended goal of quantitating terpenoids.

As a relatively new market segment, there is currently a dearth of proficiency testing and/or certified reference materials for cannabis and industrial hemp. This is especially true for terpenoids that, unlike cannabinoids or pesticides, are not faced with any regulatory restrictions regarding their absolute concentrations. The absence of these materials makes it challenging to demonstrate the accuracy of an analytical method for a broad range of terpenoids—a new program exists in the United States for hemp flower, but the number of terpene values provided is limited to 10. To provide a better assessment of method performance for all terpenes and terpenoids evaluated herein, a cross-validation approach was utilised to verify method accuracy. Since the presence of incurred terpenoids was not a concern for the cross-validation, hops (*Humulus lupulus* L. flower), a close phylogenetic relative of the family Cannabaceae (i.e. cannabis), was used.

The method used for the cross-validation is an ISO/IEC 17025 accredited method for quantitating terpenes in plant material based on a hexane extraction, followed by direct analysis of the organic extract using GC-tandem mass spectrometry (method validation parameters were consistent with those reported herein for the high-temperature headspace method; Table 7). In addition to being accredited, this method has also passed several rounds of proficiency testing for terpenes in hemp oil. While not an exact matrix match, these results do provide strong support for the accuracy of the method as a benchmark. When comparing analytes above the respective MRLs for each method, a correlation trend was obtained that demonstrated the accuracy of the high-temperature headspace method developed herein (Fig. 4A).

3.3 Method validation for e-juices and related vaping products

The original focus of this method development project was on the quantitation of terpenes in plant material. Paralleling our method development efforts in plant material were reports coming out about the safety of

Table 7 Method validation using stinging nettle as the blank matrix for quantitating terpenoids in plant materials by GC–MS/MS analysis with direct injection of liquid extracts.

ID	Analyte	Limit (ppm)		Recovery (%)			RSD (%)	
		MDL	MRL	Low (200 ppm)	Mid (500 ppm)	High (2000 ppm)	Intraday	Interday
1	α-Pinene	44.1	200	73.3	87.2	81.1	3.09	22.6
2	Camphene	51.2	200	73.2	81.2	82.9	3.09	28.7
3	β-Myrcene	52.0	200	75.2	85.8	83.8	3.55	26.4
4	(−)-β-Pinene	53.3	200	74.3	84.7	83.5	3.26	27.3
5	δ-3-Carene	57.3	200	73.2	84.4	82.1	3.19	27.9
6	α-Terpinene	46.4	200	77.2	85.6	84.9	3.37	26.8
7	Cis-β-ocimene	51.4	200	74.6	85.1	84.2	3.41	27.0
8	(+)-Limonene	38.1	200	73.3	87.2	83.1	3.66	25.9
9	p-Cymene	47.4	200	73.6	84.5	82.9	3.88	24.9
10	Trans-β-ocimene	45.0	200	75.1	87.4	83.8	3.64	26.1
11	8-Cineole (eucalyptol)	46.8	200	77.5	86.6	83.3	3.21	24.9
12	γ-Terpinene	45.2	200	75.1	87.0	84.0	3.40	23.3
13	Terpinolene	48.4	200	74.8	86.8	84.7	3.56	24.3
14	Cis-linalool oxide	49.6	200	96.6	94.9	101	3.67	22.1
15	Trans-linalool oxide	56.2	200	97.5	94.9	100	3.30	22.2
16	Linalool	38.5	200	78.1	90.5	85.2	3.39	22.0
17	(−)-Isopulegol	60.0	200	82.4	87.9	85.9	3.99	29.1

Continued

Table 7 Method validation using stinging nettle as the blank matrix for quantitating terpenoids in plant materials by GC–MS/MS analysis with direct injection of liquid extracts.—cont'd

ID	Analyte	Limit (ppm)		Recovery (%)			RSD (%)	
		MDL	MRL	Low (200 ppm)	Mid (500 ppm)	High (2000 ppm)	Intraday	Interday
18	α-Terpineol	53.3	200	95.9	94.1	102	3.97	20.6
19	Citronellol	48.4	200	94.2	87.5	98.0	4.84	27.5
20	Nerol	48.4	200	90.5	85.4	97.0	3.34	23.5
21	Geraniol	50.8	200	70.4	79.3	85.6	4.70	21.3
22	β-Damascenone	53.2	200	93.3	94.4	96.1	5.10	24.1
23	β-Caryophyllene	56.7	200	80.9	86.3	87.4	5.56	29.0
24	β-Damascone	58.6	200	81.2	94.2	83.5	5.53	27.8
25	α-Humulene	56.4	200	81.5	86.8	87.1	5.90	28.4
26	Cis-nerolidol	63.1	200	82.1	90.3	87.4	6.95	27.2
27	Trans-nerolidol	49.7	200	81.5	86.4	86.8	7.97	29.0
28	(−)-Guaiol	59.6	200	85.5	87.3	88.8	7.71	28.3
29	(−)-Caryophyllene oxide	34.7	200	85.8	78.5	86.6	10.5	28.0
30	(−)-α-Bisabolol	66.3	200	87.0	86.7	88.4	11.2	30.0

Concentrations are presented based on plant weight. MDL, method detection limit; MRL, method reporting limit. The calibration ranges were from 400 to 10,000 ppm.

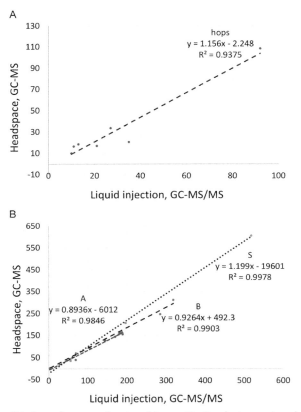

Fig. 4 Cross-validation of terpene/terpenoid quantitation between headspace GC–MS and liquid-injection GC–MS/MS. (A) Dry hop tissue (unit: ppm). (B) Commercial cannabis terpene mixtures (unit: mg/mL).

e-cigarettes and related vaping products intended to be inhaled after high-temperature vaporisation [18,20,32]. E-juices are generally formulated in a mixture of glycerol and propylene glycol, with other common ingredients including nicotine, caffeine and flavouring agents (i.e. terpenes) [33]. Given our finding that glycerol was a suitable matrix for the high-temperature headspace quantitation of terpenes, it was decided that the scope of this method should be extended to include the quantitation of terpenes in e-juices and related products. Understanding the terpene concentrations is likely to become a critical regulatory issue, as it has been documented that flavouring agents can decompose during consumer use of e-cigarettes to produce potentially harmful breakdown products [18,34–36]. Since

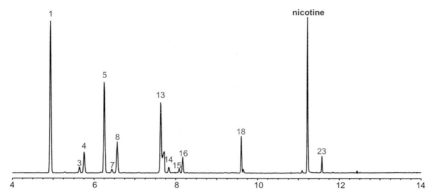

Fig. 5 Extracted ion chromatogram from the analysis of a commercial e-juice for select terpenes, terpenoids, and nicotine in the same injection. Terpene/terpenoid and nicotine peaks were extracted with m/z 93 and 84, respectively. Numbers represent structures depicted in Fig. 1.

nicotine is frequently included in e-juice formulations, it was also included in the expansion of the analytical scope of the method reported herein (Fig. 5).

A mixture of glycerol and propylene glycol (1:1 ratio) was used as a surrogate for typical e-juice matrices to develop the method for terpene/terpenoid and nicotine quantitation. Using this mixture, terpenoid spike-recoveries of at least 93% were obtained (Table 8). At the evaluated mid and high spike-recovery concentrations in e-juice (75 and 250 ppm for terpenoids, and 7.5 and 25 mg/mL for nicotine), the spike-recovery data ranged from 83% to 104% for all analytes. MDLs for all terpenoids and nicotine were less than 3 ppm for terpenoids and 0.1 mg/mL for nicotine, and MRLs were verified at 10 ppm for terpenoids and 0.5 mg/mL for nicotine (Table 8). These method performance parameters are all suitable for the intended scope, with typical nicotine concentrations in e-juices ranging mostly from 1 to 30 mg/mL [37,38].

Using this method, the terpenoids of interest were quantitated in commercially available terpenoid distillates and nicotine-containing e-juices. Owing, again, to a lack of suitable proficiency testing and/or certified reference materials, a cross-validation approach (using the same method as reported for cross-validation of terpene quantitation in plant material) was utilised to evaluate method accuracy for the terpenoid distillates. The data produced were in excellent agreement (Fig. 4B), demonstrating the applicability of this method for the quantitation of terpenoids in e-juices and other glycerol-based vaporisation products.

Quantitating terpenes/terpenoids and nicotine 163

Table 8 Method validation for quantitating terpenoids and nicotine in e-juices by headspace GC–MS.

ID	Analyte	Limit[a]		Recovery (%)[b]			RSD (%)	
		MDL	MRL	Low	Mid	High	Intraday	Interday
1	α-Pinene	0.04	5.00	105	94.6	101	2.64	4.86
2	Camphene	0.05	5.00	104	94.2	99.9	2.16	4.35
3	β-Myrcene	0.13	5.00	99.7	95.9	99.1	1.78	2.84
4	(−)-β-Pinene	0.22	5.00	105	99.0	104	2.24	4.04
5	δ-3-Carene	0.16	5.00	105	98.8	104	2.08	3.97
6	α-Terpinene	0.13	5.00	101	94.7	99.6	1.91	4.01
7	*Cis*-β-ocimene	0.77	5.00	102	97.2	99.0	2.44	2.40
8	(+)-Limonene	0.12	5.00	98.7	94.8	99.2	2.36	3.58
9	*p*-Cymene	0.10	5.00	100	94.2	99.6	2.59	4.15
10	*Trans*-β-ocimene	0.62	5.00	99.0	94.3	98.5	2.36	3.49
11	8-Cineole (eucalyptol)	0.18	5.00	101	95.3	98.6	1.81	2.91
12	γ-Terpinene	0.16	5.00	101	94.4	99.3	2.26	3.82
13	Terpinolene	0.33	5.00	105	97.8	102	2.21	3.66
14	*Cis*-linalool oxide	0.59	5.00	104	96.5	99.3	3.01	2.78
15	*Trans*-linalool oxide	1.83	5.00	105	102	104	3.02	2.49
16	Linalool	1.35	100	101	95.7	97.0	2.65	2.18
17	(−)-Isopulegol	0.35	5.00	101	96.2	98.2	3.18	2.52
18	α-Terpineol	0.53	5.00	97.3	93.2	96.7	3.81	3.65
19	Citronellol	2.54	10.0	97.8	95.3	94.7	2.69	2.83
20	Nerol	0.69	5.00	96.5	96.1	99.4	3.61	3.49
21	Geraniol	1.66	10.0	104	91.3	96.0	4.72	4.61
22	β-Damascenone	0.67	5.00	98.3	103	103	1.74	1.82
23	β-Caryophyllene	0.50	5.00	103	95.5	101	1.87	4.31
24	β-Damascone	0.63	5.00	95.7	94.2	95.6	1.54	2.30
25	α-Humulene	0.74	5.00	105	96.5	101	1.56	3.86
26	*Cis*-nerolidol	1.58	10.0	100	96.2	98.3	2.13	2.11

Continued

Table 8 Method validation for quantitating terpenoids and nicotine in e-juices by headspace GC–MS.—cont'd

ID	Analyte	Limit[a]		Recovery (%)[b]			RSD (%)	
		MDL	MRL	Low	Mid	High	Intraday	Interday
27	*Trans*-nerolidol	0.41	5.00	100	93.7	96.9	1.97	2.95
28	(−)-Guaiol	0.39	5.00	94.6	89.0	93.3	0.92	3.58
29	(−)-Caryophyllene oxide	0.33	5.00	97.8	97.8	97.3	2.08	2.14
30	(−)-α-Bisabolol	0.96	5.00	101	96.0	97.5	1.34	2.15
31	Nicotine	0.05	0.50	94.3	101	102	3.53	2.71

[a]Unit: ppm for terpenoids, and mg/mL for nicotine.
[b]Concentration: Low-, mid-, and high-level concentrations for terpenoids are 10, 75, and 250 ppm, respectively. Low-, mid-, and high-level concentrations for nicotine are 1, 7.5, and 25 mg/mL, respectively.
Concentrations are presented based on e-juice volume. MDL: method detection limit. MRL: method reporting limit. The calibration ranges for terpenoids were from 1 to 300 ppm (10–3000 ppb in vial), except *trans*-β-ocimene, linalool, citronellol, and geraniol (5–300 ppm, or 50–3000 ppb in vial). The calibration range for nicotine was from 0.25 to 30 mg/mL (2.5–300 μg/mL in vial).

In addition to terpenoids, the method reported here is suitable for the analysis of nicotine (Fig. 5 and Table 8)—a bioactive compound for which concerns about the combined neurological impact with cannabis [39,40] has been raised. In 2018, the Canadian Task Force on Cannabis Legalisation and Regulation also recommended that mixed products, including those containing both cannabis and nicotine and/or caffeine, be prohibited [41]. In this context, this method could be used as a regulatory tool to screen for nicotine, while at the same time quantitating (or profiling) the terpenoids present.

3.4 Possibility of concurrent cannabis potency test

For cannabis and related products, identification and quantitation of major cannabinoids constitute the primary analytical purpose. Liquid chromatography is routinely used for this purpose [42]. GC is less commonly used for cannabis potency test as it is not effective in differentiating acidic and neutral forms of cannabinoids, as the former would be converted by heat to the latter during analysis [43]. Nevertheless, GC analysis conditions may potentially be used to mimic those of smoking and vaping, and the analyte profiles may reflect what the users are exposed to.

Using a capillary GC column with an intermediate polarity (diphenyl: dimethylpolysiloxane stationary phase = 35:65), TG-35MS, we attempted

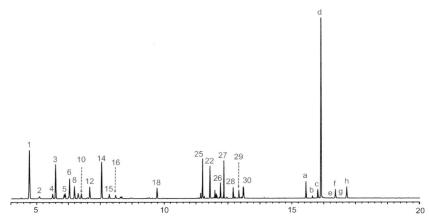

Fig. 6 Extracted ion chromatogram for select terpenes/terpenoids and cannabinoids from headspace GC–MS analysis of a cannabis sample in a single injection. Names and structures of the terpenes/terpenoids are depicted in Fig. 1. The cannabinoids are: cannabidivarin (a), tetrahydrocannabivarin (b), cannabichromene (c), cannabidiol (d), Δ^8-tetrahydrocannabinol (e), Δ^9-tetrahydrocannabinol (f), cannabigerol (g), and cannabinol (h).

to analyse cannabinoids together with terpenoids. Our results showed that we could detected eight neutral cannabinoids at 10 μg level in headspace vial (or 0.02% of plant materials if 50 mg was used). Our 20-min temperature gradient was also sufficient to resolve all 30 select terpenes/terpenoids and eight cannabinoids. The response levels of cannabinoids and terpenes/terpenoids are comparable when the former was much more abundant than the latter (e.g. 80-folds; Fig. 6). This level of difference between cannabinoids and terpenoids are close to the range often observed in cannabis materials [44]. With this method, it is entirely possible to quantitate major groups of plant bioactive compounds of interest in cannabis and related products using same-injection analyses.

4. Conclusion

The headspace GC–MS analysis method presented here facilitates the precise quantitation of select terpenes and terpenoids in plant materials and vaping liquids with minimal sample preparation. Sample matrices (plant tissues and liquid extracts) do not affect method performance and critical non-terpenoid components of e-juices, such as nicotine, can also be included in the analysis. Our method also has the potential to be optimised further for quantitative analysis of all major groups of bioactive compounds in cannabis-related and vaping products, namely, cannabinoids, terpenoids, nicotine, and caffeine in same injections.

Acknowledgement

Most of the work presented is adopted from an article by the same authors published in the journal *ACS Omega*: https://doi.org/10.1021/acsomega.0c00384. We thank Mitacs for an Elevate Fellowship to T.D.N.

References

[1] L.B. Pickens, Y. Tang, Y.-H. Chooi, Metabolic engineering for the production of natural products, Annu. Rev. Chem. Biomol. Eng. 2 (2011) 211–236.

[2] G.D. Wright, Unlocking the potential of natural products in drug discovery, J. Microbial. Biotechnol. 12 (2019) 55–57.

[3] U. Ravid, M. Elkabetz, C. Zamir, K. Cohen, O. Larkov, R. Aly, Authenticity assessment of natural fruit flavour compounds in foods and beverages by auto-HS–SPME stereoselective GC–MS, Flavour Fragr. J. 25 (2010) 20–27.

[4] N.W. Davies, T. Larkman, P.J. Marriott, I.A. Khan, Determination of enantiomeric distribution of terpenes for quality assessment of Australian tea tree oil, J. Agric. Food Chem. 64 (2016) 4817–4819.

[5] D.W. Christianson, Roots of biosynthetic diversity, Science 316 (2007) 60–61.

[6] R. Croteau, T.M. Kutchan, N.G. Lewis, Natural products (secondary metabolites), in: B. Buchanan, W. Gruissem, R.L. Jones (Eds.), Biochemistry and Molecular Biology of Plants, American Society of Plant Physiologists, Rockville, MD, 2000, pp. 1250–1318.

[7] P.K. Ajikumar, K. Tyo, S. Carlsen, O. Mucha, T.H. Phon, G. Stephanopoulos, Terpenoids: opportunities for biosynthesis of natural product drugs using engineered microorganisms, Mol. Pharm. 5 (2008) 167–190.

[8] M. Huang, J.J. Lu, M.Q. Huang, J.L. Bao, X.P. Chen, Y.T. Wang, Terpenoids: natural products for cancer therapy, Expert Opin. Investig. Drugs 21 (2012) 1801–1818.

[9] F.H. Abdel-Rahman, N.M. Alaniz, M.A. Saleh, Nematicidal activity of terpenoids, J. Environ. Sci. Health 48 (2013) 16–22.

[10] L. Caputi, E. Aprea, Use of terpenoids as natural flavouring compounds in food industry, Recent Pat. Food Nutr. Agric. 3 (2012) 9–16.

[11] J.T. Fischedick, Identification of terpenoid chemotypes among high $(-)$-trans-Δ^9-tetrahydrocannabinol-producing *Cannabis sativa* L. cultivars, Cannabis Cannabinoid Res. 2 (2017) 34–47.

[12] J. Gertsch, M. Leonti, S. Raduner, I. Racz, J.Z. Chen, X.Q. Xie, K.H. Altmann, M. Karsak, A. Zimmer, Beta-caryophyllene is a dietary cannabinoid, Proc. Natl. Acad. Sci. U. S. A. 105 (2008) 9099–9104.

[13] S.S. Dahham, Y.M. Tabana, M.A. Iqbal, M.B.K. Ahamed, M.O. Ezzat, A.S.A. Majid, A.M.S.A. Majid, The anticancer, antioxidant and antimicrobial properties of the sesquiterpene β-caryophyllene from the essential oil of *Aquilaria crassna*, Molecules 20 (2015) 11808–11829.

[14] E.B. Russo, Taming THC: potential cannabis synergy and phytocannabinoid-terpenoid entourage effects, Br. J. Pharmacol. 163 (2011) 1344–1364.

[15] J.K. Booth, J. Bohlmann, Terpenes in *Cannabis sativa*—from plant genome to humans, Plant Sci. 284 (2019) 67–72.

[16] Z. Zeng, F.T. Chau, H.Y. Chan, C.Y. Cheung, T.Y. Lau, S. Wei, D.K.W. Mok, C.O. Chan, Y. Liang, Recent advances in the compound-oriented and pattern-oriented approaches to the quality control of herbal medicines, Chinas Med. 3 (2008) 1–7.

[17] Health Canada, Quality of Natural Health Products Guide 3.1, 2015.

[18] H.C. Erythropel, S.V. Jabba, T.M. DeWinter, M. Mendizabal, P.T. Anastas, S.E. Jordt, J.B. Zimmerman, Formation of flavorant-propylene glycol adducts with novel toxicological properties in chemically unstable e-cigarette liquids, Nicotine Tob. Res. 21 (2019) 1248–1258.

[19] J. Buckell, J. Marti, J.L. Sindelar, Should flavors be banned in combustible and electronic cigarettes? Evidence on adult smokers and recent quitters from a discrete choice experiment, Tob. Control 28 (2018) 168–175.

[20] H. Ledford, Scientists chase cause of mysterious vaping illness, Nature 574 (2019) 303–304.

[21] J.M. McPartland, E.B. Russo, Cannabis and cannabis extracts, J. Cannabis Ther. 1 (2001) 103–132.

[22] Z. Piñeiro, M. Palma, C.G. Barroso, Determination of terpenoids in wines by solid phase extraction and gas chromatography, Anal. Chim. Acta 513 (2004) 209–214.

[23] Z. Jiang, C. Kempinski, J. Chappell, Extraction and analysis of terpenes/terpenoids, Curr. Protoc. Plant Biol. 1 (2016) 345–358.

[24] European Centre for Ecotoxicology and Toxicology of Chemicals, ECETOC Technical Report No. 117—Understanding the Relationship Between Extraction Technique and Bioavailability, 2013.

[25] G.A. Mills, V. Walker, Headspace solid-phase microextraction procedures for gas chromatographic analysis of biological fluids and materials, J. Chromatogr. A 902 (2000) 267–287.

[26] M.J. Sithersingh, N.H. Snow, Headspace-gas chromatography, in: C.F. Pool (Ed.), Gas Chromatography, Elsevier Inc, 2012, pp. 221–233.

[27] A. Shapira, P. Berman, K. Futoran, O. Guberman, D. Meiri, Tandem mass spectrometric quantification of 93 terpenoids in Cannabis using static headspace injections, Anal. Chem. 91 (2019) 11425–11432.

[28] N.C. Bouvier-Brown, R. Holzinger, K. Palitzsch, A.H. Goldstein, Quantifying sesquiterpene and oxygenated terpene emissions from live vegetation using solid-phase microextraction fibers, J. Chromatogr. A 1161 (2007) 113–120.

[29] M. Dziadas, H.H. Jeleń, Analysis of terpenes in white wines using SPE-SPME-GC/MS approach, Anal. Chim. Acta 677 (2010) 43–49.

[30] T.F. Jorge, A.T. Mata, C. António, Mass spectrometry as a quantitative tool in plant metabolomics, Philos. Trans. A Math. Phys. Eng. Sci. 374 (2016) 20150370.

[31] United States Environmental Protection Agency, Definition and procedure for the determination of the method detection limit, Revision 2, December 2016, Retrieved: April 2020 from https://www.epa.gov/sites/production/files/2016-12/documents/mdl-procedure_rev2_12-13-2016.pdf.

[32] T. Basáñez, A. Majmundar, T.B. Cruz, J.P. Allem, J.B. Unger, E-cigarettes are being marketed as "vitamin delivery" devices, Am. J. Public Health 109 (2019) 194–196.

[33] National Academies of Sciences, Engineering, and Medicine of the United States of America, Public Health Consequences of E-cigarettes, The National Academies Press, Washington, DC, 2018.

[34] S. Klager, J. Vallarino, P. MacNaughton, D.C. Christiani, Q. Lu, J.G. Allen, Flavoring chemicals and aldehydes in e-cigarette emissions, Environ. Sci. Technol. 51 (2017) 10806–10813.

[35] Z.T. Bitzer, R. Goel, S.M. Reilly, R.J. Elias, A. Silakov, J. Foulds, J. Muscat, J.P. Richie, Effect of flavoring chemicals on free radical formation in electronic cigarette aerosols, Free Radic. Biol. Med. 120 (2018) 72–79.

[36] Y. Son, V. Mishin, J.D. Laskin, G. Mainelis, O.A. Wackowski, C. Delnevo, S. Schwander, A. Khlystov, V. Samburova, Q. Meng, Hydroxyl radicals in e-cigarette vapor and e-vapor oxidative potentials under different vaping patterns, Chem. Res. Toxicol. 32 (2019) 1087–1095.

[37] J.F. Etter, E. Zäther, S. Svensson, Analysis of refill liquids for electronic cigarettes, Addiction 108 (2013) 1671–1679.

[38] J.M. Cameron, D.N. Howell, J.R. White, D.M. Andrenyak, M.E. Layton, J.M. Roll, Variable and potentially fatal amounts of nicotine in e-cigarette nicotine solutions, Tob. Control 23 (2014) 77–78.

[39] T.R. Gray, R.D. Eiden, K.E. Leonard, G.J. Connors, S. Shisler, M.A. Huestis, Identifying prenatal cannabis exposure and effects of concurrent tobacco exposure on neonatal growth, Clin. Chem. 56 (2010) 1442–1450.

[40] F.M. Filbey, T. McQueeny, S. Kadamangudi, C. Bice, A. Ketcherside, Combined effects of marijuana and nicotine on memory performance and hippocampal volume, Behav. Brain Res. 293 (2015) 46–53.

[41] Government of Canada, Regulations Amending the Cannabis Regulations (New classes of Cannabis): SOR/2019-206, 153, Canada Gazette, 2018, pp. 3558–3728.

[42] M. Wang, Y.H. Wang, B. Avula, M.M. Radwan, A.S. Wanas, J. Van Antwerp, J.F. Parcher, M.A. Elsohly, I.A. Khan, Decarboxylation study of acidic cannabinoids: a novel approach using ultra-high-performance supercritical fluid chromatography/photodiode array-mass spectrometry, Cannabis Cannabinoid Res. 1 (2016) 262–271.

[43] E.A. Ibrahim, W. Gul, S.W. Gul, B.J. Stamper, G.M. Hadad, R.A. Abdel Salam, A.K. Ibrahim, S.A. Ahmed, S. Chandra, H. Lata, M.M. Radwan, M.A. Elsohly, Determination of acid and neutral cannabinoids in extracts of different strains of *Cannabis sativa* using GC-FID, Planta Med. 84 (2018) 250–259.

[44] D. Namdar, H. Voet, V. Ajjampura, S. Nadarajan, E. Mayzlish-Gati, M. Mazuz, N. Shalev, H. Koltai, Terpenoids and phytocannabinoids co-produced in *Cannabis sativa* strains show specific interaction for ell cytotoxic activity, Molecules 24 (2019) 3031.

CHAPTER SIX

Improving cannabis differentiation by expanding coverage of the chemical profile with GCxGC-TOFMS

Elizabeth M. Humston-Fulmer*, David E. Alonso, Joseph E. Binkley*

LECO Corporation, Saint Joseph, MI, United States
*Corresponding author: e-mail address: joe_binkley@leco.com

Contents

1.	Classifications of cannabis and the need for nontargeted chemical analyses	169
2.	Analytical tools and methods provide nontargeted chemical profiling for chemovar classifications	172
	2.1 Chromatography	173
	2.2 Mass spectrometry	175
	2.3 Multidimensional chromatography	178
3.	Application of GCxGC-TOFMS for nontargeted chemical analysis and chemovar classifications	184
4.	Methods	194
	4.1 Definition	194
	4.2 Rationale	194
	4.3 Materials, equipment, and reagents	194
	4.4 Protocols	195
	4.5 Analysis and statistics	195
	4.6 Summary	195
References		196

1. Classifications of cannabis and the need for nontargeted chemical analyses

Cannabis is a complex botanical consisting of a diverse group of chemicals in a wide concentration range. The plant has been cultivated in Asia since the beginning of recorded history and used for a wide range of purposes. In China, cannabis was used for the production of various goods including strings, ropes, textiles, and paper [1]. It was used for both medicinal and ceremonial purposes in India, and listed as a sacred plant associated with happiness, joy, and liberation in ancient Hindu writings (Atharva Veda) [2]. Cannabis reached the Western world in the 19th Century and was one of the three most prescribed medicines in the United States from the mid-to-late 1800s. In 1843, the Irish physician William O'Shaughnessy called cannabis an "extraordinary agent," and described the broad medical benefits in the *Provincial Medical Journal* and *Retrospect of the Medical Sciences* [3]. Its therapeutic utility as a sedative, hypnotic, and analgesic was detailed in Sajous's *Analytic Cyclopedia of Practical Medicine* in the 1920s [4]. Unfortunately, there were significant inconsistencies in the cannabis preparations of the period, which resulted in unacceptable clinical variability of cannabis as medication [5]. It was during this time that countries around the world initiated the process of legally banning or severely restricting cannabis and its products. The United States eventually passed the Controlled Substances Act in 1970 prohibiting the use of cannabis altogether. Today, cannabis is the most commonly cultivated, trafficked, and abused illicit drug [6].

In recent years, the discovery of cannabinoid receptors, together with the synthesis and structural elucidation of cannabinoids has led to a renewed interest in cannabis, its physiological effects, and potential therapeutic applications. With these advancements, there is a growing trend toward legalization of cannabis for both medical and recreational use around the world. In light of this increasing global demand for cannabis, there is an urgent need for establishment of standardized regulations and testing protocols. To this end, robust analytical methods are needed to support this active area of research. Accurate documentation relating to the cannabis products, their classification, chemical composition and safety should be readily available to growers, distributors, medical providers, researchers, and consumers. Implementation of novel analytical methods and improved classification strategies are vital for the development of well-defined and reproducible cannabis materials.

Historically cannabis was classified into three species or "strains": (1) *C. indica* (medicinal or drug-type), (2) *C. sativa* (fibre or hemp type) and (3) *C. ruderalis* (intermediate, wild-type) [7]. These classifications are based on plant morphology and place of origin. *Indica* plants are relatively short and densely branched with wide leaves. In contrast, *sativa* plants are tall with long and narrow leaves. The *indica* strain is associated with relaxing or calming properties, appetite stimulation, sleep promotion, and pain relief. *Sativa* strains have a reputation of being stimulating, uplifting, and are recommended for daytime activities. This cultivar method of classification has been used as a predictor of physiological effects, but it is highly contested in the scientific community. Decades of cannabis hybridization have resulted in hundreds of different strains with prescribed effects that are ambiguous or worse yet, inaccurate. Underground growers with the goal of producing hybrids with increasing amounts of Δ9-tetrahydrocannabinol (Δ9-THC), the psychoactive component in cannabis, have further increased the variability of strains. Recently, McPartland has argued that "Categorizing cannabis as either '*Sativa*' or '*Indica*' has become an exercise in futility. Ubiquitous interbreeding and hybridization renders their distinctions meaningless" [8]. In fact, some scientists have concluded that there is actually only one highly variable species or strain of cannabis.

As cultivar classification has been deemed less reliable, there has been a shift toward classifying cannabis into chemical types or chemotypes (I–V). This classification is based on cannabinoid ratios of Δ9-THC, cannabidiol (CBD), and cannabigerol (CBG) in plants [9]. Type I cannabis (Δ9-THC dominant) and Type II (Δ9-THC/CBD codominant) describe a large portion of the recreational and medical cannabis used globally today. Type III (CBD dominant) is essentially the fibre-type cannabis (Hemp) used in formulation of various over the counter products (e.g., CBD oils and beverages) and consumer goods such as textiles, ropes, and building materials. Type III, along with Type IV (CBD/CBG codominant) cannabis, also has enormous potential for the formulation of new pharmaceutical products. Type V class cannabis has minimal or nonexistent cannabinoid content. Because cannabinoid levels enable some predictions of therapeutic potential, this chemically-based classification system is an improvement over cultivar classification. For example, cannabinoids such as CBD and CBG modulate many of the adverse effects associated with Δ9-THC (e.g., tachycardia and anxiety). Unfortunately, this classification focuses only on a very limited group of cannabinoids and ignores many of the other biologically active cannabis constituents that contribute to the plants overall value. Several cannabinoids and terpenes have healing properties associated with the reduction of pain, inflammation, and nausea. Additionally, the beneficial attributes of the plant are enhanced through the interaction of cannabinoids with each other (i.e., Δ9-THC and CBD) or with terpenes (i.e., Δ9-THC and limonene). It should therefore be no surprise that it is cannabis as a whole, not the individual constituents, which results in greater medical efficacy for the treatment of various chronic diseases. This synergistic relationship between different cannabis compounds is termed the "Entourage Effect" [10].

Noncannabinoid compounds, such as terpenes, not only contribute to the different aroma and flavour characteristics of cannabis, but also to its overall biological activity [11].

For these reasons, there has been a move in recent years to classify cannabis into chemovars based on the overall chemical content. This includes cannabinoids, terpenes, and other biologically active components in the plant [12]. These efforts focus on the identification of biochemical markers for differentiation of cannabis varieties. The overall goal is to obtain deeper insight into the relationship between chemical composition and desirable cannabis attributes [13]. Reputable laboratories have developed advanced breeding programs utilizing markers to improve the safety and therapeutic benefits of their scientifically derived, well-documented strains [14]. While this is a marked improvement, many of these studies still focus on a relatively small and targeted set of cannabinoids and terpenes. Expanding these analyses towards a more comprehensive chemical characterization helps researchers better understand cannabis chemical composition, variability, and may improve chemovar classifications. New analytical methodology has the potential to properly differentiate cannabis products through compositional analyses and to help in the development of well-defined, reproducible strains [15].

Expanding chemical profiles can be accomplished with nontargeted analytical tools. These extended profiles may provide better chemovar classifications and improved links with therapeutic efficacy. In this chapter, we explore analytical technologies that may be useful for expanding coverage of the chemical profile in Section 2 and we apply these techniques to improve the characterization of a set of cannabis samples in Section 3.

2. Analytical tools and methods provide nontargeted chemical profiling for chemovar classifications

Chemical analysis tools lead to chemical profiling information by facilitating the determination and comparison of individual analyte components within a complex sample like cannabis. In these types of analyses, where the experimental goals include determining a more complete chemical profile, nontargeted analytical tools are particularly useful because analytes of interest that may be important for chemovar classification are not always known ahead of data acquisition. As nontargeted analytical capabilities are extended with more powerful instrumentation, the coverage of the chemical profile grows so that additional information about a sample of interest can be obtained: better characterizations, improved comparisons, and a deeper understanding of the samples. In this section, we describe how chromatography, multidimensional chromatography, and time-of-flight mass spectrometry combine to provide a more thorough chemical profile of cannabis samples and may lead to improved chemovar classification.

2.1 Chromatography

Chromatographic methods are standard for determining information about individual analytes within complex samples. This analytical technology has been applied broadly across a range of industries and sample types for a variety of experimental goals. The fundamental concept of chromatography is that individual chemical analytes separate from each other and from sample matrix based on how the chemicals partition between a mobile and stationary phase. With gas chromatography (GC), for example, an analyte in the gas phase moves through a capillary column partitioning between the gaseous mobile phase that flows through the column and the stationary phase that coats the inside of the column. An individual chemical's relative affinity for each phase determines the time required for the analyte to elute, travelling through the column and reaching the detector. Analytes that interact with the stationary phase less travel through the column in less time, while analytes that interact with the stationary phase more require more time to travel through the column. These interactions are based on the chemical and physical properties of each analyte. Variation in these properties per analyte cause different elution times, which results in the separation of analytes from each other.

The elution times for an analyte are reproducible for a given mobile and stationary phase combination, so many chromatography applications are routine targeted analyses. In these cases, analytes known to be of interest prior to analysis are analysed as standards, which provides information to then identify and potentially quantify the same analytes in samples of interest. This targeting can be done during data acquisition, data analysis, or both. The target analyte is identified in the sample by matching the retention times of the known standard to retention times of observed analytes in the sample. Quantitative information (relative or absolute in the absence or presence of calibration, respectively) is then determined from the integration of chromatographic peaks.

An example of targeted analysis of CBD in a cannabis sample that was analysed with GC coupled to time-of-flight mass spectrometry (TOFMS)

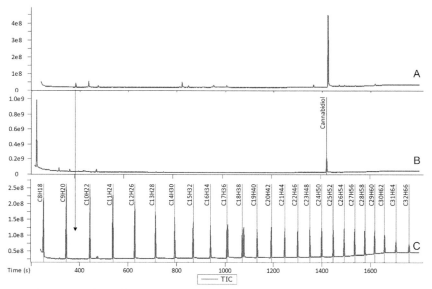

Fig. 1 GC-TOFMS data of a Type III cannabis sample (A), a CBD standard (B), and a linear alkane standard (C) is shown.

is shown in Fig. 1. The total ion chromatogram (TIC) that sums all mass to charge (m/z) fragments collected from the GC-TOFMS data for a Type III cannabis sample is shown in Fig. 1A and data for a CBD standard is shown in Fig. 1B (Sample preparation and method details for this analysis are provided in Section 4). As described in Section 1, CBD is one of the cannabinoids that is important for distinguishing Type classifications (I–V) and it was identified in this sample by comparing the retention time of candidate peaks to the retention time of the CBD peak in the standard. Mass spectral data verifies this identification, but retention matching alone is a common mode of identification in the absence of MS data. As expected for a Type III cannabis sample (CBD dominant), the largest peak in this sample was determined to be CBD. Several other analyte peaks can be observed in this sample and could be identified in a targeted way by analysing additional standards to verify identifications in the samples.

Additionally, the use of retention order extends GC applicability to nontargeted analyses, where target analytes of interest are not analysed as standards, but determined through analysis. While absolute retention time values depend on the separation conditions as well as the dimensions of the column, the retention order is reproducible for a given stationary phase material. Adjusting retention times to Retention Index (RI) values allows for comparison of observed elution to database values independent of experimental conditions. To standardize elution order with RI, a known mixture (linear alkanes is common) can be used to create a calibration to adjust the observed retention times to standard RI values that are contained in library databases. [16] Data for a linear alkane standard containing C_8 through C_{32} was used to standardize retention time to RI to support analyte identifications, as shown in Fig. 1C. For example, a peak in the cannabis sample shown in Fig. 1A can be observed at a retention time of 380 s. Compared to the alkane standard, this peak elutes between nonane (RI = 900) at 344 s and decane (RI = 1000) at 442 s. Using the entire alkane series to adjust retention times to retention index, an RI value of 938 was calculated for the peak at 380 s in the sample. Analytes with retention index close to 938 could be considered as candidates for this observed analyte. The power to determine the identification of this analyte and other nontargeted peaks in a sample is greatly improved when the GC separation is coupled to a multichannel detector, like a mass spectrometer.

2.2 Mass spectrometry

Coupling GC with mass spectrometry (MS) increases the capabilities of the analytical analysis by extending the information that can be determined about a sample into a nontargeted analysis. Mass spectrometry maintains the ability to perform either relative or absolute quantitative work, depending on the presence or absence of calibration, while improving the capabilities for compound identification, leading to better characterization of samples. This is particularly true when the entire m/z range of data is acquired for the duration of the chromatographic separation, as with scanning quadrupole instruments or TOFMS instruments. In both instances, the full m/z range of data is collected and the observed m/z and fragmentation patterns of the analytes eluting from the column can be searched against library databases to return potential analyte identifications. For example, the TOFMS data that corresponds with the peak with a RI value of 938 in Fig. 1A is shown in Fig. 2. On the right side of Fig. 2, the acquired spectral information is shown on the top and the NIST library database spectrum is shown on the bottom. The observed peak matched to α-pinene with a spectral similarity score of 924 out of 1000, suggesting very good agreement. This identification is also supported by RI as α-pinene has a retention index of 937 [16]. α-pinene is a monoterpene that is routinely encountered in cannabis samples and may help with chemovar distinction. While RI on its own

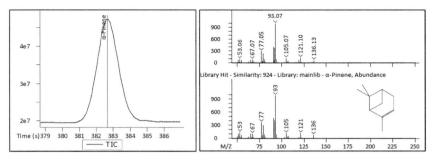

Fig. 2 A zoom-in on the chromatographic peak for α-pinene is shown along with the corresponding spectral information. The observed spectrum matched to α-pinene in the NIST library with a similarity score of 924.

could have suggested α-pinene was a potential candidate for this peak, many other analytes are likely to have similar interactions with the mobile phase and stationary phase and could have similar RI values. Having MS information to library match, along with the RI verification, adds greater confidence that the chromatographic peak observed in the cannabis sample was indeed α-pinene. This nontargeted capability allows for adding tentatively identified peaks (through mass spectral and RI matching) to better classify the samples by chemovar without requiring the analysis of all analytes as standards.

While both scanning MS and TOFMS instruments acquire the full m/z range of information and can be used for library matching, there is a crucial difference in the way the information is acquired that impacts the analytical capabilities for both targeted and nontargeted analyses. A TOFMS instrument acquires the complete mass spectrum simultaneously, while a scanning instrument collects the entire spectrum by scanning through the m/z range and collecting one m/z at a time. One of the implications of this difference is that intensity changes over the rise and fall of the chromatographic peak can affect the intensity of the m/z fragments with a scanning instrument and skew the spectral ratios, as diagramed in Fig. 3 [17]. TOFMS spectral information does not have skew since the complete m/z range is acquired simultaneously, so consistent spectral information is observed across the entire width of the chromatographic peak.

When spectral patterns are consistent across the width of the chromatographic peak as with TOFMS, additional mathematical analysis options become available for interpreting the data. In complex samples such as these, it is common to observe chemical constituents that interact similarly with the chromatographic stationary phase, causing them to elute together.

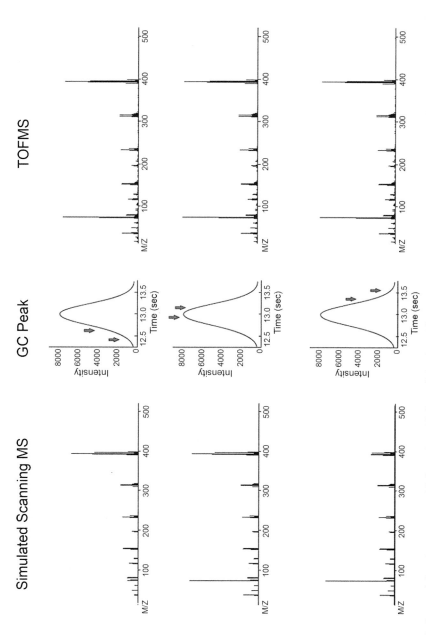

Fig. 3 Spectral skew is common with Scanning MS data while TOFMS has more consistent m/z ratios across the width of the chromatographic peak [17].

These chromatographic coelutions often obscure the individual analyte information and cause error in qualitative and quantitative interpretations because they combine into a single apparent peak in the TIC view. Peak areas are combined and the spectral information from multiple analytes are merged together. When the MS data does not have skew, deconvolution algorithms can often separate chromatographic coelutions by mathematically resolving the combined spectral information to return deconvoluted spectra and pure chromatographic data for each of the coeluting analytes, addressing both the potential misidentification and quantitative error. An example of mathematical deconvolution of two coeluting analytes in the cannabis sample is shown in Fig. 4. The TIC view suggests that one analyte elutes at 364 s. With deconvolution, though, two unique analytes are mathematically separated and pure spectral information is returned. By plotting extracted ion chromatograms (XICs) for an m/z associated with each analyte, distinct chromatographic peak shapes can be observed. These unique m/z can be integrated for reliable peak areas and the resolved spectral data can be compared to library databases for reliable identifications. In this case, the deconvoluted spectra matched to butyrolactone and ethyl pyrazine with similarity scores of 909 and 873, respectively. Both identifications are also supported by RI with database values of 915 and 921 compared to the observed RI of 918. Although neither of these analytes are routinely included in chemical screening of cannabis samples, adding their information may provide a better understanding of the samples and chemovar classifications.

Mathematical deconvolution that is possible with the non-skewed TOFMS data extends the characterization capabilities of an analysis by providing individual analyte information, even in instances of chromatographic coelution. This leads to an increased number of identified and reliably quantified analytes, which facilitates better characterization of samples without the analysis of chemical standards. With complex samples, however, there can be chromatographic coelution present that exceeds the capabilities of deconvolution and information within this portion of the data may remain obscured. In these cases, improving the chromatographic resolution with an extra dimension of separation can lead to an extension of chemical information and further improve the chemovar classification capabilities.

2.3 Multidimensional chromatography

Comprehensive two-dimensional GC (GCxGC) extends the chromatographic separation by pairing two columns with complementary stationary

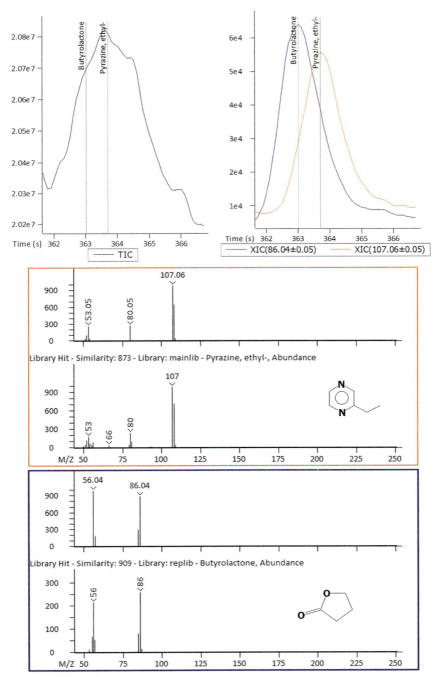

Fig. 4 Ethyl-pyrazine and butyrolactone coelute in the GC separation. The TOFMS data is deconvoluted to determine identification and quantification information for each analyte.

phases. Since chromatographic separations are based on analyte interactions with the stationary phase, GCxGC takes advantage of the fact that analyte interactions differ depending on the properties of the phase. Analytes that coelute with one type of stationary phase do not necessarily coelute on a different type of stationary phase, for example. The concept of GCxGC is to pair columns with complementary stationary phases in series so that each sample is separated by both separation mechanisms at essentially the same time [18]. Analytes may coelute on one phase or the other, but fewer will coelute on both. As diagramed in Fig. 5, the heart of GCxGC is a modulating device that connects the primary column to the secondary column. The modulator also collects the primary column effluent, refocuses the effluent, and reinjects it to the second dimension column. This is performed with a frequency that balances maintaining the primary column separation with sufficiently resolving analytes on the complementary stationary phase in the short second separation dimension. Careful consideration of detector is also crucial as the peak widths from the second dimension separation are expected to be much narrower than is typical with one-dimensional GC. Because of its fast acquisition rate capabilities, TOFMS is ideal for proper delineation of these narrow second dimension peaks. Commercially available hardware and method development software tools [19] have the capacity to make this type of analysis routine.

A GCxGC separation of the cannabis sample shown in Fig. 1A is shown in Fig. 6. The method details for the GCxGC separation are provided in Section 4, but the primary column separation conditions were maintained with both analyses. The retention times in the first dimension are consistent and analytes from the GC separation can be visually connected to those in the GCxGC separation by matching first dimension retention times. For example, CBD and α-pinene, shown in Figs. 1 and 2, are also indicated

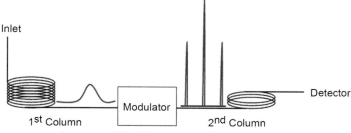

Fig. 5 Instrument schematic of GCxGC showing effluent peak from primary separation, modulator, and narrower peaks in the second dimension separation.

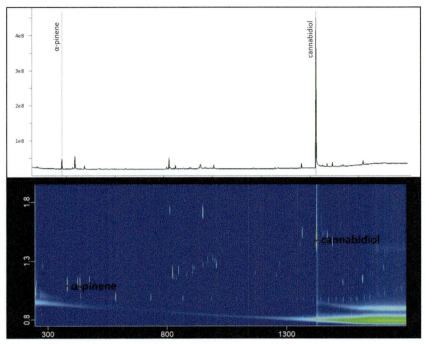

Fig. 6 The Type III cannabis sample shown in Fig. 1 was also analysed with GCxGC. CBD and α-pinene are indicated.

on the plots in Fig. 6. GCxGC effectively spreads the analytes out into two-dimensional space, as demonstrated with this top-down contour plot, shown with log colour scale.

One of the most often discussed benefits of GCxGC is the increase in peak capacity, which allows for separating more analytes in chromatographic space. For example, a GCxGC separation of the coelution that required deconvolution in Fig. 4 is shown in Fig. 7. Butyrolactone and ethyl pyrazine have similar retention index values for the primary separation (a semi-standard non polar column), but have different retention index values on a polar column (1632 and 1337, respectively), which was used for the second dimension separation. When separated with both mechanisms in GCxGC, what had required deconvolution to separate in one dimension is chromatographically resolved in the second. Analytes that are vertically aligned in the contour plot in Fig. 6 are other examples of coelutions in GC separation that have benefitted from this increase in peak capacity.

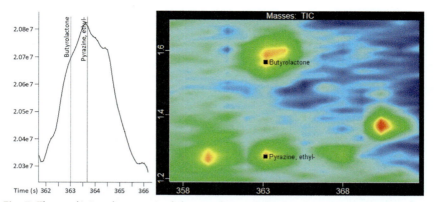

Fig. 7 The coelution that required deconvolution (left, also shown in Fig. 4) chromatographically separates in the second dimension with GCxGC (right).

This benefit is even more important when the additional peak capacity is able to separate compounds that were not efficiently deconvoluted in the GC data. For example, an apparent single peak from the GC-TOFMS data that was not reliably identified is shown in Fig. 8. The best library match for this analyte was a monoterpene, bicyco[2.1.1.]hexane, 5,5-dimethyl-1-vinyl with a similarity score of 767. While retention index supported this identification (observed = 923 and library = 921), there are notable discrepancies in the spectral information. Both m/z 67.06 and 108.07 are present higher than the expected ratios in the observed data. These discrepancies would have mostly likely left this analyte an unknown if additional targeted experimentation were not performed. However, the GCxGC data shown in Fig. 8 reveals that a perfect first dimension coelution of 2,3-dimethyl pyrazine was overlapped with the terpene. The spectrum observed from the GC separation that did not match well to library databases is the merged spectra of these two coeluting analytes. With GCxGC and chromatographic separation in the second dimension, the similarity score for the terpene improved to 914 and the previously not found compound matched to 2,3-dimethyl pyrazine with a similarity score of 862. Retention index information further supported the identification of the newly detected pyrazine compound (observed = 922 and library = 926). In this case, GCxGC was able to turn one unreliably identified analyte into two analytes with confident identifications, improving the chemical understanding of this cannabis sample.

The increased peak capacity and TOFMS detection improves sample characterization by providing additional analyte information. GCxGC further

Improving cannabis differentiation

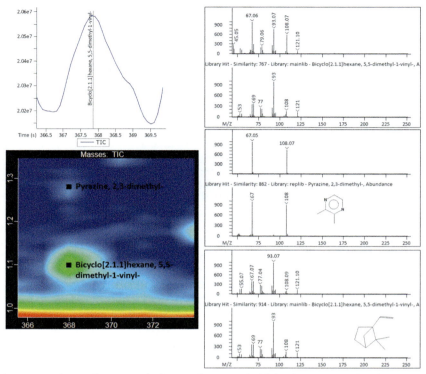

Fig. 8 A coelution that exceeds deconvolution with GC (top) is separated in the second dimension with GCxGC (bottom).

improves characterization by also producing structured chromatograms. Because there is some consistency in the types of interactions that analytes with the same functional groups have with each stationary phase, analytes of the same compound class tend to elute in structured bands across the GCxGC separation space, as demonstrated in Fig. 9. For example, bands of terpenes (monoterpenes and sesquiterpenes), terpenoids (monoterpenoids and sesquiterpenoids), and cannabinoids are all readily apparent in the contour plot displayed below. This aspect of GCxGC provides better visual characterization of a sample and leads to improved chemical profiling.

The combined analytical capabilities of GCxGC-TOFMS leads to additional analyte information for better characterization of complex samples, which facilitates improved differentiation of cannabis samples and chemovar classifications, as described in the next section.

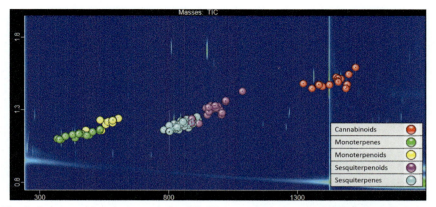

Fig. 9 Structured bands through the GCxGC separation space highlight compound classes of interest.

3. Application of GCxGC-TOFMS for nontargeted chemical analysis and chemovar classifications

Seven different cannabis botanical samples were analysed with GCxGC-TOFMS (see Section 4 for method details). The goals of this work were to both understand the chemical composition of these samples, and to explore how a more complete chemical profile may improve the ability to characterize and distinguish the samples by chemovar. The samples were expected to differ based on their Type, growth conditions, storage, and age. All of these factors have the potential to impact chemical profiles, so variation and distinct groups were expected within this set of samples. Extending the chemical analysis from what is typically targeted to what can be revealed in this nontargeted analysis resulted in more information and a better ability to characterize these samples by chemovar.

Representative contour plots for all samples are shown in Fig. 10. These enhanced chromatographic plots demonstrate the structured GCxGC separations and are backed by high performance TOFMS, facilitating preliminary visual review and differentiation of the cannabis samples. In a sense, they serve as maps of the chemical profile and help to reveal the chemovar for these samples. For example, differences in the cannabinoid profile can be observed by comparing the cannabinoid band. The terpene bands are also apparent, and general distinctions can be visually discerned. This differentiation of samples is enhanced by further exploring individual analyte content and trends.

One of the primary ways to distinguish cannabis samples is to target the key cannabinoids that are used for Type determinations. This Type distinction is based on the relative amount of CBD and Δ9-THC. It is one of the

Improving cannabis differentiation 185

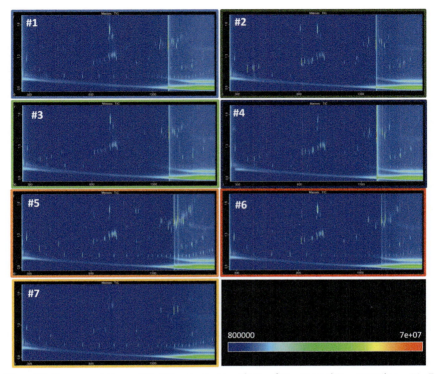

Fig. 10 Representative GCxGC contour plots for cannabis samples #1–7 (log colour scale).

more common ways of predicting therapeutic potential of products, as the ratio of these cannabinoids has implications in the physiological and therapeutic effects of consumption. A zoomed-in region of the chromatogram that contains CBD and Δ9-THC for two of the seven samples is shown in Fig. 11. CBD was determined in these samples by comparing to the standard, as illustrated in Fig. 1, and through spectral matching to library databases (similarity score = 912). The presence of Δ9-THC was confirmed through library matching (similarity score = 921) and by relative retention order to CBD in both separation dimensions. Sample #2 was expected to be CBD dominant (Type III) and Sample #6 was anticipated to be Δ9-THC dominant (Type I). The relative peak heights match expectation with CBD higher in the Type III sample and Δ9-THC higher in the Type I sample. A plot of the peak areas of CBD and Δ9-THC in all seven samples is also shown in Fig. 12. By this classification, it appears that there are two Type groups of samples. Samples #1–#4 have higher amounts of CBD and are Type III and Samples #5–#7 have higher amounts of Δ9-THC and are Type I.

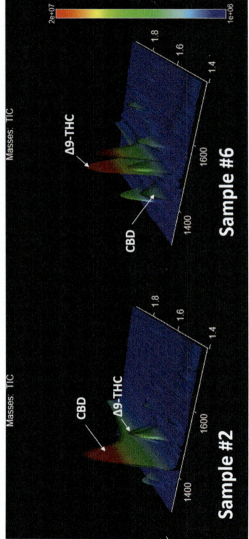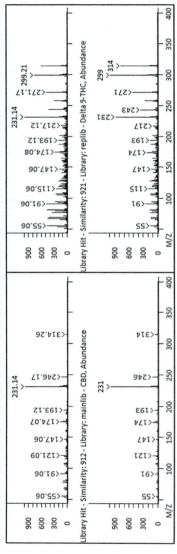

Fig. 11 A zoom-in on the cannabinoid band, highlighting differences in CBD and Δ9-THC in a Type III (Sample #2) and Type I (Sample #6) cannabis sample.

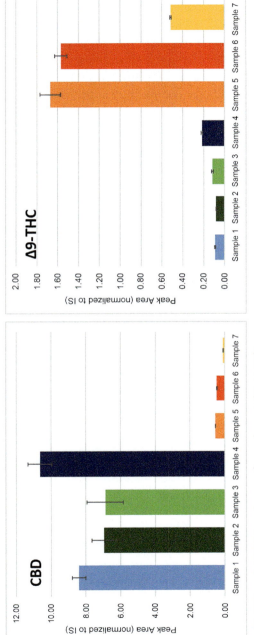

Fig. 12 CBD and Δ9-THC peak areas for each cannabis sample.

This Type distinction is important and was readily determined with the analysis. GCxGC can also add information about the cannabinoid content of the samples beyond CBD and Δ9-THC. This extended list of cannabinoids may lead to a better understanding of the samples, their properties, and how they should be classified. As this is an active area of research, not all cannabinoids have been definitively determined, nor are they present in mass spectral databases for library comparisons at this time. This would prohibit the determination of cannabinoids in many cases, but the structured nature of the GCxGC space can be leveraged to tentatively assign more analyte peaks as cannabinoids. These analytes that make up a "cannabinoid collective" have similar structure, functionality, and properties and can be observed in the band shown in Fig. 9. This band encompasses the peaks surrounding CBD and Δ9-THC in Fig. 11 and can be observed for all the samples in Fig. 10. Tentative identification of these cannabinoids is supported by retention order and also by the underlying mass spectral data. While some of the analytes in this band are unknowns, they have spectral patterns similar to known related cannabinoids. This spectral information taken along with the retention patterns of the structured chromatograms allows for tentative assignments to the cannabinoid collective, even in the absence of a definitive identification. The exact identifications could be determined with further investigation and research, but the ability to distinguish and classify the samples can still be explored prior to absolute identifications.

GCxGC-TOFMS extended the cannabinoid coverage to 26 analytes and a summary of the samples based on their cannabinoid profile was done with Principal Component Analysis (PCA), as shown in Fig. 13. PCA is a data analysis tool that captures the inherent variation in the data and describes the similarities and differences between the samples related to this variation. The output of PCA shown in Fig. 13 is a scores plot where each sample is represented as a data point in principal component space. The proximity of the samples in this plot indicates their similarity to each other. Samples that cluster closer to each other are more similar while samples that are further from each other in the scores plot are less similar. Based on CBD and Δ9-THC, shown in Figs. 11 and 12, there appeared to be two cannabinoid profiles present in these samples. Several distinct cannabinoid profiles were revealed, however, when a more complete cannabinoid profile was considered. The samples do separate by Type on the x-axis (Type $I > 0$ and Type $III < 0$), but there are additional distinctions within each Type that are apparent by the different scores on the y-axis. With the exception of the Samples #2 and #4, each type of cannabis has a distinct cannabinoid profile when the extended cannabinoid profile is considered.

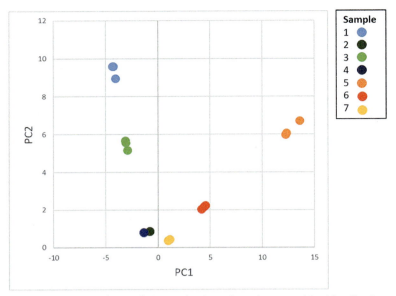

Fig. 13 PCA summary of cannabis samples based on the cannabinoid collective.

Whether the additional cannabinoids prove to be helpful in predicting physiological effects or therapeutic potential is yet to be determined, but having this additional information to include in the chemovar distinction allows for exploring these questions and may lead to a better understanding of the properties of the samples.

Terpenes and terpenoids are other important classes of compounds that are of interest in cannabis samples. Cannabis terpene content has been hypothesized to work in concert with the cannabinoid profile to impact the plant's physiological effects. Seventeen terpenes are commonly targeted for cannabis characterization and these are described in Table 1. Each terpene was identified with spectral matching to NIST library databases with similarity scores indicated. The analyte identifications were also supported by RI with observed and library values listed in the table, and by their elution order within the structured GCxGC space. The relative amount of each terpene in each sample is indicated in the heat map (normalized to mean per analyte to see trends across sample types) with colour scale indicating relative intensity.

Several observations about the samples can be made from these 17 terpenes and some distinction between the samples is apparent. For example, many of the terpenes are observed at their highest levels in Sample #2 (including α-pinene, β-pinene, β-myrcene, limonene, β-ocimene, etc.). Another set of terpenes were observed at their highest levels in Sample

Table 1 Routinely screened terpenes and terpenoids.

Name	R.T. (s)	Formula	CAS	Similarity	RI	Lib. RI	Sample 1	Sample 2	Sample 3	Sample 4	Sample 5	Sample 6	Sample 7
α-pinene	381.611, 1.099	$C_{10}H_{16}$	80-56-8	937	937.2	937							
camphene	397.213, 1.121	$C_{10}H_{16}$	79-92-5	931	953.3	952							
β-pinene	424.815, 1.129	$C_{10}H_{16}$	127-91-3	944	981.8	979							
β-myrcene	435.616, 1.106	$C_{10}H_{16}$	123-35-5	954	992.9	991							
α-phellandrene	450.017, 1.134	$C_{10}H_{16}$	99-83-2	848	1007.9	1005							
3-carene	457.217, 1.126	$C_{10}H_{16}$	13466-78-9	821	1015.5	1011							
α-terpinene	462.018, 1.133	$C_{10}H_{16}$	99-86-5	829	1020.6	1017							
limonene	474.019, 1.132	$C_{10}H_{16}$	138-86-3	923	1033.2	1030							
β-ocimene	480.019, 1.125	$C_{10}H_{16}$	13877-91-3	790	1039.5	1037							
γ-terpinene	502.821, 1.148	$C_{10}H_{16}$	99-85-4	811	1063.5	1060							
α-terpinolene	531.623, 1.156	$C_{10}H_{16}$	586-62-9	849	1093.8	1088							
caryophyllene	820.846, 1.206	$C_{15}H_{24}$	87-44-5	947	1437.2	1419							
humulene	847.249, 1.225	$C_{15}H_{24}$	6753-98-6	944	1472.1	1454							
linalool	537.624, 1.161	$C_{10}H_{18}O$	78-70-6	892	1100.2	1099							
fenchol	555.625, 1.201	$C_{10}H_{18}O$	1632-73-1	910	1120.1	1113							
α-terpineol	625.231, 1.252	$C_{10}H_{18}O$	98-55-5	859	1197.3	1189							
caryophyllene oxide	943.256, 1.316	$C_{15}H_{24}O$	1139-30-6	910	1605.1	1581							

#6 (including carene, α-terpinene, γ-terpinene, and α-terpinolene). Linalool is distinctly higher in Sample #5 compared to the other samples. The other terpenes are elevated in other miscellaneous samples with less distinct patterns.

While differences in individual terpenes in each sample can be observed, the overall trends in the samples related to these 17 terpenes can also be summarized with PCA, as shown in Fig. 14. This plot suggests that there are three terpenes profiles and groups of related samples. One group has a terpene profile that is represented by Sample #6 (with the elevated levels of carene, α-terpinolene, γ-terpinene, and α-terpinene) and another profile is represented by Sample #2 (with elevated levels of α-pinene, β-pinene, and β-myrcene). The remaining five samples are more similar to each other than either the Sample #6 or Sample #2, as they all cluster together and away from the other two.

GCxGC-TOFMS not only extended the cannabinoid profile, but also increased the number of terpenes and terpenoids that were identified and measured in these samples. In some instances, the increase in GCxGC peak capacity was required to determine the additional terpenes. For example, the terpene highlighted in Fig. 8, Bicyclo[2.1.1]hexane, 5,5-dimethyl-1-vinyl-, coeluted with another compound in the 1D separation, but was chromatographically resolved in the second dimension with GCxGC. The relative trends of this terpene between the samples are shown in Fig. 15. This particular terpene was not reliably determined without GCxGC and

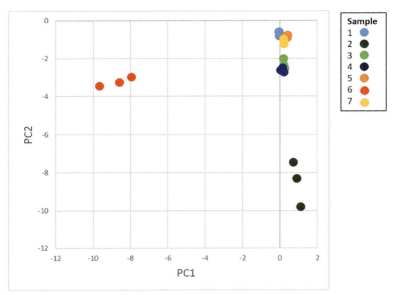

Fig. 14 PCA summary of cannabis samples based on routinely targeted terpenes and terpenoids.

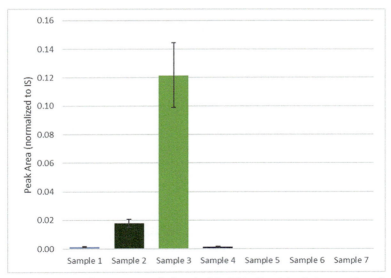

Fig. 15 The terpene that was revealed with GCxGC in Fig. 8 is distinct to Sample #3.

appears to be distinct to Sample #3, a cannabis sample that did not cluster independently with just the 17 routinely screened terpenes and terpenoids.

The GCxGC-TOFMS technology allowed for an extension of terpene and terpenoid profile to 73 analytes. These identifications were determined

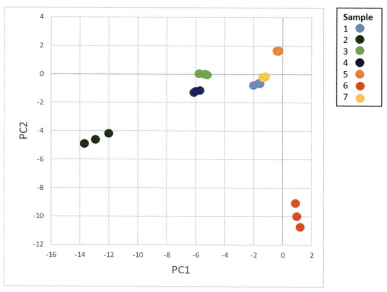

Fig. 16 PCA summary of cannabis samples based on the extended terpenes and terpenoids.

with mass spectral matching, retention index matching in the first dimension, and by their elution within the appropriate band in the structured GCxGC separation space. PCA was used again to summarize the samples with this extended list, as shown in Fig. 16. Better distinction between the samples was observed with almost all sample types clustering independently, except for Samples #1 and #7 that clustered together. As with the extended cannabinoid profile, whether these terpenes prove to be crucial in predicting and understanding the samples is to be determined, but having this additional information does allow for continuing study on how terpenes may work synergistically with other cannabis constituents.

The targeted cannabinoids suggested two profiles while the targeted terpenes suggested three profiles. Extending each group independently led to more distinction between the samples. The cannabinoid collective on its own provided pronounced information on all of the samples except for Samples #2 and #4 and the extended terpene profile on its own provided distinct information on all of the samples except Sample #1 and #7. When the extended cannabinoid and extended terpene profiles are taken together, though, a more complete view is gained and a unique and distinct grouping by chemical profile is determined for each sample, as shown with the PCA scores plot in Fig. 17. This extended chemical profile allowed for better characterization and classification of these cannabis samples.

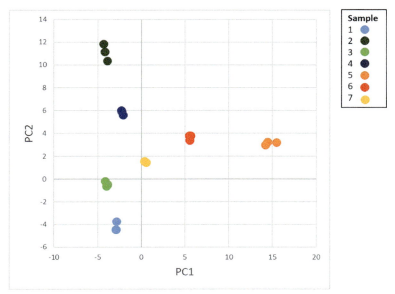

Fig. 17 PCA summary of cannabis samples based on the cannabinoid collective and the extended terpenes and terpenoids.

As described in Section 2 and demonstrated with the characterization of these seven samples, there are numerous benefits that can be attributed to the analytical capabilities of GCxGC-TOFMS. TOFMS provides full m/z range nonskewed spectral information that turns chromatographic peaks and intensity into chemical identification information, even in instances of coelution. GCxGC extends the list of identified analytes with an increased peak capacity, better isolation of individual analytes, and additional retention context from structured chromatograms. This ultimately provides better chemical coverage, and the specific impact on the ability to characterize and differentiate these seven samples was demonstrated. Classification by chemovar has been enhanced with the overall improved coverage of the chemical profile.

The GCxGC data is rich, and it contains even more information about these samples. The focus of this work has been profiling the cannabinoids, terpenes, and terpenoids. Many other compound classes are present including esters, aldehydes, ketones, furans, lactones, pyrazine, pyridine, and aromatics, among others. The chemical profile could be further extended, and even more about these samples could be understood by probing deeper into these other compound classes. The power of GCxGC-TOFMS is that all of this information is determined from one analysis on one instrument platform.

4. Methods

4.1 Definition

The determination of chemical profiles to serve as Chemovar Maps for description of cannabis botanicals through the combination of enhanced chromatographic resolution and underlying spectral information provided by high performance GCxGC-TOFMS.

4.2 Rationale

A large portion of analytical methods used for cannabis analysis target a small subset of the numerous compounds in these samples. A more comprehensive analytical approach that combines GCxGC-TOFMS was implemented for improved characterization and differentiation of cannabis botanicals.

4.3 Materials, equipment, and reagents

An assortment of cannabis samples were acquired from different sources including a collaborating forensic laboratory, a local vape shop, and a certified reference material standard (Part No. 54999C) purchased from Emerald Scientific (San Luis Obispo, CA, USA). Anhydrous ethanol (200 proof, >99.5%, Cat. No. 459836) and a C_8-C_{40} linear alkane standard (Cat. No. 40147-U) were purchased from Sigma-Aldrich (St. Louis, MO, USA). A CBD standard was purchased from RESTEK. A Viking Ace Tobacco Grinder (Item No. GV002-63) was used for sample milling. Microwave-Assisted Extraction and Cannabis Activation (MAECA) were conducted using an Anton Paar Monowave 50+P synthesis reactor (Vernon Hills, IL 60061, USA). MAECA extracts were transferred to 2 mL GC vials (Cat. No. 03-377-298, Fisher Scientific, Batavia, IL, USA) and diluted with ethanol prior to analysis.

A Pegasus® BT 4D mass spectrometer (LECO Corp., St. Joseph, MI, USA) with dual stage quad jet thermal modulator and L-PAL3 auto sampler was used for sample analysis. ChromaTOF® 5.5 software was used for instrument control, data acquisition, and processing. The GC column flow rate was 1.4 mL/min with constant GC inlet and transfer line temperatures of 250 and 300 °C, respectively. Split 20:1 injections of 1 μL were made onto an R*xi*-5 ms primary column (30 m × 0.25 mm i.d. x0.25 μm) coupled to an R*xi*-17sil ms secondary column (0.45 m × 0.25 mm i.d. x0.25 μm; Restek, Bellefonte, PA, USA). The primary oven temperature was held at 40 °C for 1 min, ramped 10 °C/min to 300 °C and held for 3 min. The secondary oven temperature offset was +20 °C relative to the primary oven. A modulation period of 1.2 s with hot pulse duration of 360 ms and cooling time between stages of 240 ms was used for GCxGC-TOFMS data acquisition (Liq. N_2, thermal modulation). The modulator temperature was maintained +15 °C relative to the secondary oven. The MS ion source temperature was 250 °C and MS data were collected at spectral acquisition rates of 10 spectra/s for GC-TOFMS and 200 spectra/s for GCxGC-TOFMS. Data acquisition with a m/z range of 45–600 started after a solvent delay of 240 s.

4.4 Protocols

1. Mill cannabis sample bud(s) with a tobacco grinder
2. Weigh 0.05 g of ground sample into a tared 10 mL reaction vessel with a magnetic stir bar
3. Add 5 mL of absolute ethanol
4. Purge the vessel using a steady stream of gaseous N_2 for 5 s
5. Cap the vessel, place it in the microwave reactor and lock the reactor cavity
6. Extract and Activate the sample for 30 min at 180 °C
7. Remove the vessel from the reactor and let it cool to room temperature
8. Remove 0.50 mL of the supernatant and transfer to a 2 mL gas chromatography vial
9. Dilute with 1.0 mL of additional ethanol
10. Acquire data using the Pegasus BT 4D GCxGC-TOFMS system

4.5 Analysis and statistics

ChromaTOF Software (LECO Corp., St. Joseph, MI, USA) was used for nontargeted data processing with automated peak deconvolution, spectral similarity comparisons (NIST 17 & Wiley 11) and retention index calculation and filtering. Peak areas were determined through automated peak find and normalized to an internal standard. Compiled peak areas were the variables for PCA and pre-processing included normalization to each variable's mean (MatLab).

4.6 Summary

MAECA was used to prepare cannabis samples for analysis. GCxGC-TOFMS and software tools were used for the characterization and differentiation of cannabis botanicals. Extending the chemical profile of the cannabis sample led to better differentiation of the sample types.

References

[1] A.W. Zuardi, History of cannabis as a medicine: a review, Rev. Bras. Psiquiatr. 28 (2) (2006) 153–157.

[2] J. Gumbiner, History of Cannabis in India, in: Psychology Today, 2011 June 16, https://www.psychologytoday.com/us/blog/the-teenage-mind/201106/history-cannabis-in-india. Accessed 2-12-20.

[3] W.B. O'Shaughnessy, On the preparations of the Indian hemp, or Gunjah, (Cannabis Indica), Prov. Med. J. Retrosp. Med. Sci. 123 (1843) 363–368.

[4] S. Kumar, P. Sarma, H. Jumar, A. Prakash, B. Medhi, Modulation of endocannabinoid system: success lies in the failures, Indian J. Pharmacol. 50 (4) (2018) 155–158.

[5] Jin D., Jin S. and Chen J., "Cannabis classification systems and growth trends of the North American Medical Cannabis Industry", The Second Annual Academic Conference of the Specialty Committee of TCM Pharmacognosy of the World Federation of Chinese Medicine Societies, pg. 23–27, October 2015.

[6] World Health Organization, "Management of Substance Abuse"; https://www.who.int/substance_abuse/facts/cannabis/en/; Accessed 2-12-2020, 2020.

[7] S. Elzinga, J. Fischedick, R. Podkolinski, J.C. Raber, Cannabinoids and terpenes as chemotaxonomic markers in Cannabis, Nat. Prod. Chem. Res. 3 (4) (2015) 1–9.

[8] J.M. McPartland, Cannabis systematics at the levels of family, genus, and species, Cannabis Cannabinoid Res. 3 (1) (2018) 203–212.

[9] E. Small, H.D. Beckstead, Common cannabinoid phenotypes in 350 stocks of Cannabis, Lloydia 36 (2) (1973) 144–165.

[10] (a) E.B. Russo, Taming THC: potential cannabis synergy and phytocannabinoid-terpenoid entourage effects, Br. J. Pharmacol. 163 (2011) 1344–1364. (b) E.B. Russo, The case for the entourage effect and conventional breeding of clinical cannabis: no "strain," no gain, Front. Plant Sci. 9 (2019) 1–8.

[11] J.T. Fischedick, Identification of terpenoid chemotypes among high (−)-trans-Δ9-tetrahydrocannabinol-producing *Cannabis sativa* L. cultivars, Cannabis Cannabinoid Res. 2 (1) (2017) 34–47.

[12] A. Hazekamp, J.T. Fischedick, Cannabis—from cultivar to chemovar. Drug Test. Anal. 4 (7–8) (2012) 660–667, https://doi.org/10.1002/dta.407. wileyonlinelibrary.com.

[13] A. Hazekamp, K. Tejkalova, S. Papadimitriou, Cannabis: from cultivar to Chemovar II—a metabolomics approach to cannabis classification, Cannabis Cannabinoid Res. 1 (1) (2016) 202–215.

[14] M.A. Lewis, E.B. Russo, K.M. Smith, Pharmacological foundations of cannabis chemovars, Planta Med. 84 (2018) 225–233.

[15] J.K. Booth, J. Bohlmann, Terpenes in *Cannabis sativa*—from plant genome to humans, Plant Sci. 284 (2019) 67–72.

[16] NIST Library Databases, 2017.

[17] J. Binkley, M. Libardoni, Comparing the capabilities of time-of-flight and quadrupole mass spectrometers, LCGC Spec. Issue 8 (3) (2010) 28–33.

[18] Z. Liu, J.B. Phillips, Comprehensive two-dimensional gas chromatography using an on-column thermal modulator interface, J. Chromatogr. Sci. 29 (1991) 227–231.

[19] www.simplyGCxGC.com.

CHAPTER SEVEN

Analysis of terpenes in hemp (*Cannabis sativa*) by gas chromatography/mass spectrometry: Isomer identification analysis

E. Michael Thurman*

Center for Environmental Mass Spectrometry, Department of Environmental Engineering, University of Colorado, Boulder, CO, United States
*Corresponding author: e-mail address: michael.thurman@colorado.edu

Contents

1. Introduction — 198
2. Major terpenes and terpenoids — 198
   2.1 α-Pinene and β-pinene — 200
   2.2 δ-Limonene — 200
   2.3 β-Myrcene — 200
   2.4 Terpinolene — 201
   2.5 Linalool — 201
   2.6 Terpineol — 201
   2.7 α-Humulene — 201
   2.8 β-Caryophyllene — 202
   2.9 β-Caryophyllene-oxide — 202
3. Methods and materials — 202
4. Results and discussion — 203
   4.1 The *Cannabis sativa* plant — 203
   4.2 Chromatographic separation of the terpenes — 205
   4.3 Mass spectral fragmentation of the monoterpenes, monoterpenoids, and sesquiterpenes — 205
   4.4 Mass spectral identification for terpenes and terpenoids in Cannabis flower — 215
   4.5 Mass spectral identification of two major cannabinoids in Cannabis flower — 228
5. Conclusions and major takeaway ideas — 232
Acknowledgements — 232
References — 232

1. Introduction

Terpenes are a varied and significant class of plant organic compounds (hydrocarbons) that occur particularly in evergreens or conifers and make up a major component of the resin of these plants, which is called turpentine. They are simple hydrocarbons, made up only of hydrogen and carbon atoms. They are volatile compounds and give odour that discourage herbivores, appeal to their predators, or discourage insects that attack the plants. The essential oils of many plants, flowers, and trees, are made up of terpenes and their related compounds. The terpene molecule consists of an isoprene unit that is linked to the next isoprene unit, head to tail, so to speak. This chemical structure was realized by Leopold Ruzicka in 1953 and was named the isoprene rule or the C-3 rule [1]. The isoprene units may be either straight chain, branched, or cyclic structures. It was Kekule who gave terpenes their name because they were all hydrocarbons with the same molecular formula of $C_{10}H_{16}$ [2]. Kekule used this name to remove the confusion associated with the many isomers of the same formula, $C_{10}H_{16}$, which previously had all been called camphene [2].

Terpenes are sometimes called terpenoids, but there is a chemical difference in that terpenoids contain oxygen, while terpenes do not. Terpenes are grouped by the number of isoprene units in the molecule. They include: hemiterpenes (a single isoprene unit), monoterpenes (two isoprene units), sesquiterpenes (three isoprene units), diterpenes (four isoprene units), sesterterpenes (five isoprene units), triterpenes (six isoprene units), sesquarterpenes (seven isoprene units), tetraterpenes (eight isoprene units), and polyterpenes (many isoprene units). In this chapter, we will discuss only monoterpenes, sesquiterpenes, and their related terpenoids, which are monoterpenoids and sesquiterpenoids, all of which are either major or minor components of the hemp plant, *Cannabis sativa*. In that regard, Table 1 shows the 10 major terpenes found in *Cannabis sativa* based on current research of *Cannabis sativa* [3].

2. Major terpenes and terpenoids

The following sections discuss the importance of these terpenes and terpenoids in the *Cannabis sativa* plant. The discussion follows the order of the compounds shown in Table 1.

Table 1 Chemical name, formula, molecular ion mass, and molecular ion structure of 10 important terpenes and terpenoids in *Cannabis sativa* L [3].

Chemical name	Formula	Molecular ion mass	Molecular ion structure
Monoterpenes and monoterpenoids			
α-Pinene	$C_{10}H_{16}$	136.1	
β-Pinene	$C_{10}H_{16}$	136.1	
δ-Limonene	$C_{10}H_{16}$	136.1	
β-Myrcene	$C_{10}H_{16}$	136.1	
Terpinolene	$C_{10}H_{16}$	136.1	
Linalool	$C_{10}H_{18}O$	154.1	
α-Terpineol	$C_{10}H_{16}O$	152.1	
Sesquiterpenes and sesquiterpenoids			
α-Humulene	$C_{15}H_{24}$	204.2	
α-Caryophyllene	$C_{15}H_{24}$	204.2	
Caryophyllene oxide	$C_{15}H_{24}O$	220.2	

2.1 α-Pinene and β-pinene

These two cyclic monoterpene isomers are the main constituents of turpentine and come from the resin of conifer trees (e.g., pine tree, genus *Pinus*). Not only do they occur in conifers they are also found in sagebrush and many other plants. Both α-pinene and β-pinene are produced from geranyl pyrophosphate following cyclization followed by the subsequent loss of a proton from the carbocation [4]. The role of geranyl pyrophosphate will be seen in many of the terpenes found in the *Cannabis sativa* plant.

2.2 δ-Limonene

δ-Limonene is a monoterpene found not only in *Cannabis sativa*, but also in many citrus plants, such as lemons, limes, and oranges. It is a major component of aromatic scents characteristic of both conifers and broad leaf trees. Furthermore, limonene is a dietary supplement, and is a fragrance ingredient in cosmetics and cleaning solutions. It is an anti-oxidant and has properties that may include anticancer and other properties [4]. The δ-limonene structure occurs as part of the structure of cannabidiol (CBD) but not as part of the structure of tetrahydrocannabinol (THC). δ-Limonene is synthesized from geranylpyrophosphate as shown below [4] in Fig. 1.

2.3 β-Myrcene

β-Myrcene is a branched-chain alkene hydrocarbon and is also classified as a monoterpene with the same formula of $C_{10}H_{16}$ as the other monoterpenes in *Cannabis sativa*. Myrcene is a major component of essential oils of many

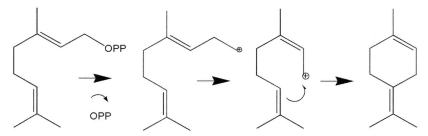

Fig. 1 Shows the synthesis of δ-limonene from geranyl pyrophosphate, where OPP is pyrophosphate.

plants including both hops and cannabis. Myrcene is associated with an herbal essence, as compared to the pine and citrus odours of the previous terpenes. Myrcene is a key component to the synthesis of the other monoterpenes as shown *supra previa* in geranylpyrophosate (OPP). See Ref. [4].

2.4 Terpinolene

Terpinolene is one of four isomers of terpinene (δ-terpinene) with the double bond in a different location for each isomer. Each isomer has the same carbon skeleton but with the double bonds in different locations. They exist as colourless liquids with a turpentine odour.

2.5 Linalool

Linalool is a monoterpenoid with the formula $C_{10}H_{18}O$, which occurs in hundreds of species of plants, including many flowers. This terpenoid has many commercial uses based on its floral scent. The structure of linalool looks quite similar to the monoterpene, myrcene, but with the addition of an OH group and loss of the double bond at the C-3 position (see Table 1). Linalool is used in the majority of scented hygiene products [5] and can even be used as an insecticide for fleas and cockroaches.

2.6 Terpineol

Terpineol is a monoterpenoid of terpinene. Similar to its monoterpene it has four isomers. The α-terpineol isomer is present in cannabis. It has a pleasant odour similar to the lilac flower and, thus, is a common ingredient in perfumes and cosmetics.

2.7 α-Humulene

Humulene is isomeric with caryophyllene, both are naturally occurring sesquiterpenes that are found in hops (*Humulus lupulus*) and cannabis [6,7]. Its' name is derived from its common and major occurrence in hops, where it can be present at as much as 40% of the essential oil. The hoppy aroma of beer derives from this sesquiterpene. The many flavours of beers being derived from epoxides of humulene. Humulene occurs not only in hops and cannabis but also in cloves, basil, oregano, black pepper, rosemary and many other herbs and spices. Humulene often co-occurs with its isomer β-caryophyllene in many plants and flowers. Together these two isomers give the characteristic odour of cannabis, the smell of "pot".

2.8 β-Caryophyllene

β-Caryophyllene is the beta isomer of humulene [7] and is also a bicyclic sesquiterpene that is a major constituent of many essential oils, such as clove oil, oil of rosemary, cannabis oils, and hops. It occurs typically as a mixture of isomers. The aroma of black pepper comes from β-caryophyllene.

2.9 β-Caryophyllene-oxide

This compound is simply β-caryophyllene epoxide (see structure in Table 1), which is also associated with the aroma of cannabis.

3. Methods and materials

The Intuvo 9000 GC comes with an Agilent HP-5MS Ultra Inert column along with a guard chip installed for column protection. The GC column is 30 m long with an inner diameter of 0.25 mm and a film thickness of 0.25 μm. While the oven was ramped from 30 °C to 275 °C over a 20-min period, the guard chip was held at 55 °C until the oven reached 55 °C, and then the two were ramped to the final temperature of 275 °C. The guard chip was recommended to be run at about 25 °C higher than that of the oven. The Intuvo comes with a *Cycle Time Optimization* feature that has "fast cooling" option set by default to minimize the delay between runs using a rapid heating and cooling technology. Table 2 shows the details used in the GC/MS analysis.

Standard mixes terpenes were purchased from Sigma Aldrich, Pennsylvania.

Table 2 GC–MS system parameters.

Parameter	Value
Agilent Intuvo 9000 GC	
Inlet	250 °C (Splitless, 10.549 psi)
Column	Agilent HP-5MS Ultra Inert (30 m × 250 μm × 0.25 μm)
Column Flow	1.2 mL/min (Helium)

Table 2 GC–MS system parameters.—cont'd

Parameter	Value
Oven	30 °C (hold: 1 min) 10 °C/min to 40 °C (hold: 5 min) 40 °C/min to 250 °C (hold: 5 min) 40 °C/min to 275 °C (hold: 10 min) *Total Run Time: 27.875 min*
Guard chip	55 °C (hold: 7.5 min) 40 °C/min to 250 °C (till the end of the run)
Purge flow to split vent	100 mL/min at 0.25 min
Injection volume	1 μL
Bus temperature	275 °C
MSD connector	275 °C
MSD transfer line	275 °C
Agilent 5977B Mass Spectrometer	
Operating mode	Scan (normal scanning)
MS source temperature	230 °C
MS quad temperature	200 °C
Solvent delay (MS start time)	2.90 min
Low mass	50 m/z
High mass	500 m/z
Scan speed	1562 μ/s

4. Results and discussion

4.1 The *Cannabis sativa* plant

The *Cannabis sativa* plant is the source for both hemp and contains the CBD and THC isomers. The definition for hemp is that the plant has less than 0.3% THC. In this study we are focussing on hemp, which by definition has much higher levels of CBD than THC. The monterpenes and sesquiterpenes, discussed above (the 10 major terpenes and terpenoids in Table 1) are the important compounds in cannabis that impart the characteristic odour and flavours of the cannabis plant [8,9]. The 10 major terpenes present include: alpha and beta pinene, beta-myrcene, δ-limonene, terpinolene, linalool, α-terpineol, β-caryophyllene, α-humulene, and caryophyllene oxide. Interestingly all of the monoterpenes have the same formula (thus, they are isomers) and the two sesquiterpenes have the same formula. Thus, identification should be based on more than GC retention time and selected ion monitoring, but also it is important to have the mass spectrum and to know how the molecule fragments. Furthermore, when possible, it is valuable to have structures for fragment ions of the terpene in order to aid in the identification of these compounds, as well as to aid in the elucidation of new isomers and their terpene structures. The goal of this chapter is to interpret the fragmentation of these 10 compounds (Table 1) and other terpenes and terpenoids that have been found in the hemp plant.

The literature shows that the cannabis plant has a family of terpenes, with as many as 120 sesquinoid and terpenoid compounds [9], which are produced in the trichomes of the cannabis plant. Slight differences in the genetics of the cannabis plant may give rise to the varying concentrations of the mono- and sesquiterpenes, as well as the concentrations of CBD, THC, and their other isomers.

There is considerable discussion at various web sites, and in both the scientific and nonscientific literature, concerning the health benefits of the terpenes when taken in conjunction with CBD and THC [10,11]. There are reports that caryophyllene interacts with cannabis receptor type 2, which gives an anti-inflammatory effect [11]. The importance of terpenes as a pharmacological component of cannabis is still in need of further research and study.

Thus, a valuable tool for this work is a reliable GC/MS method for these important terpenes. Such a method was recently developed by Ibrahim et al. [3] for the 10 common terpenes. However, when one analyses the cannabis extracts by GC/MS there are many more than 10 terpenes that are present and many are isomeric with similar if not identical fragmentation. Thus, there is a need for a detailed study and examination of the fragmentation pathways of the major terpenes. This chapter looks at both major and minor terpenes and terpenoids found in *Cannabis sativa.*

The chapter consists of four sections. First is the chromatographic separation of the major terpenes that occur in *Cannabis sativa* as found in the standards provided by Restek Corporation (Section 3). The second section deals with the mass spectral fragmentation and identification of the major terpenes in cannabis. The third section looks at several examples of *Cannabis sativa* that is rich in CBD with regard to the identification and quantification of the terpenes in the plant extract. The fourth section discusses the fragmentation of both CBD and THC and how they differ when using electron ionization. The combination of this three prong approach will give the reader a good idea of how to tackle the analysis of terpenes in cannabis, as well as giving the reader tools for further study of the mono and sesquiterpenes and their terpenoids.

4.2 Chromatographic separation of the terpenes

Fig. 2 shows the chromatographic separation of 20 monoterpenes and sesquiterpenes (and several of their isomers) that occur in *Cannabis sativa*. This shows the separation of the standard mix currently available from Restek that was used in this study of terpenes and terpenoids in cannabis. The details of the separation of these compounds are found in Section 3. Both CBD and THC are also separated in this chromatogram and will be shown later in this chapter.

Table 3 shown below is the detailed information on the separation of these compounds by GC/MS. The important takeaway from Table 3 is that for the most part the order of retention time of the monoterpenes and monoterpenoids follows the boiling point of the compound. The exception to this rule is ocimene, which boils at a very low temperature but is retained by hydrophobicity on the GC column giving it a longer retention time than might be expected. The sesquiterpenes also have a low boiling point and yet show longer retention on the GC column due to their larger molecular weight and hydrophobicity.

4.3 Mass spectral fragmentation of the monoterpenes, monoterpenoids, and sesquiterpenes

4.3.1 Monoterpenes

The mass spectral fragmentation of major terpenes present in *Cannabis sativa* are shown in this section (Fig. 3).

The molecular ion is m/z 136 and is a small but recognizable peak in both mass spectra. The molecular ion forms by the loss of a single electron forming an odd electron (OE) ion, since the electron lost is the bonding electron in the outer shell. This OE ion is unstable and immediately loses a radical molecule, a methyl group ($-CH_3$), which results in the loss of 15 mass units and forms the m/z 121 ion ($136-15=121$). The base peak ion, which is m/z 93, forms from the loss of an isopropene radical molecule

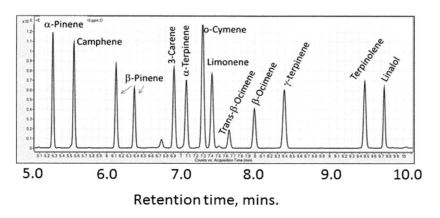

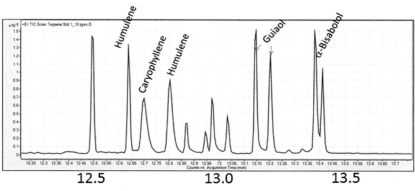

Fig. 2 Separation of 20 standard mix of mono and sesquiterpenes using gas chromatography/mass spectrometry, using the gradient described in Section 3. Separation is on a DB-5 column, polysiloxane 5% phenylmethyl, which is the most commonly used column for GC/MS analysis.

Table 3 Chemical characteristics of the terpenes and terpenoids found in the standard Mix #1 and 2 from Restek, a commonly available standard for these compounds, 20 compounds in total.

Compound	Molecular ion	Boiling point	Retention time	Frag. ion 1	Frag. ion 2	Frag. ion 3
Standard #1 Mix						
α-Pinene	136.1	157	5.3	121	105	93
Camphene	136.1	159	5.6	121	107	93

Table 3 Chemical characteristics of the terpenes and terpenoids found in the standard Mix #1 and 2 from Restek, a commonly available standard for these compounds, 20 compounds in total.—cont'd

Compound	Molecular ion	Boiling point	Retention time	Frag. ion 1	Frag. ion 2	Frag. ion 3
β-Pinene (2 isomers)	136.1	166	6.1 and 6.3	121	107	93
Delta-3-carene	136.1	171	6.9	121	105	93
Alpha-terpinene	136.1	174	7.1	121	105	93
p-cymene	134.1	177	7.3	119	103	91
D-Limonene	136.1	176	7.4	121	107	93
Ocimene (2 isomers)	136.1	100	7.65 and 8.0	121	105	93
Gamma-terpinene	136.1	183	8.4	121	105	93
Terpinolene	136.1		9.5	121	105	93
Linalool	154.1	198	9.7	136	121	107
Isopulegol	154.1		10.5	136	121	111
Geraniol	154.1	230	11.5	136	123	111
Beta-caryophyllene (2 isomers)	204.2	127	12.5 and 12.7	189	175	161
Alpha-humulene (2 isomers)	204.2	106	12.6 and 12.8	189	161	147
Nerolidol (4 isomers)	222.2	122	12.8–13.0	204	189	161
(−)-Guaiol (2 isomers)	222.2		13.1–13.2	204	189	161
(−)-Alpha-Bisabolol (2 isomers)	222.2	153	13.4	204.2	189.1	161.1
Standard #2 Mix						
1,8-Cineole (eucalyptol)	154.1	176	7.5	139	125	108
(−)-Caryophyllene oxide (4 isomers)	220.2	127	13.1–13.4	205	177	161

(−43) to give rise to the m/z 93 ion. The loss of 43 mass units could come from many different positions on the pinene isomers and without isotopic labelling it is not possible to predict which is the favourable loss. The ion that forms at m/z 93 is an even electron ion and forms the stable base peak.

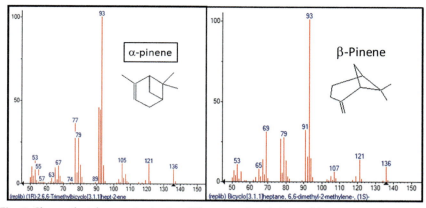

Fig. 3 Shows the electron ionization (EI) spectra for alpha and beta pinene.

We will see in the mass spectra of the monoterpenes of cannabis that the m/z 93 ion is an important and common base peak ion, or in many cases, forms a major ion in the mass spectrum. Finally, the difference in the mass spectra of the two isomers of pinene is minimal but slightly different in the mass range of m/z 68–93. However, the two isomers are readily separated by gas chromatography with retention times of 5.3 and 6.1 min, respectively (Table 3). The two peaks at 6.1 and 6.3 min had identical spectra, both of which matched the library for β-pinene; we suggest that these isomers are conformational, not structural isomers.

The two isomers of δ-limonene and δ-terpinene (terpinolene), although quite similar in chemical structure (Fig. 4), have strikingly different mass spectra. Delta-limonene loses a methyl radical ($-CH_3$) to form the m/z 121 ion, or it may lose an ethyl radical (-29 mass units) to form the even electron ion at m/z 107. The m/z 107 ion is still a minor fragment ion as is the m/z 93 ion, which was a base peak for the two pinene isomers shown in Fig. 3. The base peak ion for δ-limonene is the m/z 68 ion, which presumably forms from the loss of the isoprene unit as a neutral molecular loss (136–68=68 mass units). This fragmentation proceeds from the OE molecular ion to another OE ion at m/z 68 by loss of an isoprene molecule. This is a distinguishing feature of the mass spectrum of δ-limonene and is not found in any of the other monoterpenes associated with cannabis, which makes the spectrum of δ-limonene unique. The m/z 68 ion is diagnostic for δ-limonene and is exactly half of the molecular ion of 136.

Terpinolene shows two major fragment ions at m/z 121 and 93, both of which were ions found in the pinene isomers. Whenever an ion has a major

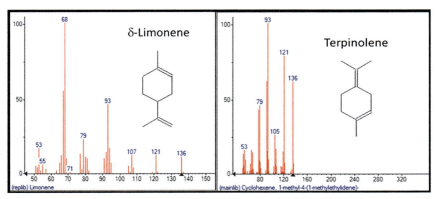

Fig. 4 Comparison of mass spectra of the monoterpenes, delta-limonene and delta terpinene (also known as terpinolene).

intensity relative to the base peak ion (the base peak ion being the largest ion in the mass spectrum) then it means that the ion is relatively stable and can be used for identification or quantification in the mass spectrum. Thus, terpinolene loses a methyl radical to form the m/z 121 ion or it loses an isopropyl radical to form the m/z 93 ion, which is the base peak ion of the mass spectrum. It is interesting to note that these two different mass spectra are also reflected in the chemistry of these two monoterpene isomers. For example, δ-limonene is in the citrus family of compounds, while terpinolene is in the turpentine family of compounds with a fresh smell and is commonly used in soaps for this reason. No doubt that the fragmentation mass spectra of these two isomers are related to their chemistry and function. This inference, albeit, will be called upon throughout the discussion of mono and sesquiterpene spectral interpretation. Perhaps fragmentation will be a useful tool to add some insight into the curious properties of these terpene compounds in cannabis.

4.3.2 Monoterpenoids

The monoterpenoids are formed by the addition of an alcoholic group to the monoterpene structure. Fig. 5 shows the mass spectra of two isomers, linalool and isopulegol. Linalool has a tiny, but barely recognizable, molecular ion at m/z 154. A characteristic loss of the terpenoids is the loss of a water molecule, mass loss of 18 mass units. This shows as an ion with m/z 136 (154–18=136). One should be careful here since the monoterpenes have a molecular ion of m/z 136 and the mass spectrum could easily be misinterpreted as a monoterpene rather than a monoterpenoid. Thus, the

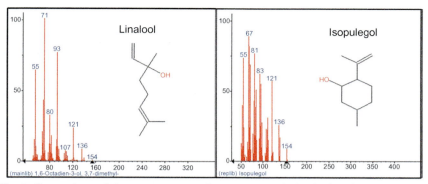

Fig. 5 Comparison of the mass spectra of two terpenoids, linalool and isopulegol.

small but important ion at m/z 154 gives one a clue that the structure is a monoterpenoid rather than a monoterpene. This fact can be very important in the identification of unknowns in cannabis, as will be shown in the following section where we identify other monoterpenoids in the flower of *Cannabis sativa*.

The base peak ion for linalool is the m/z 71 ion, which forms from the fragmentation that occurs at the C-3 position, which is substituted by both the hydroxyl group and a methyl group. The ion that forms is the epoxide fragment ion of m/z 71, which is characteristic of the linalool structure and is not seen in any of the other monoterpenoids. The other fragment ions in the mass spectrum of linalool (m/z 121, 107, and 93) form as direct losses from the dehydrated ion of m/z 136 and are analogous to the losses seen for the monoterpenes above, i.e., terpinolene and pinene.

The mass spectrum for isopulegol, an isomer of linalool, is quite different. The molecular ion at m/z 154 is definitely more visible in the mass spectrum. Again there is an immediate water loss of 18 mass units to give the m/z 136 ion. This fragment ion at m/z 136 is quickly seen as an isomer of terpinolene (spectrum above in Fig. 4). Notice that the m/z 121, 105 (not labelled), and 93 ion are present. The base peak ion is m/z 67 but there are also strong signals from the fragment ions at m/z 55, 81, and 93. Basically the fragmentation pathway involves first the loss of water to form the m/z 136 ion, followed by a branched pathway, meaning that from the m/z 136 ion, there is the possibility to form the m/z 93, 81, 67, and 55 ion directly by independent losses from the m/z 136 ion. This is interpreted from the fact that forming the other cascade of ions requires forbidden losses. That means basically that the ions cannot form from the m/z 93 ion (i.e., the m/z 81, 67, and

55 ions) since these are forbidden losses (i.e., either 12 or 14 mass units). Again this mass spectrum is unique for isopulegol and is not seen in other monoterpenoid mass spectra.

4.3.3 Sesquiterpenes

The sesquiterpene structure consists of three isoprene units, which give rise to a molecular mass of m/z 204 from the monoterpenes (i.e., $136 + 58 = 204$ mass units). Fig. 6 shows the mass spectra of the two important sesquiterpenes in cannabis, that is humulene and caryophyllene, which are isomers.

Inspection of the spectra in Fig. 6 shows that they are quite different. Humulene shows a relatively simple mass spectrum with the basic peak ion of m/z 93, which is again the same common fragment ion seen in all of the monoterpenes. The m/z 136 ion is present, which suggests that the fragmentation pathway proceeds to the base peak ion, m/z 93, by first a loss of isoprene neutral molecule (68 mass units) to give the m/z 136 ion. Once the m/z 136 ion forms, it continues to fragment to give the m/z 93 ion, similar to the fragmentation of the pinene isomers.

A unique fragment ion of humulene is the m/z 80 ion, which also has an intensity of approximately 40% relative to the base peak ion. This ion is an OE ion forming, most likely, by the neutral loss of a larger fragment, shown below, the 124 neutral molecule (Fig. 7).

Thus, this fragment ion of m/z 80 is considered diagnostic for the humulene structure and will be used in the next section on the identification of terpenes in the flower of the cannabis plant.

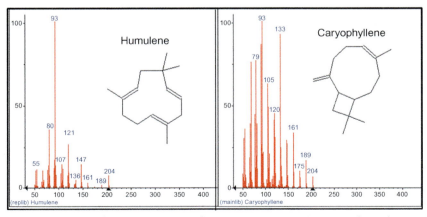

Fig. 6 Comparison of the mass spectra of two important sesquiterpenes, humulene and caryophyllene.

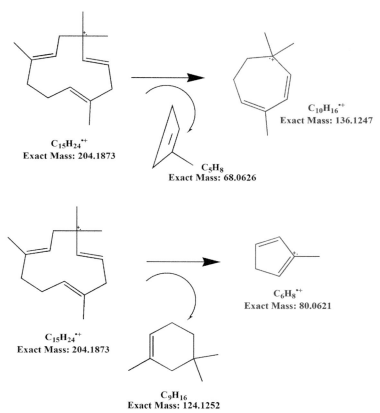

Fig. 7 Fragmentation pathway for humulene to give two important fragment ions, one of which is diagnostic for humulene, the m/z 80 ion.

The mass spectrum for caryophyllene is shown in Fig. 6. The molecular ion, m/z 204, is approximately the same intensity as its isomer, humulene, and is an important ion for the identification of both of these sesquiterpene isomers. The base peak ion is m/z 93 but it also has a large ion of 95% intensity at m/z 133. This ion forms from the loss of an isoprene radical with a mass of 71 mass units (see Fig. 8 below). This is also a unique fragmentation that is diagnostic of the caryophyllene family of isomers and will be used in the following section for identification of unknown terpenes in cannabis flower. Another diagnostic ion is the m/z 120 ion, which originates from the loss of 84 mass units (see Fig. 9) below.

Presumably the base peak ion of m/z 93 forms from a pathway similar to humulene in that there is a loss of 68 mass units via a neutral isoprene unit

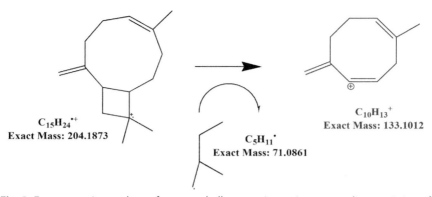

Fig. 8 Fragmentation pathway for caryophyllene to give an important diagnostic ion of m/z 133.

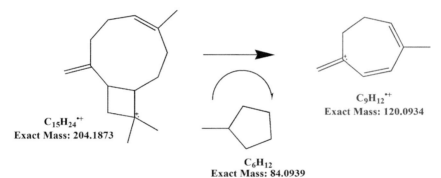

Fig. 9 Fragmentation pathway for the OE ion, m/z 120, also a diagnostic ion fragmentation.

(see Fig. 7) to give the m/z 136 ion that subsequently fragments to give a base peak ion of m/z 93, similar to its isomer, humulene.

Another major ion in the caryophyllene mass spectrum is the m/z 161 ion, which is an even electron ion that results from the loss of a propene radical of 43 mass units. This loss was common for the monoterpenes, such as pinene, and gave rise to the base peak ion in the pinene spectrum at m/z 93 (see Fig. 3). Caryophyllene also gives radical losses of 15 mass units (CH$_3$ radical) and 29 mass units (CH$_3$CH$_2$ radical) to give ions at m/z 189 and 175, respectively. Again this fragmentation pathway is identical to monoterpene fragmentations.

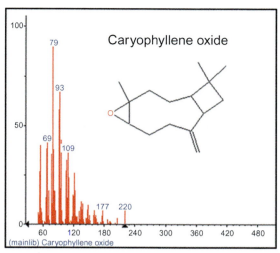

Fig. 10 Mass spectrum of caryophyllene oxide, a major sesquiterpenoid in cannabis.

The last compound to consider is the sesquiterpenoid, caryophyllene oxide, which consists of the addition of an oxygen atom to the caryophyllene structure. Fig. 10 shows the mass spectrum for this compound. The molecular ion is present with a small but recognizable ion at m/z 220. Major ions occur at m/z 79, 93, and 109. The m/z 93 ion is an important ion in all of the monoterpenes and sesquiterpenes, as shown in the preceding Figs. 3–6. The presumable pathway for formation most likely passes through the m/z 136 ion, followed by a loss of a propene radical with mass of 43 units to give the m/z 93 ion. What is more interesting, a diagnostic ion for this spectrum, is the ion at m/z 109, which is not seen in any of the other mass spectra shown in Figs. 3–6. In particular the m/z 109 ion does not occur in the monoterpenoids, linalool or isopulegol, which both contain oxygen. Presumably the m/z 109 ion is 16 mass units larger than the m/z 93 ion, which is associated with the same structure plus an oxygen atom. This proposed fragmentation pathway is shown in Fig. 11.

The fragmentation involves the loss of a radical fragment of 121 mass units (Fig. 11). Finally, the base peak ion is m/z 79, which is basically the loss of a methylene group from the m/z 93 ions and results in a simple five-membered ring ion, shown below in Fig. 12.

This fragment ion of m/z 79 is seen in the fragmentation of monoterpenes, such as pinene and terpinolene. It is also found in caryophyllene as

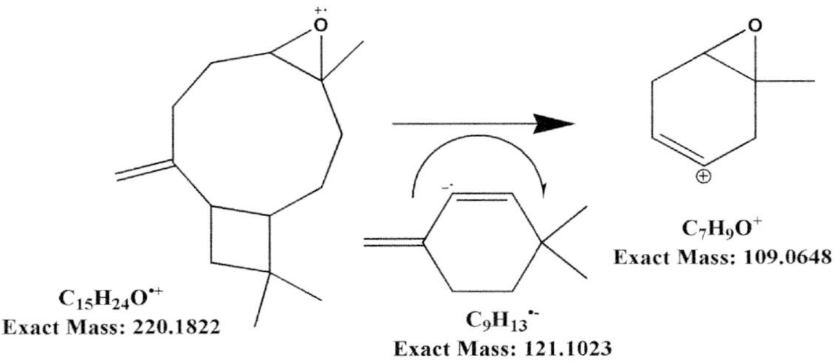

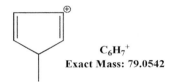

Fig. 11 Proposed fragmentation of caryophyllene oxide to give the diagnostic fragment ion of m/z 109.

Fig. 12 Structure of the five-membered ring ion, m/z 79, which is the base peak ion.

a major ion (see Fig. 6). These are the major fragmentation studies of the important terpenes and terpenoids in cannabis, as reported in the recent literature [3]. Table 4 shows the summary of diagnostic ions for the compounds studied thus far.

In the next section, the major terpene and terpenoids found in cannabis flower will be examined and the use of diagnostic ions will be used along with the library of electron ionization (EI) spectra to identify the major compounds.

4.4 Mass spectral identification for terpenes and terpenoids in Cannabis flower

This section deals with the extraction of several *Cannabis sativa* cultivars, named A and B. These plants were gifts from local growers of *Cannabis sativa*, which contained less than 0.3% THC. Samples of the plant were processed without drying and extracted with a simple procedure. Basically the plant is weighed in its green state so that critical terpenes and terpenoids are not lost by volatilization. They were ground in a mortar and pestle and immediately extracted with methanol. Then they were centrifuged and filtered through 0.45 μm PTFE filters and analysed directly by the same GC/MS procedure described in Section 3. Fig. 13 shows

Table 4 Base peak and diagnostic ions for monoterpenes, monoterpenoids, sesquiterpenes, and sesquiterpenoids.

Compound	Base peak (m/z)	Diagnostic ion (m/z)	Molecular ion (m/z)
α-Pinene and β-Pinene	93	None	136
δ-Limonene	68	68	136
Terpinolene	93	None	136
Linalool	71	71, 80	154
Isopulegol	67	67, 81	154
Humulene	93	80	204
Caryophyllene	93	133	204
Caryophyllene oxide	79	109	220
Sesquiterpenoids, *In General*	N/A	69	N/A

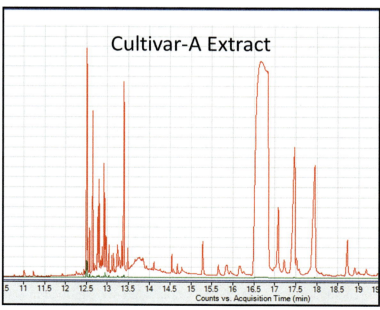

Fig. 13 Total ion chromatogram of A-Cultivar extract by GC/MS.

Analysis of terpenes in hemp by gas chromatography/mass spectrometry   217

the chromatogram for the A cultivar. The total ion chromatogram is shown in Fig. 13. There are several important things to note in the chromatogram. Firstly, the very large peak at 16.5 min, which is the retention time for cannabidiol (CBD), which is a major component of the plant extract. Given that this is an extract of *Cannabis sativa*, which furthermore is a cultivar that contains greater than 10% CBD and less than 0.3% THC, it is not surprising that the major peak in the chromatogram corresponds to CBD. The evidence for this identification will be shown in the following section that deals with the analysis of the cannabinoids by GC/MS. In fact, the majority of the peaks after the elution of CBD, at 16.5 min, are related directly to the cannabinoids.

Fig. 14 shows an enlarged view of the components eluting before CBD, from 5 to 16.5 min. The large cluster of peaks from 12.5 to 13.5 min corresponds to the retention time of sesquiterpenes. The monoterpenes elute much earlier from 5 to 10 min based on the standard chromatogram shown in Fig. 1. Careful examination of the chromatogram in Fig. 14 shows that there are trace levels of α-pinene, β-pinene, and δ-limonene, which are trace amounts of monoterpenes present in the extract. There are several reasons for the trace levels of monoterpenes present.

Firstly, they are actually present at trace levels in the plant and are being used as part of the synthesis of CBD and THC. Secondly, they were present at higher levels but were volatilized during transport of the plant material to the laboratory. Thirdly, they were lost during sample preparation. Both of these two previous hypotheses seem unlikely based on the analysis and our experience. The monoterpenoids were present at trace levels also, mainly linalool was identified at trace amounts in the chromatogram shown in Figs. 13 and 14, which was based on retention time matching and matching of the EI spectrum to the GC/MS library.

What are overwhelmingly present in the chromatogram are the peaks from 12.5 to 16.5 min shown in Figs. 13 and 14. These peaks corresponded to a series of isomers matching both in retention time and mass spectra to humulene and caryophyllene and their isomers. The concentrations, too were quite high, in the 100–ppm range based on standards analysed at the same time. Thus, this analysis of the A cultivar of *Cannabis sativa* suggests that the monoterpenes and monoterpenoids play a minor role (concentration wise) in the terpene profile of this cultivar; while the sesquiterpenes play the major role (concentration wise) in the terpene and terpenoid profile of the A cultivar.

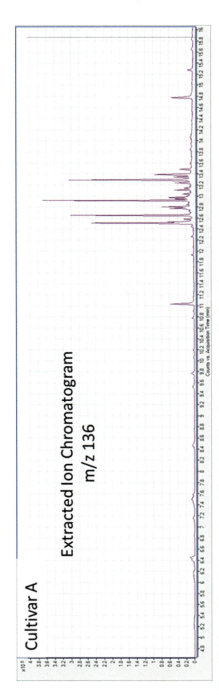

Fig. 14 Extracted ion chromatogram of the *m/z* 136 ion for terpenes and terpenoids.

Analysis of terpenes in hemp by gas chromatography/mass spectrometry 219

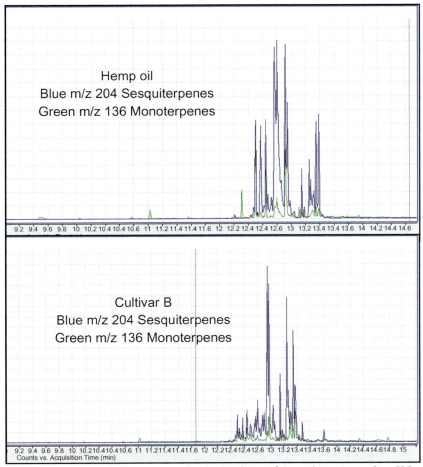

Fig. 15 Extracted ion chromatograms for the B cultivar of *Cannabis sativa* and an CBD oil sample.

We tested this finding with two other samples, one a different cultivar called, B cultivar, and an oil extract that was purchased at a local health store. The chromatograms for those two samples are shown in Fig. 15 below. The blue trace in the chromatogram corresponds to the m/z 204, which are sesquiterpenes that elute from 12 to 16 min at the end of the chromatogram on the far right of the figures. Both a CBD oil sample and the B cultivar sample had similar chromatograms with the sesquiterpenes being the major terpenes present. The green trace was the m/z 136, which corresponds to the monoterpenes. Only trace amounts of pinene and limonene were present in the samples.

4.4.1 Ocimene—A minor monoterpene in cannabis

The minor terpene, ocimene, was present in all of the three samples. This monoterpene has not been reported as a major terpene [3], although in these cultivars it was an important monoterpene and many web sites report this monoterpene. It is discussed as a folk medicine remedy [11] and in the scientific literature, [12]. The structure is shown in Fig. 16 along with its' mass spectrum. The mass spectrum is similar to other monoterpenes such as alpha and beta pinene. The base peak ion is m/z 93; however, the molecular ion is barely seen in the mass spectrum because of the lability of the ion. The ocimenes are a group of monoterpenes found not only cannabis but also in plants and fruits and have been known to give mild antimicrobial activity as an oil mixture [12]. The name comes from an ancient Greek word for basil, ocimum. They have a pleasant odour and are used in perfumes for their herbal scent. Similar to myrcene, the ocimenes are unstable and are easily oxidized in air.

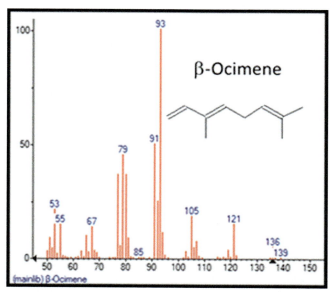

Fig. 16 Mass spectrum and structure of β-ocimene.

4.4.2 Camphene—A minor monoterpene in cannabis

Camphene is a monoterpene with a bicyclic structure (Fig. 17) and a pungent odour. It is used as an additive in foods and as a fragrance additive. When present in cannabis it does contribute an earthy or musky smell to the oil, as well as a somewhat piney like odour. In fact, camphene is sometimes synthesized from pinene. It was present in trace amounts in both the A and B cultivars but not in the oil that was analysed. Camphene has been noted to have some potential ability to treat skin conditions, such as eczema and psoriasis [10,11]. Thus, camphene may be one of the more useful monoterpenes in cannabis creams and worthy of more research efforts.

The mass spectral fragmentation of camphene shows that the base peak ion is m/z 93, the typical loss of a propyl radical to give a stable even electron ion (EE). It is worth noting that camphene does give a relatively large molecular ion at m/z 136, with an intensity of approximately 20% of the base peak ion. Also worthy of note in the identification of camphene by mass spectrum is the large ion at m/z 121, which is the loss of a methyl radical from the molecular ion. Given that there are two methyl groups available for this loss, the odds are increased for the formation of this ion, along

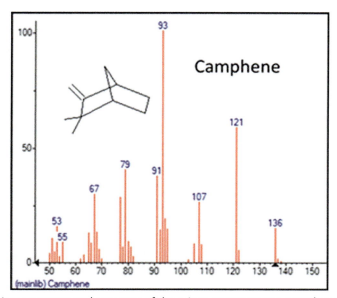

Fig. 17 Mass spectrum and structure of the minor monoterpene, camphene.

with the fact that the resulting ion would be stabilized on the substituted bicyclic ring, further increasing the probability of a stable ion. The intensity of the m/z 121 ion was 60%. Furthermore, camphene has a low boiling point of 159°C and a retention time similar to that of pinene, eluting at 5.6 min (Table 3).

4.4.3 Delta-3-carene—A minor monoterpene in cannabis

Delta-3-carene is also a bicyclic monoterpene but with a cyclopropane ring rather than a five-membered ring structure of camphene. Carene is a major constituent of turpentine and comes from the resin of conifer trees as a major source. Carene has a sweet and pungent odour resembling fir needles, musky, earthy, and a damp woodlands-like odour. Among the natural health products, δ-3-carene claims to have properties that promote brain health and is an antihistamine [10,11].

The mass spectrum is similar to the bicyclic monoterpenes in that it has a molecular ion at m/z 136 with an intensity of approximately 20% (Fig. 18). It has a base peak ion of m/z 93 again showing the common loss of a propyl radical loss of 43 mass units. The m/z 121 ion is not as intense as that of camphene. It too has a low boiling point of 171°C and a retention time of 6.9 min (Table 3).

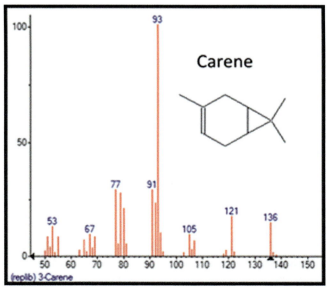

Fig. 18 Mass spectrum and structure of the minor monoterpene, delta-3-carene.

4.4.4 p-Cymene—A minor alkylbenzene-terpene in cannabis

Para-cymene is an aromatic organic compound and thus is not an isomer of the monoterpenes, discussed thus far. It contains one more double bond than the monoterpenes of mass 136 mass units, which lowers the mass to 134 mass units. It has a much different mass spectrum than the monoterpenes discussed thus far. It is a naturally occurring compound formed in plants such as cumin and thyme. Its structure is similar to that of terpinene, with a methyl and isopropyl group substituting the aromatic ring at the para position (Fig. 19). The mass spectrum shows a molecular ion at m/z 134 with an intensity of 25%, showing some stability for the OE ion. The molecular ion loses a methyl radical (15 mass units) to give the base peak ion at m/z 119. The spectrum also shows the defining fragment ion of m/z 91 of the aromatic ring, called a tropylium ion, which is a well-known ion in electron ionization and a classic ion for showing the aromatic ring structure. The other two geometric isomers of cymene, that is ortho- and meta-cymene, do not occur naturally, only p-cymene is a natural terpene.

4.4.5 Eucalyptol—A minor monoterpenoid in cannabis

Eucalyptol is a monoterpenoid that has a spicy aroma and is a well-known smell associated with the eucalyptus tree, which is native to Australia, but is found worldwide. The oil of the eucalyptus tree is from the leaf of the tree and contains about 90% eucalyptol. The eucalyptus oil is used as a flavouring

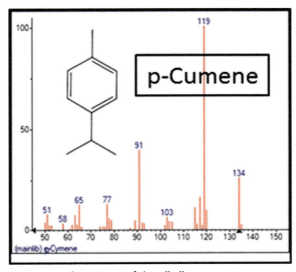

Fig. 19 Mass spectrum and structure of the alkylbenzene-terpene, p-cymene.

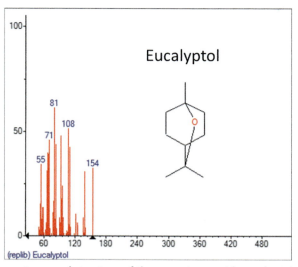

Fig. 20 Mass spectrum and structure of the monoterpenoid, eucalyptol.

agent in many products, including: mouthwash, baked goods, meat products, beverages, and even cigarettes. It is present in minor amounts in cannabis, but may give both odour and flavour to the cannabis oil. Fig. 20 shows the mass spectrum of eucalyptol. The molecular ion is m/z 154 with a large intensity of 60% relative to the base peak ion of mass m/z 81. The characteristic and distinctive mass spectrum of eucalyptol makes it an easy identification by library matching using mass spectrometry.

4.4.6 Geraniol—A minor monoterpenoid in cannabis

Geraniol is a monoterpenoid and is a major component of rose oil and citronella oil [11]. It also occurs in geranium (hence, its name), lemon, and other essential oils. It is commonly used in perfumes. Interestingly, geraniol is also produced by honeybees to mark nectar-bearing flowers and to locate the entrances to their hives. Geraniol is used in the biosynthesis of other terpenes, such as myrcene and ocimene, both of which occur in cannabis. Its' mass spectrum is shown in Fig. 21. The molecular ion is m/z 154, which for all practical purposes is not present. There is an ion at m/z 139 ion, which is the loss of a methyl radical to give rise to the EE ion at m/z 139. The base peak ion is m/z 69. Geraniol was present in both cultivars, A and B cultivars, at low concentrations of approximately 1–10 ppm.

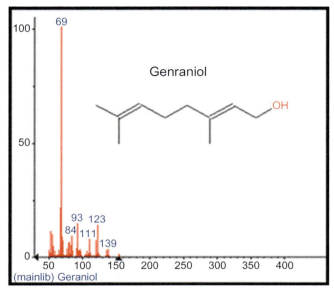

Fig. 21 Mass spectrum and structure of the monoterpenoid, geraniol.

4.4.7 Nerolidol—A minor sesquiterpenoid in cannabis

Next is to consider the less abundant sesquiterpenoids in cannabis, starting with nerolidol, which is shown in Fig. 22 along with its mass spectrum. Nerolidol occurs naturally in many essential oils of plants and flower [11].

Nerolidol is present in ginger, jasmine, lavender, lemon grass, and of course cannabis. The aroma of nerolidol is woody and reminds one of fresh bark odour. It has various biological activities such as antioxidant, antifungal, and antimicrobial activity. The molecular ion is not present in the mass spectrum, but would be an ion with mass of m/z 222, which is 18 mass units larger than its sesquiterpenoid mass of 204 mass units. Thus, the molecule loses 33 mass units, which is actually two losses. First is the loss of water, 18 mass units to give an ion of mass m/z 204 ion, followed by the loss of a methyl radical, 15 mass units, to give the m/z 189 ion shown in the spectrum. This interpretation was worked out from looking at the NIST [13] spectrum on line, which does show very small ions at m/z 222 and 204.

The base peak ion is m/z 69. The m/z 121 and 107 ions are reminiscent of monoterpenoids, such as geraniol and caryophyllene oxide, both of which have the m/z 69 ion. None of the monoterpenoids or monoterpenes

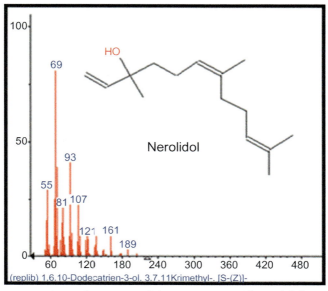

Fig. 22 Mass spectrum and structure of the monoterpenoid, nerolidol.

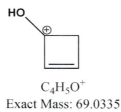

$C_4H_5O^+$
Exact Mass: 69.0335

Fig. 23 Shows the structure of the *m/z* 69 ion, which is the base peak for nerolidol.

have the *m/z* 69 ion as a prominent ion in the mass spectrum; thus, the *m/z* 69 ion is diagnostic of the alcoholic group of the sesquiterpenes. The hypothesized structure of the *m/z* 69 ion is shown below in Fig. 23, which could be confirmed with accurate mass.

4.4.8 Bisabolol—A minor sesquiterpenoid in cannabis

Bisabolol is a natural monocyclic sesquiterpenoid. It is a colourless viscous oil that is the primary constituent of some plant oils [11]. Bisabolol has a weak, sweet, floral aroma and is used in various fragrances. It has also been used for hundreds of years in cosmetics because of its perceived skin healing properties. Bisabolol is known to have anti-irritant, anti-inflammatory, and

Analysis of terpenes in hemp by gas chromatography/mass spectrometry

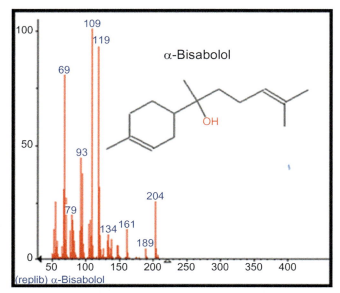

Fig. 24 Chemical structure and mass spectrum of the sesquiterpenoid, α-bisabolol.

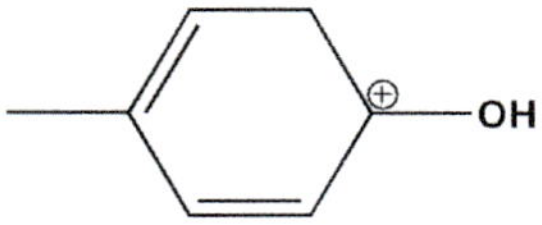

$C_7H_9O^+$
Exact Mass: 109.0648

Fig. 25 Shows the structure of the base peak ion of *m/z* 109 in α-bisabolol.

anti-microbial properties. The chemical structure and mass spectrum of bisabolol is shown in Fig. 24.

The molecular ion for α-bisabolol is not present at *m/z* 222, but the *m/z* 204 ion is a relatively large ion, at 20% intensity. This spectrum could be mis-interpreted as a sesquiterpene if it were not for the presence of the *m/z* 69 ion that is characteristic of sesquiterpenoids, as previously discussed. The base peak ion is *m/z* 109 and also shows the presence of an oxygen atom as shown below in Fig. 25.

Caryophyllene oxide also has the *m/z* 109, so again this ion is diagnostic for oxygen containing sesquiterpenoids (Fig. 11). Like nerolidol, a-bisabolol also has the *m/z* 69 ion, again diagnostic of the sesquiterpenoids.

4.4.9 Guaiol—A minor sesquiterpenoid in cannabis

Guaiol is a sesquiterpenoid found in the oil of the guaiacum tree, cypress pine, and in the cannabis plant. It has been associated with decreasing anxiety, a feature that is known in the field of homeopathy [10,11,14]. Guaiol has a history of use as a treatment for arthritis and gout. It has also been used as a diuretic and to lower blood pressure. It is known to have anti-inflammatory and analgesic properties. The structure and mass spectrum for guaiol is shown in Fig. 26.

The molecular ion for guaiol is *m/z* 222 and shows prominently in the mass spectrum with an intensity of about 15%. The presence of oxygen is shown by the water loss to give the *m/z* 204 ion, which then forms a sesquiterpene. This water loss then controls the remainder of the fragmentation giving major ions at *m/z* 189 (methyl radical loss of 15 mass units), *m/z* 161, which is the base peak ion. The *m/z* 161 ion is a loss of 28 mass units from the *m/z* 189 ion, which is an ethylene molecular loss, which gives the stable even electron ion at *m/z* 161 (base peak ion).

4.5 Mass spectral identification of two major cannabinoids in Cannabis flower

This section looks at the mass spectra of two major cannabinoids in *Cannabis sativa*, cannabidiol (CBD, the major component) and tetrahydrocannabinol

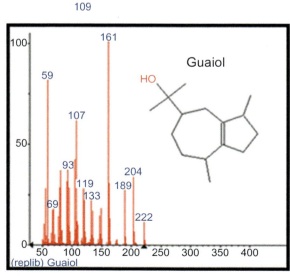

Fig. 26 Chemical structure and mass spectrum of the sesquiterpenoid, guaiol.

Analysis of terpenes in hemp by gas chromatography/mass spectrometry

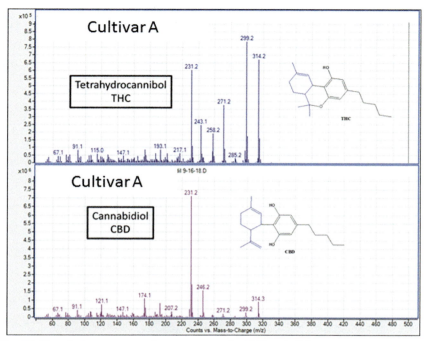

Fig. 27 Chemical structure and mass spectra for THC and CBD.

(THC, a less abundant component, which by definition, is less than 0.3% THC, in *Cannabis sativa*). The A cultivar extract is shown in Fig. 13 with a large (overloaded peak) at 16.5–16.8 min in the chromatogram. This peak is CBD and its mass spectrum is shown along with THC in Fig. 27.

What is obvious from the comparison of the two mass spectra for THC and CBD is that the mass spectra are quite different, which is not true for the electrospray spectra using liquid chromatography/mass spectrometry that show identical spectra, with both the same ions and same ion intensity (see chapter "Analyses of cannabinoids in hemp oils by LC/Q-TOF-MS" by Ferrer). Thus, EI spectra for these two compounds and their related isomers are an important tool for understanding their structures, and perhaps their function. The fragmentation pathway for THC is shown in Fig. 28.

There are six major ions in the mass spectrum with masses of m/z 314 (molecular ion), 299 (base peak ion), 271, 258, 243, and 231. The fragmentation of the molecular ion proceeds by three pathways with losses of methyl radical (-15 mass units), loss of ethyl radical (-29 mass units), or the loss of

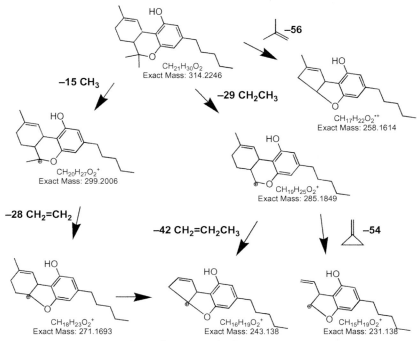

Fig. 28 Fragmentation pathway for THC with electron ionization.

an isoprene neutral molecule to give the m/z 258 ion. Subsequent losses of neutral molecules give the other fragment ions. The third ring structure containing the ether oxygen apparently stabilizes the even electron ions (the ions with odd masses) and gives many stable ion fragments as shown in Fig. 28.

This EI fragmentation of THC looks considerably different than the fragmentation of CBD, which gives all of the same ions as THC but with much different intensities, with only the m/z 231 ion being stable as the base peak ion.

The synthesis of both THC and CBD involves the molecules, olivetol and a monoterpene, such as myrcene, which is cyclized to form either THC with three rings, or CBD, which has two rings, one of which has the structure of limonene, a monoterpene. Fig. 13 shows the total ion chromatogram for the A cultivar, which is predominated by a large peak of CBD at 16.6 min, as discussed above. The molar concentration of the limonene fragment of CBD is equal to all of the monoterpenes and sesquiterpenes in the chromatogram! This surprising result suggests that the cannabis plant, in this case the A cultivar, has spent all of its energy synthesizing several simple monoterpenes, such as myrcene and limonene, which is then consumed in synthesizing CBD. Thus, it is not surprising then to see a chromatogram for the A cultivar that is depleted in monoterpenes relative to both sesquiterpenes and CBD. The sesquiterpenes have an extra isoprene unit so they are not part of the synthesis pathway for CBD. Thus, the odour from the A cultivar is coming chiefly from the sesquiterpenes, since CBD has no odour in its pure form. Thus, it appears that the plant uses its monoterpenes (myrcene) to make the high yields of CBD. In the A cultivar, the CBD content was greater than 10%.

4.5.1 The entourage effect

The entourage effect is a recent concept (1999, Ref. 15) that there is a synergy among the cannabinoids, terpenes, and terpenoids that display as a cooperative effect in eliciting different cellular responses [16]. The term has continued to evolve into the phrase now meaning that the combined effect of these three major components of the cannabis plant work together to create an effect that is "greater than the sum of its parts" to give almost magical, or at least, potent positive effects on the health of one taking this mixture. At least one study has found that the common terpenes in cannabis do not directly activate the CB1 and CB2 receptors, while sesquiterpenes, THC, and CBD do activate the receptors [11].

The terpenes and terpenoids do have medicinal effects as discussed earlier in this chapter, such as powerful antioxidants, which is created by the conjugated double bonds present in many of the monoterpenes, which have the ability to scavenge oxygen radicals and form terpenoids or epoxides. This is especially true for compounds like myrcene and ocimene, and the terpenoids, linalool and geraniol. There is also the possibility of toxic effects of some of the monoterpenes and terpenoids in higher doses. For example, eucalyptol is an example with a LC50 of 2480 mg/kg in rats and is classified as a reproductive toxin for females.

There is the practical side from producers of these products to tout the entourage effect, which is that it is much easier and less expensive to produce an oil that is a mixture of terpenes, terpenoids, and cannabinoids, rather than to make a pure product. As the study of *Cannabis sativa*, CBD, and THC proceed, much more will be learned about their role in human medicine and its value as a medicinal plant.

5. Conclusions and major takeaway ideas

1. The major terpenes in *Cannabis sativa* are the sesquiterpenes and sesquiterpenoids, which give the characteristic odours of each of the cultivars of the cannabis plant.
2. Monoterpenes, although important in the cannabis plant, are mainly found as part of the structure of CBD or THC, at least based on the study herein, where two cultivars of *Cannabis sativa* were studied with greater than 10% CBD and less than 0.3% THC.
3. Monoterpenes have the same formula and therefore by definition are isomers. They commonly give similar mass spectra (often nearly identical spectra to the human eye). There are some diagnostic ions being found, which are helpful in the identification of monoterpene isomers by mass spectrometry. See Tables 1, 3 and, especially, Table 4.
4. Fragmentation studies of THC and CBD by electron ionization reveal a quite different stability pattern of fragment ions, with the third ring of THC giving stability to the ions. CBD, on the other hand, fragments chiefly to one base peak ion, m/z 231.
5. The entourage effect, an interesting concept, maybe supported by the findings herein. That is the important terpenes in the two cultivars studied yield mainly sesquiterpenes (caryophyllene, e.g.), which do activate the CB1 and CB2 receptors, based on the literature. The monoterpenes are in low concentration relative to the sesquiterpenes and cannabinoids. An analysis of at least one oil from the *Cannabis sativa* plant show that the monoterpenes are present at low concentrations but the sesquiterpenes are a major component.

Acknowledgements

Our appreciation is given to the Writers' Cabin, where much of this chapter was written. Our chapter is dedicated to our Beloved Kitty, "Gato Negro", who was a patient partner in many of our former papers and chapters. He loved to sit beneath our desk, or behind the computer, on a cold winter's night and listen to the clicking of the keys as words appeared on the page. We feel his presence today as we finish up this book on cannabis analysis. Hopefully, many good medicinal treatments will come from this plant to help both animal and man.

References

[1] L. Ružička, The isoprene rule and the biogenesis of terpenic compounds, Cell. Mol. Life Sci. 9 (10) (1953) 357–367, https://doi.org/10.1007/BF02167631. 13116962.
[2] A. Kekulé, Lehrbuch der organischen Chemie [Textbook of Organic Chemistry] (in German), vol. 2, Ferdinand Enke, Erlangen, (Germany), 1866, pp. 464–465. Mit dem Namen Terpene bezeichnen wir ... unter verschiedenen Namen aufgeführt werden.

[3] E.A. Ibrahim, M. Wang, M.M. Radwan, A.S. Wanas, C.G. Majumdar, B. Avula, Y.H. Wang, I.A. Khan, S. Chandra, H. Lata, G.M. Hadad, R.A.A. Salam, A.K. Ibrahim, S.A. Ahmed, M.A. ElSohly, Analysis of terpenes in *Cannabis sativa* L. using GC/MS: method development, validation, and application. Planta Med. 85 (2019) 431–438, https://doi.org/10.1055/a-0828-8387.

[4] D. Tholl, Biosynthesis and biological functions of terpenoids in plants. Adv. Biochem. Eng. Biotechnol. 148 (2015) 63–0106. https://doi.org/10.1007/10_2014_295.

[5] US National Library of Medicine, Linalool, PubChem, US National Library of Medicine, 2017. Retrieved 14 February 2017.

[6] G. Tinseth, The Essential Oil of Hops: Hop Aroma and Flavor in Hops and Beer, https://realbeer.com/hops/aroma.html, 2020. (accessed April 8, 2020).

[7] S.T. Katsiotis, C.R. Langezaal, J.J.C. Scheffe, Analysis of the volatile compounds from cones of ten Humulus lupulus cultivars. Planta Med. 55 (7) (1989) 634, https://doi.org/10.1055/s-2006-962205.

[8] A. Bertoli, S. Tozzi, L. Pistelli, L.G. Angelini, Fibre hemp inflorescences: from crop-residues to essential oil production, Ind. Crop. Prod. 32 (2010) 329–337.

[9] L.O. Hanus, S.M. Meyer, E. Munoz, O. Taglialatela-Scafati, G. Appendino, Phytocannabinoids: a unified critical inventory, Nat. Prod. Rep. 12 (2016) 1357–1392.

[10] B. Rahn, What are Cannabis Terpenes and What Do They Do?, https://www.leafly.com/news/cannabis-101/terpenes-the-flavors-of-cannabis-aromatherapy, 2014 (accessed April 8, 2020).

[11] J.K. Booth, J. Bohlmann, Terpenes in *Cannabis sativa*—from plant genome to humans, Plant Sci. 284 (2019) 67–72.

[12] J. Novak, K. Zitteri-Egiseer, S.G. Deans, C.M. Franz, Essential oils of different cultivars of *Cannabis sativa* L. and their antimicrobial activity, Flavour Fragr. J. 16 (2001) 259–262, https://doi.org/10.1002/ffj.993.

[13] Webbook, https://webbook.nist.gov/cgi/cbook.cgi?ID=C142507&Mask=200. (accessed April 8, 2020).

[14] K.W. Hillig, A chemotaxonomic analysis of terpenoid variation in cannabis. Biochem. Syst. Ecol. *32* (2004) 875–891, https://doi.org/10.1016/j.bse.2004.04.004.

[15] E.B. Russo, Taming THC: potential cannabis synergy and phytocannabinoid-terpenoid entourage effects, Br. J. Pharmacol. 163 (2011) 1344–1364.

[16] D.W. Christianson, Roots of biosynthetic diversity, Science 316 (2007) 60–61.

CHAPTER EIGHT

Gas chromatography/electron ionization mass spectrometry (GC/EI-MS) for the characterization of phytocannabinoids in *Cannabis sativa*

Jodie V. Johnson[a],*, Adam Christensen[b,c], Daniel Morgan[c], Kari B. Basso[a]

[a]Department of Chemistry, University of Florida, Gainesville, FL, United States
[b]Essential Validation Services (EVS), Gainesville, FL, United States
[c]Botanica Testing, Inc., Gainesville, FL, United States
*Corresponding author: e-mail address: jvj@chem.ufl.edu

Contents

1. Introduction	235
2. Experimental	239
3. Results and discussion	241
3.1 Extraction	241
3.2 Maintenance	242
3.3 EI-MS characterization of cannabinoid standards	242
3.4 GC/EI-MS of extracts of THC-type and CBD-type cannabis botanicals	246
4. Concluding remarks	272
References	272

1. Introduction

Cannabis sativa has been used by humans for centuries for a variety of purposes. Extensive research efforts in the 1940s thru 1971 led to the chemical characterization and synthesis of many of the phytocannabinoids [1–3]. These lead to the identification of the phytocannabinoids and in particular delta-9-*trans*-tetrahydrocannabidiol (d9-*trans*-THC) as being responsible for the psychoactivity of Cannabis [4]. When Cannabis was listed as a Schedule 1 drug in 1971 here in the United States and the "war on drugs" was waged,

Comprehensive Analytical Chemistry, Volume 90
ISSN 0166-526X
https://doi.org/10.1016/bs.coac.2020.05.003

© 2020 Elsevier B.V.
All rights reserved.

research on *Cannabis sativa* was sharply curtailed. Instead, most of the analytical efforts were devoted to developing methods to detect THC in seized "drugs" and in biological matrices (urine, blood, plasma, hair) for forensic and judicial reasons. However, research continued on the chemical characterization and pharmacology of Cannabis in a few labs, most notably with Mahmoud Elsohly, Turner and many others at the University of Mississippi, USA [5] and Raphael Mechoulam and others in Israel [6]. Almost 50 years later, there have been dramatic changes in the world's attitude towards Cannabis. With the continuing legalization of hemp- and drug-type *Cannabis sativa* around the world, there has also been a resurgence in research into the characterization and medical effects of the various chemical constituents of cannabis. As for almost any plant, there are hundreds of compounds present in Cannabis flowers, leaves, stems and roots. Most of the interest continues to be with the identification of and pharmacological characterization of the phytocannabinoids. Hartsel reviewed the history, biology, biochemistry and pharmacology of the major phytocannabinoids [7] while the many different chemical constituents were updated by ElSohy and Gul [8]. Hanus et al. provided an extensive review of all of the phytocannabinoids discovered to date including their structures and pharmacology [9]. They indicated there are almost 200 cannabinoids known now, not only from Cannabis but also in several other plant species. In addition to the delta-9-*trans*-THC, the most interest in the last few years have been cannabidiol (CBD). The major phytocannabinoids are present in plants as acids as shown in Fig. 1. However, the biological activities of the phytocannabinoids are produced by their decarboxylated forms, which are generated via gradual degradation following harvest or from the specific application of heat to promote the conversion of acid to neutral form [9].

Analytical efforts to identify and characterize the phytocannabinoids have involved numerous analytical methods almost always with some chemical separation involved. In early research, thin layer chromatography (TLC), low pressure or gravity column chromatography have provided for chemical separation. Collected spots and fractions were then characterized by the classic analytical methods still in use today: ultra-violet/visible spectroscopy (UV–vis), infrared spectroscopy (IR), nuclear magnetic resonance (NMR), gas chromatography (GC) usually with flame ionization detectors (FID) and mass spectrometry (MS). The analytical methods used for the determination of phytocannabinoids have been reviewed [10] with a specific review of the use of GC [11]. GC/FID and GC/MS are relatively inexpensive analytical instruments and can be used for a wide variety of

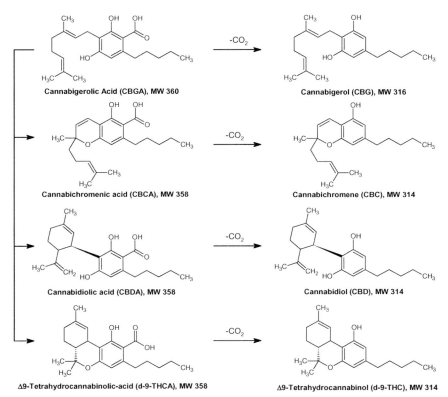

Fig. 1 The major phytocannabinoids are present in the plant as acids with CBGA being the precursor of the other three (left hand side). With heat and other environmental conditions, the acids are decarboxylated to form the corresponding neutral phytocannabinoid (right hand side).

analyses. This likely accounts for the fact that the GC/FID and GC/MS are still one of the most used analytical technique for analysis of cannabis and cannabis-derived products [10–12]. It was recognized in the 1960s that the phytocannabinoids in Cannabis have a carboxylic acid function while the pharmacological active form is the decarboxylated version (see reviews above and references therein). Decarboxylation is promoted by heating of the samples; 15 min at 150 °C results in almost complete conversion of THCA to THC and several other products [13]. As the injection port of the GC is typically around 250 °C, the cannabinoid acids are converted into the neutral forms in the injection process and only the neutral forms are detected with GC. It is also suspected that other degradants may be formed in the hot injection port. In order to observe the cannabinoid acids and other polar cannabinoids, derivatization is required [10–12,14].

Berés et al. compared GC/FID, GC/EI-MS with and without derivatization and UHPLC/ESI-MS/MS for quantitative determination of the major phytocannabinoids [12]. Their conclusion was the GC/EI-MS of the derivatized plant material was the "method of choice because of its accuracy, robustness, and versatility as it allows analysis of terpenoids using the same instrument". As many of the cannabis products have been decarboxylated, i.e. the cannabinoids are in their neutral forms, the less expensive, simple to use, robust GC/FID can readily separate and detect the major cannabinoids of interest in less than 3 min with baseline resolution [15].

Liquid chromatography connected to UV detectors or mass spectrometers enables the ready detection of the acid and neutral forms of the cannabinoids without derivatization. Due to the relatively large concentrations of the major cannabinoids, their good UV sensitivity due to the resorcinol portion of the cannabinoid structure and the good chromatographic separation possible with modern analytical columns, HPLC/UV is an accepted analytical method for the determination of the major cannabinoids in Cannabis and cannabis-related products [16,17].

While HPLC/UV may work for the major cannabinoids, when attempting to monitor the more trace cannabinoids, HPLC/MS and especially with tandem mass spectrometry can give added specificity, sensitivity and structure elucidation capabilities. The main ionization being used with HPLC/MS today is electrospray ionization (ESI), which can produce both positive and negative ions. While there are several different gas-phase chemistry and physic explanations of how this works, the end result is usually a protonated molecule $[M+H]^+$ in the positive ion mode and an $[M-H]^-$ ion in the negative ion mode with usually minimal fragmentation. To obtain structural information on these ions, the most common method is the use of collision-induced dissociation (CID) of the ions with some collision gas and multiple stages of mass spectrometry, i.e. tandem mass spectrometry. The process of tandem mass spectrometry involves the mass selection of a precursor ion, CID of the precursor ion to form product ions and then mass analysis of these product ions. There are two basic types of tandem mass spectrometers, tandem–in–space and tandem–in–time [18]. With tandem–in–space instruments (e.g. triple quadruple, QQ-TOF), these processes occur in separate portions of the instruments: mass spectrometer-collision cell-mass spectrometer. With a tandem–in–time instrument (3-D and linear quadrupole ion trap mass spectrometers), all the processes happen in the same part of the instrument but the functions change with time. There are a number of parameters which determine how much energy goes into the CID event but generally the tandem-in-space instruments produce more energetic spectra than the tandem-in-time instruments. From the standpoint of structure elucidation, the tandem-in-time instruments are capable of performing MSn ($n = 1-10$), that is one can continue investigating the structure of ions produced by the preceding CID event. While several tandem-in-space instruments were built to do MS^3, their expense and complexity did not result in very many of these being built. Finally, the use of high resolution, high mass accuracy mass spectrometers for the final mass spectrometers have greatly aided the structural elucidation capabilities and enhanced the specificity of analysis utilizing both GC and HPLC [19–22]. Importantly, high resolution MS/MS spectra and data have been provided either in their supplementary material or upon request [20–22]. In an excellent and I would say must-read manuscript, Berman et al. describe their use of HPLC/ESI-MS/MS HRMS work and demonstrate why this type of research is necessary [22].

While much of the analytical work today still revolves around the analysis of the major phytocannabinoids for legal requirements of plant and products, it was our desire to continue to characterize the numerous minor compounds observed in our GC/MS and HPLC/MS chromatograms. Quantitative determination of the major cannabinoids (neutral and acidic forms) was done via HPLC/UV and will not be discussed here. Our GC/MS analyses of the cannabis originated with the determination of the monoterpenes and sesquiterpenes and their oxygenated forms in essential oils and solvent extractions of cannabis botanicals and cannabis-derived products. It was obvious that the phytocannabinoids were readily extracted with the 1:1 hexane: ethanol we used for extraction. This is another advantage of the GC/MS method in that the volatiles and semivolatile cannabinoids can be analysed in one analysis.

2. Experimental

Standards and reagents: An 11 cannabinoid standard mix was purchased from Shimadzu (Kyoto, Japan) (Cayman Chemicals (Ann Arbor, MI); part # 220-92329-21) containing 250 µg/mL of each cannabinoid: THCA, THCV, d8-THC, d9-THC, CBD, CBDA, CBDV, CBC, CBN, CBG and CBGA. Appropriate dilutions were made to produce abundant GC/MS and HPLC/MS peaks (ca. 50 µg/mL). Water, methanol and formic acid were Optima, LC/MS-grade from Fisher Scientific.

Ethanol was molecular-biology grade from Fisher Scientific. Hexane was ACS certified from Fisher Scientific.

Samples and sample preparation: Cannabis and cannabis-derived products were received from various sources for various analysis. They are referred to here by generic labels. Often little was known about the characterization of Cannabis plant and/or method of production. For the work presented here, botanical material from a THC-type cannabis and a CBD-type cannabis were prepared manually just prior to their analysis. Known amounts of botanicals (184 mg CBD-type; 297 mg THC type) were placed in a glass vial and 2 mL (CBD-type) and 1.5 mL (THC-type) solvent were added. Solvent was usually some ratio of hexane: ethanol (1:1, 2:1, 3:1, v/v). With experience, for HPLC analysis, often only ethanol was used, as hexane is not miscible with the water and methanol used as mobile phases. The botanicals were macerated with a glass rod briefly (ca. 1 min) and then placed in an ultrasonic bath for 10 min. The liquid of each sample was transferred via pipette to smaller glass vial which was centrifuged for 10 min. This resulted in a clear, bright green (i.e. contained chlorophyll) supernatant. The supernatant was transferred to an auto-sampler vial for GC/MS and HPLC/MS. For Cannabis-derived tinctures, distillates, pastes, tars, crude extracts, ca. 100 mg were dissolved in 1–3 mL solvent; this usually resulted in a clear solution of various colours and opaqueness. These were generally amenable to analysis without further treatment except perhaps further dilution. CBD-oils and tinctures were often produced with various vegetable oils. These were problematic due to the lipid content. In general, they were treated like the other products but often with a "freezing" out of the lipids which was only marginally successful.

GC/EI-MS: GC/EI-MS data were acquired on a ThermoScientific (San Jose, CA) DSQ II: DSQ II 2.0.1, SP1 version. This is a traditional low resolution linear quadrupole mass spectrometer. Ionization was via 70 eV electron ionization with an ion source temperature of 250 °C. Typically, the instrument was scanned from m/z 35 to m/z 600 or m/z 700 at 1500 u/s. Acquisition of spectra was delayed for 6 min after sample injection to permit the solvent to elute; no compounds of interest eluted in this time period. Chromatographic separation was controlled with a ThermoScientific (San Jose, CA) Trace GC Ultra with 2.0 version software operated with an injection port temperature of 300 °C and a GC/MS transfer line temperature of 300 °C. A Restek Corp (Bellefonte, PA) Rxi-5MS (30 m × 0.25 mm i.d. and 0.25 μm df) GC column was operated with a constant 1 mL/min flow of helium as a carrier gas with a split flow rate of

10 mL/min (split ratio = 10:1). Two slightly different temperature programs were used for the data presented here:
Temperature program for CBD-type: 40 °C (0–3 min) > 300 °C at 4 °C/min > 350 °C at 10 °C/min; hold 5 min.
Temperature program for THC-type: 40 °C (0–3 min) > 350 °C at 4 °C/min; hold 5 min.
A ThermoScientific AI/AS3000 Auto-Sampler (2.0 version) controlled sampling, sample injections and syringe rinsing. Injection volume was 1.0 µL split injection.

EI Mass Spectral Library: GC/EI-MS spectra were searched against the NIST/EPA/NIH Mass Spectral Library, Version 2.3, build May 4, 2017 using the NIST Mass Spectral Search program.

Data Reduction: Data reduction for both GC/MS and HPLC/MS was done with ThermoScientific Xcalibur software, Version 2.2 SP1.48.

3. Results and discussion

3.1 Extraction

Our entry into the analysis of Cannabis-related samples was via analysis of Cannabis essential oils. 10 µL of the essential oil in 1 mL 1:1 (v:v) hexane: ethanol provided a good concentration for GC/EI-MS of the essential oil components. From experience, the hexane: ethanol would also extract lipophilic material from plant extracts and did so quite nicely with the phytocannabinoids. Please note again, that we were not trying to be quantitative here in the sense of determining the absolute concentration of the cannabinoids. The intent was to have a simple, quick extraction which gave us plenty of cannabinoids that were readily detected. The extraction procedure described above did so; in fact, the resulting concentrations were usually too high and overloaded the GC and MS for the major cannabinoids. For quantitative work, optimum extraction would be required and an assessment of the recovery. Béres et al. performed an extraction study with 50 mg of pulverized botanical material extracted with 2 mL of 95% ethanol with ultrasonication for 15 min at room temperature and subsequent centrifugation [12]. This process was repeated twice more and each individual extraction was assayed for the major cannabinoids. The result was that 93% of THCA and more than 95% of the CBDA, CBD and d9-THC were obtained in the first extraction. They deemed that a single ethanol extraction was more than adequate for quantitative purposes. Cannabis-derived material, such as tinctures, pastes, tars, resins, usually were readily dissolved with ethanol to give clear solutions which often required no further sample preparation except dilution to prevent excessive overloading of the instruments. CBD-oils, some of the tinctures and pastes contain various vegetable oils and/or mid-chain triglycerides (MCTs) which can definitely be a problem for GC/MS. We have tried using methanol and ethanol with "freezing" out the lipids with marginal success. For HPLC/UV, this is not particularly problematic, but for GC/MS some of the MCTs and/or their degradation products elute in the region of the phytocannabinoids and may obscure some of them.

3.2 Maintenance

The injection of extracts of cannabis botanicals has obvious maintenance issues. The GC injection port acts as another step in sample cleanup; only compounds volatile at the injection port temperature have a chance of entering the GC and being detected. The solvent-extracts were bright green, indicating high concentrations of chlorophyll. There would also be various high MW lipids (e.g. triglycerides) extracted from the rupturing of membranes and other non-volatiles present in these extracts. While these higher MW and non-volatile compounds are not detected, they definitely lead to severe contamination of the GC injection port liners. As the contamination builds up, peak tailing and other chromatographic effects may be seen. Prior to analysing cannabis, the injection port liner was either cleaned or replace monthly, after about 150 analyses of mostly essential oils. More frequent maintenance was required when cannabis botanicals were analysed. The cannabis-derived products (pastes, tars, oils, distillates) produced much less contamination than the botanical extracts but more than the essential oils.

The non-volatiles can be carried into the GC column itself. As this contamination builds up, chromatographic separation and peak shapes will start to suffer. If the first 2–3 ft of column are removed, an improvement in chromatographic behaviour can be obtained for a short while. Eventually, we decided to just install a new column, usually every 6 months or so.

3.3 EI-MS characterization of cannabinoid standards

Fig. 2 shows the electron ionization (EI) mass spectra of four of the major molecular weight (MW) 314 u phytocannabinoids. EI ionizes by bombarding neutral gaseous compounds with 70 eV electrons which knocks out an electron leaving a positively-charge ion with a lot of internal energy which results in extensive fragmentation characteristic of EI mass spectra.

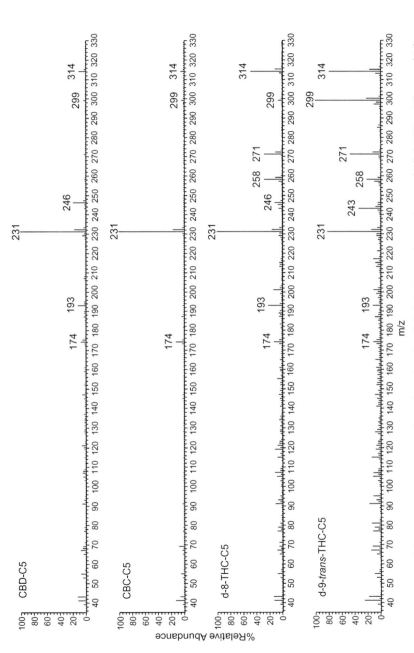

Fig. 2 EI mass spectra of the major MW 314 cannabinoids with a C5-alkyl chain (top to bottom): cannabidiol (CBD—C5), cannabichromene (CBC-C5), delta-8-*trans*-tetrahydrocannabinol (d-8-THC-C5) and delta-9-*trans*-tetrahydrocannabinol (d-9-*trans*-THC-C5).

With the aromatic resorcinol, typically relative abundant molecular ions ($M^{+\cdot}$) are observed at m/z 314 for each of the MW 314 isomers. Almost all the MW 314 isomers produce the same m/z ions which have varying relative intensities related to their structures with some ions being distinctive for some isomers. In particular, CBD produces an m/z 246 fragment ion which is distinctive. The mechanisms of the fragmentation to produce the ions above were initially determined by Budzikiewicz in 1965 [23]. Belgado–Povedano et al. have provided high resolution EI mass spectra in their supplementary material along with more fragmentation Schemes [19].

Electron ionization (70 eV) spectra are very reproducible from laboratory to laboratory and between instrument types. EI mass spectra obtained today can be compared to spectra obtained years ago. Hazekamp et al. [24] showed EI mass spectra of the major cannabinoids which are nearly identical to those obtained in our lab and aided us in identification of some cannabinoids. This feature of EI mass spectra led to the development of compilation of EI mass spectra, which are commercially available, and permit the tentative identification of compounds via very rapid searching (less than 5 s per search) of thousands of spectra [25,26]. This is one of the major advantages of GC/EI-MS. HPLC/MSn spectral libraries are also available; the number of entries are steadily growing but are still relatively limited.

Note in the structures of the four major MW 314 cannabinoids of Fig. 1 that they all have a 5-carbon alkyl chain (C5) attached to the resorcinol. The C3-homologues (with a 3-carbon alky chain) of these and other cannabinoids are also well known [9]. The EI mass spectra of the C3 versions of CBD and d-9-THC are shown in Fig. 3. The correct nomenclature and abbreviations for these are cannabidivarin (CBDV) and delta-9-*trans*-tetrahydrocannabivarin (d-9-THCV), respectively. For clarification in this document, the C3 versions and other homologues of the cannabinoids will use the C5-version abbreviation with an appended carbon number. In this case, CBD-C3 for cannabidivarin and d-9-*trans*-THC-C3 for delta-9-*trans*-tetrahydrocannabivarin.

If the spectra of the CBD-C3 and CBD-C5 are compared and the d-9-*trans*-THC-C3 and d-9-*trans*-THC-C5 spectra are compared, one will note they reflect very similar spectra with an offset of 28 u for the major ions, e.g., the m/z 231 of the C5-cannabinods are equivalent to the m/z 203 of the C3-cannabinoids. Also note that the CBD-C3 has a prominent m/z 218 ion equivalent to the m/z 246 of CBD—C5. This characteristic was evident in the spectra of all of the compounds related to the CBD family of compounds.

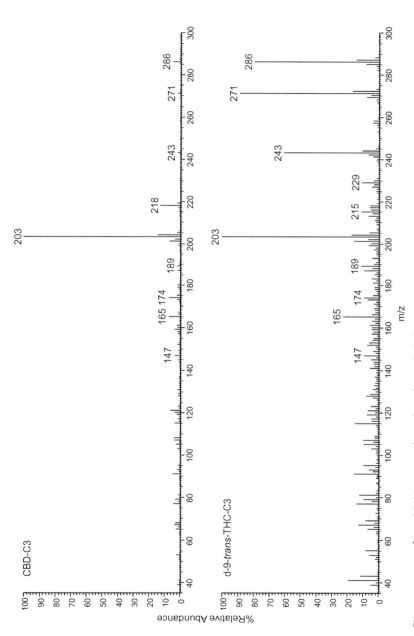

Fig. 3 EI mass spectra of two MW 286 cannabinoids with C3-alkyl chain: (top) Cannabidivarin (CBDV; CBD—C3) and (bottom) delta-9-*trans*-tetrahydrocannabivarin (d-9-THCV; d-9-THC-C3).

Another aspect of the EI mass spectra of the cannabinoids, is that the major ions are in the m/z 170–350 range. Many of the potential interfering compounds in extracts of cannabis produce major ions below this range, making the detection of cannabinoids more apparent and chances of interference less likely.

3.4 GC/EI-MS of extracts of THC-type and CBD-type cannabis botanicals

Fig. 4 shows the GC/EI-MS chromatograms of the hexane: ethanol extracts of a THC-type cannabis (low CBD, high THC) and a CBD-type cannabis (high CBD, low THC). The most abundant peak by a wide margin in each extract is the cannabinoid which describes the cannabis type: d-9-*trans*-THC at RT 58.69 in the THC-type and CBD at RT 56.76 in the CBD-type. With the major cannabinoids being the predominant peaks and eluting in a retention time window without significant interferences, permits GC/FID to be a cost-effective and reliable and accurate method for quantitative determination of the major cannabinoids. In addition, if the terpenes, sesquiterpenes and diterpenes are of no interest, the analysis time can be significantly shortened with a higher starting temperature and rapid heating to the elution temperature range of the cannabinoids.

The early elution portion of each of the chromatograms is dominated by monoterpenes (RT 12–18), monoterpenoids (RT 19–27), sesquiterpenes (RT 29–36) and sesquiterpenoids (RT 36–43). Not as noticeable, but diterpenes are usually detected (RT 44–48), which overlap with some of the C1-cannabinoids discussed later. We have compared the GC/MS of solvent (1:1 hexane: ethanol) extracted fresh, not dried, cannabis botanicals to that of an essential oil of cannabis and they agreed very well (data not shown or published). However, during the drying of cannabis botanicals after harvest or during sample preparation to obtain a dry weight, the ratio of sesquiterpenes to monoterpenes increases due to losses of the more volatile monoterpenes.

The first significant cannabinoid eluted at ca. 48 min which was equivalent to an elution temperature of 212 °C and the last significant cannabinoid eluted at RT 62 min, equivalent to an elution temperature of 228 °C. As the boiling points of the cannabinoids are a physical characteristic and as gas chromatographic separation depends in large part upon the differences in boiling points of compounds, the relative elution of cannabinoids in gas chromatography has not changed significantly since GC was first used. Thus, if one knows the temperature gradient of an analysis, one can predict

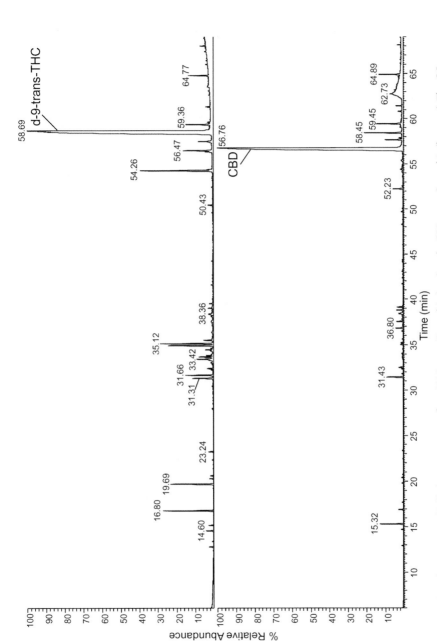

Fig. 4 GC/EI-MS total ion current (TIC) chromatograms for a THC-type Cannabis (top) and a CBD-type Cannabis (bottom). The GC temperature program was: 40 °C (0- min) heated linearly to 300 °C at 4 °C/min; then heated to 350 °C at different heat rates and then held at 350 °C for 5 min. These later sections did not contain any obvious cannabinoids and are not shown.

where one might expect to see certain cannabinoids elute. Relative retention times are often listed in references to aid in identifying cannabinoids.

In general, the elution of the cannabinoids occurs in a region of the chromatogram relatively free of other compounds. Their mass spectra, as discussed above, have major ions in the m/z 170–370 range while most compounds eluting in this retention time window, e.g., alkanes, have spectra dominated by low m/z ions (m/z 41–150). Thus, the potential interference is much reduced, especially when using ions characteristic of specific cannabinoids.

These analyses were obtained with low polarity Rxi-5MS GC column and a single linear temperature gradient. In a number of instances discussed below, the co-elution of some cannabinoids, most notably CBD and CBC, would be problematic from a quantitative standpoint. With multi-step temperature programs, we could separate the CBD/CBC but often still with a significant valley between the two. A longer GC column of the same stationary phase may or may not improve the CBD/CBC separation. There are a wide variety of stationary phases available from column manufacturers. The selection of column with a more polar stationary phase may improve the chromatographic separation and behaviour of the cannabinoids [15].

CBD-C5 (RT 56.34 min) and CBC-C5 (RT 56.47) elute closely together (Fig. 5). When their concentrations are relatively low, as in the THC-type cannabis, they are chromatographically separated. In the EI mass spectra shown previously, CBD-C5 is characterized by a prominent m/z 246 ion which is almost absent in CBC-C5 and indeed in the other MW 314 isomers. CBC-C5 has a relatively high abundance of m/z 174 compared to CBD-C5. These attributes are reflected in their mass chromatograms of Fig. 5. Indeed, these differences between CBD-C5 and CBC-C5 appear to exist in all of the compounds related to them.

In the CBD-type extract, the concentration of CBD-C5 was too high and it overloaded the column, creating a broad, badly shaped peak which merged with those of CBC-C5. Sometimes, the m/z 174 ion-peak of CBC-C5 could be distinguished. Dilution of the sample should have given peaks like those demonstrated in Fig. 5.

For the THC-cannabis, the d-9-*trans*-THC-C5 overloaded the GC producing a broad, poorly shaped GC peak at ca. 58 min (top trace in chromatogram of Fig. 6). There was a very minor peak for the expected d-8-*trans*-THC-C5 at RT 57.83 min in the THC-cannabis. In the analyses of CBD-type cannabis extracts, there was almost always a MW 314 compound which eluted earlier than d-9-*trans*-THC-C5 and actually matched

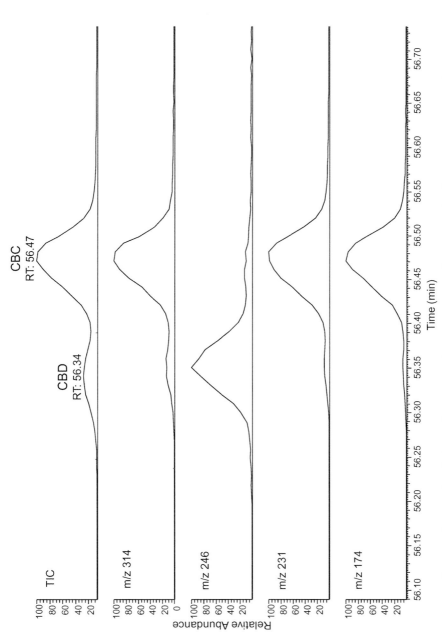

Fig. 5 GC/EI-MS of CBD-C5 (RT 56.34) and CBC-C5 (RT 56.47) in the THC-type cannabis extract. The chromatograms (top to bottom) were TIC, m/z 314, m/z 246, m/z 231 and m/z 174.

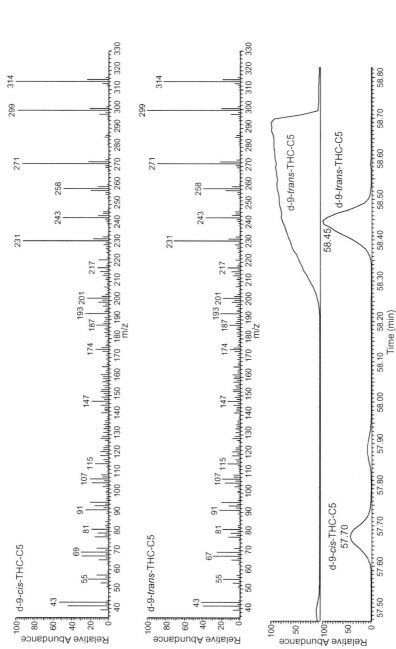

Fig. 6 EI mass spectra of (top) d-9-*cis*-THC-C5 and (middle) d-9-*trans*-THC-C5 from the CBD-type cannabis. In the bottom panel are the TIC chromatograms of the THC-cannabis (top trace) showing the overloaded d-9-*trans*-THC-C5 and the TIC chromatogram of the CBD-cannabis (bottom trace) showing d-9-*cis*-THC-C5 peak at RT 57.70 and the d-9-*trans*-THC-C5 peak at RT 58.45.

well with it from a library search. As seen by the two mass spectra of Fig. 6, there were only minor differences in the spectra of these two compounds. Eventually, the earlier eluting peak was identified as d-9-*cis*-THC-C5 based upon the work of Smith and Kempfert [27]. They identified the d-9-*cis*-THC-C5 via synthesis and chemical characterization. They analysed a number of different cannabis plants and came to the following conclusion. If the ratio of the amount of CBD to total amount of THC (CBD:THC) was greater than 16:1, then the *trans*-THC to *cis*-THC ratio was between 1:1 and 2:1. If the CBD:THC ratio was less than 1, then the *trans*-THC:*cis*-THC was much greater than 10:1 [27]. In our samples, the d-9-*trans*-THC-C5 to d-9-*cis*-THC-C5 was 357:1 for the THC-type cannabis and 2.4:1 for the CBD-type cannabis. Smith and Kempfert's description of the relative retention time, mass spectra and survey results agree well with those shown above.

The remainder of this document will discuss the detection or and relative abundances of some of the other cannabinoids which were detected. As the concentration and peaks of individual compounds decreased, the use of characteristic ion-peaks instead of the TIC peaks made detection of compounds more evident. Also in complex samples, often there are common ions either in the background and/or from overlapping and co-eluting compounds. In the Tables below, the areas of specific ion-peaks were determined and used to calculate the percent relative abundance (%RA) of ions instead of using the actual mass spectra. For minor peaks, this is much more accurate. For major peaks, one gets about the same %RA between the mass-intensity listing of the spectra and the use of peak area ratios. For semi-quantitation, the specific ion-peak areas were summed to give an overall abundance of the individual compounds instead of using the peak areas from the total ion current (TIC) chromatograms. This is again more accurate within a set group of compounds (e.g. the MW 314 isomers) but clearly under-reports the areas compared to the TIC peaks. For the compounds with obvious TIC peaks, the ratio of the (sum of the ion-peak areas) to the (TIC peak areas) varied from 0.2 to 0.5. A number of the very minor compounds which produced obvious specific ion-peaks could not be detected in the TIC traces.

GC/EI-MS chromatograms of individual ions characteristic of the MW 314 cannabinoid-C5s were used to look for and identify MW 314 compounds. Four other isomers were detected eluting prior to CBD and indicated by MW 314-A to MW 314-D (Table 1). None of these matched well with the NIST EI mass spectral library, if their relative retention times were

Table 1 GC/EI-MS characteristics of the MW 314 cannabinoid-C5 isomers.

MW 314 cannabinoid-C5	RT (min)		%Total area		% Relative abundance									
	THC-type	CBD-type	THC-type	CBD-type	174	187	193	231	243	246	258	271	299	314
MW 314-A	53.44	53.57	0.12	0.06	4.73	5.44	8.84	100.00	11.89	4.64	10.99	11.81	9.00	28.01
MW 314-B	54.46	54.57	0.01	0.05	18.80	9.47	20.24	3.51	3.47	1.31	6.70	100.00	28.13	43.38
MW 314-C	54.58	54.71	0.02	0.08	9.73	3.76	5.53	100.00	4.29	0.46	4.10	3.70	3.64	17.94
MW 314-D	55.28	55.42	0.01	0.08	interfer	interfer	2.28	100.00	0.34	0.00	0.88	1.58	3.38	6.66
CBD-C5[a]	56.36	56.62	0.47	93.27	8.48	1.78	6.18	100.00	1.32	10.74	0.87	1.21	2.27	5.40
CBC	56.47	nd	5.02	0.00	9.26	4.12	1.36	100.00	0.37	0.39	0.37	0.85	3.39	4.65
d-9-*cis*-THC	57.53	57.67	0.26	1.92	13.31	12.91	25.25	91.39	49.98	7.72	48.87	89.39	95.64	100.00
d-8-*trans*-THC	57.83	nd	0.02	0.00	4.26	11.16	41.50	76.04	21.75	22.68	79.13	49.66	33.84	100.00
d-9-*trans*-THC	58.66	58.45	94.07	4.54	7.38	5.52	10.66	60.15	28.10	28.13	25.30	56.01	100.00	79.42

[a]CBD from averaged spectra.
GC retention times (RT, min) and percent relative abundances calculated from measurement of the areas of the individual ion-peaks with the exception of CBD, which was taken from the mass-listing of its averaged spectrum. The sum of all the individual ion-peak areas was used to calculate the %Total Area.
nd, not detected; interfer, interference from a co-eluting/overlapping compound.

taken into account. There are several MW 314 isomers indicated to elute prior to CBD in several references but we have been unable to assign them to any of these peaks at this time.

The MW 286 CBD-C3, CBC-C3, and d-9-*trans*-THC-C3 were readily detected in the two cannabis extracts while the d-9-*cis*-THC-C3 was detected in the CBD-type cannabis (Fig. 7). Table 2 contains a summary of their GC/EI-MS characteristics. The characteristic of CBD-C3 and d-9-*trans*-THC-C3 matched their corresponding standards described in the earlier section. While we had no standard for CBC-C3, its relative RT to CBD-C3 compared well with those of the CBD-C5 and CBC-C5 (of the THC-cannabis) and the spectra patterns of the CBC-C3 matched those of the CBC-C5. Similarly, the d-9-*cis*-THC-C3 was identified. It was advantageous doing the analyses of the two different types of cannabis together as it aided identification by comparison of peak areas between the extracts. As an additional evidence for their identification, one can see the relative peak areas and relative retention times of the C3-homologues match those of the C5-homologues.

As for the MW 286 C3 isomers, the CBD-C1 and CBC-C1 homologues were readily detected in the CBD-type cannabis and the CBC-C1 and the d-9-*trans*-THC-C1 were readily detected in the THC-type cannabis (Fig. 8, Table 3). The other C1-homologues were not relatively abundant but could be readily detected with the use of ion-peaks of the expected MW 258 characteristic ions. No standards were available for any of these MW 258 compounds and their spectra were not in the NIST EI mass spectral library. Identification was based upon their characteristic ion abundances matching their corresponding C5-homologues and the good correlation of the relative amounts and relative retention times.

With a single temperature gradient, the retention times of homologous series usually show a linear relationship with carbon number. When the retention time of the C1-, C3- and C5-homologues of the major cannabinoids are plotted against the number of carbons in the alkyl chain of the resorcinol, a linear relationship was indeed observed for each cannabinoid series (Fig. 9). Vree et al. did an extensive study showing linear relationships between a number of chemical features of the homologues of natural and synthetic cannabinoids and their retention times [28]. They advocated this as a tool to aid in finding and identifying other members of a homologous series. The lead author of this document has done this numerous times over

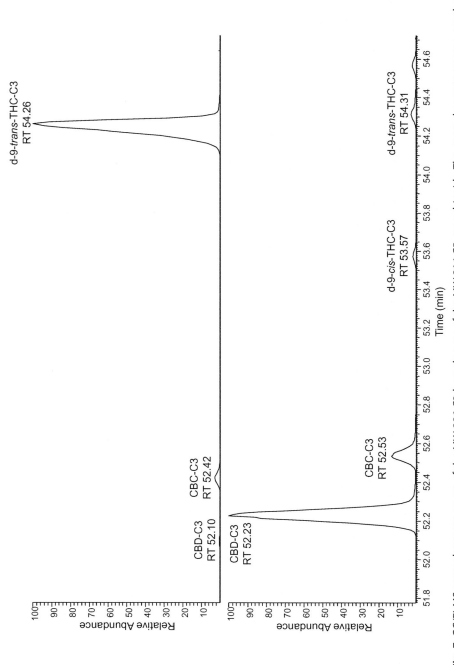

Fig. 7 GC/EI-MS mass chromatograms of the MW 286 C3-homologues of the MW 314 C5-cannabinoids. The mass chromatograms were the sum of the intensities of *m/z* 203, 271, and 286 ions for the THC-cannabis (top) and the CBD-cannabis (bottom).

Table 2 GC/EI-MS characteristics of the MW 286 cannabinoid-C3 homologues.

MW 286 cannabinoid-C3	RT (min)		%Total areas		% Relative abundance							
	THC-type	CBD-type	THC-type	CBD-type	174	203	215	218	229	243	271	286
CBD–C3	52.10	52.23	0.30	84.57	3.89	100.00	0.59	6.01	0.30	1.24	1.72	3.12
CBC–C3	52.41	52.55	2.06	11.26	6.24	100.00	0.20	0.00	0.21	0.49	2.20	1.66
d-9-*cis*-THC–C3[a]	53.46	53.57	0.00	1.45	12.56	100.00	6.74	0.78	4.68	32.47	84.66	35.48
d-9-*trans*-THC–C3	54.26	54.30	97.65	2.72	4.22	99.34	5.83	2.74	5.77	67.07	100.00	44.52

[a]d-9-cis-THC–C3 co-eluted with a MW 314 compound in the CBD–type cannabis.

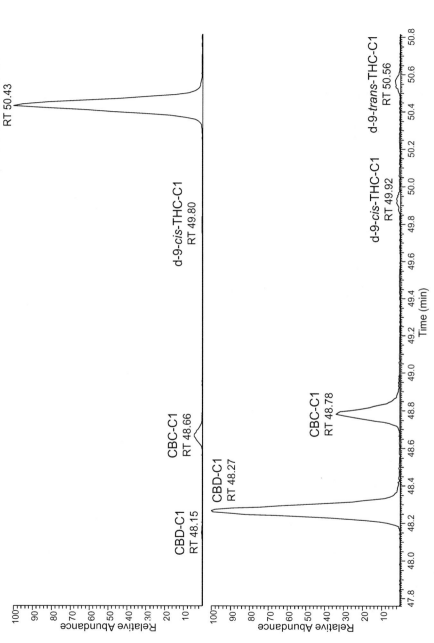

Fig. 8 GC/EI-MS mass chromatograms of the MW 258 C1-homologues of the MW 314 C5-cannabinoids. The mass chromatograms were the sum of the intensities of m/z 175, 243 and 258 ions for the THC-cannabis (top) and the CBD-cannabis (bottom).

Table 3 GC/EI-MS characteristics of the MW 258 cannabinoid-C1 homologues.

MW 258 cannabinoid-C1	RT (min)		%Total area		% Relative abundances					
	THC-type	CBD-type	THC-type	CBD-type	175	190	201	215	243	258
CBD–C1	48.15	48.27	0.34	75.12	100.00	22.66	2.03	2.11	2.27	10.27
CBC–C1	48.66	48.79	3.75	21.30	100.00	0.00	0.72	0.93	2.88	6.31
d-9-*cis*-THC–C1	49.80	49.94	0.48	1.41	62.46	0.10	19.08	41.19	61.95	100.00
d-9-*trans*-THC–C1	50.43	50.56	95.42	2.16	46.52	6.56	11.89	27.29	86.04	100.00

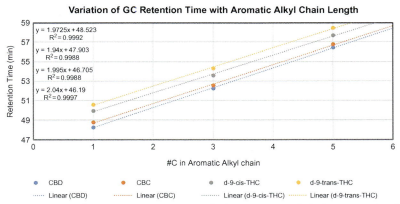

Fig. 9 Variation in the GC Retention Time with the Aromatic Alkyl Chain Length of the C1-, C3- and C5-homologues of the major cannabinoids.

his career. Usually, the C3-isomers are very evident but the C1-isomers and others may not be so evident in cannabis extracts. The lines obtained with just the C3- and C5-homologues can be used to find the C1-homologues. With this linear relationship, the retention times of the other isomers can be calculated in order to aid in their detection.

The linear relationships shown in Fig. 9 were used to predict the retention time of other members of the homologous series. The mass chromatograms of the expected ions were then used to look for correlating ion-peaks of the ions expected of a particular cannabinoid. In this manner, additional homologous members of the major C5-cannbinnoids were tentatively identified. The GC/EI-MS characteristic ions are shown for the C4-cannabinoids in Table 4, the C2-cannabinoids in Table 5 and the C7-cannabinoids in Table 6. The CBD-C6 was detected at RT 58.45 in the CBD-cannabis and d-9-*trans*-THC-C6 was detected at RT 60.25 in the THC-cannabis. Both C6-isomers co-eluted with other compounds which made it difficult to obtain reliable ion ratios. The C4-homologues were actually relatively abundant for the CBD-C4 and d-9-*trans*-THC-C4 in their respective cannabis types. In the CBD-type cannabis extract, the d-9-*trans*-THC-C4 eluted under the CBD-C5 and was not detected. The C2-homologues were very minor but detectable for the two shown. Citti et al. recently isolated and did extensive characterization of the CBD-C4 named cannabidibutol [29] and d-9-*trans*-THC-C7 named (−)-*trans*-Δ^9-tetrahydrocannabiphorol

GC/EI-MS of phytocannabinoids 259

Table 4 GC/EI-MS characteristics of the MW 300 cannabinoid-C4 homologues.

MW 300 cannabinoid-C4	Type	RT	%Relative abundances						
			217	229	232	244	257	285	300
CBD-C4	CBD	54.41	100.00	0.62	9.40	0.32	2.62	1.79	4.61
CBC-C4	CBD	54.57	100.00	1.46	0.63	6.98	5.95	3.31	2.34
d-9-*cis*-THC-C4	CBD	55.68	100.00	8.09	0.00	6.71	60.64	71.94	41.38
d-9-*trans*-THC-C4	THC	56.32	60.13	interfer	interfer	11.81	100.00	78.39	88.55

The "intefer" in the table indicates interference from a co-eluting/overlapping compound.

Table 5 GC/EI-MS characteristics of the MW 272 cannabinoid-C2 homologues.

MW 272 cannabinoid-C2	Type	RT(min)	% Relative abundances					
			189	204	215	229	257	272
CBD-C2	CBD	50.24	100	0.00	0.00	0.00	0.00	15.30
d-9-*trans*-THC-C2	THC	52.36	100	0.00	0.00	26.73	89.33	62.74

Table 6 GC/EI-MS characteristics of the MW 342 cannabinoid-C7 homologues.

MW 342 cannabinoid-C7	Type	RT	% Relative abundance					
			243	259	274	299	327	342
CBD-C7	CBD	60.52	0.00	100.00	15.23	0.00	0.00	3.71
d-9-*trans*-THC-C7	THC	62.12	36.48	87.40	0.00	49.69	100.00	67.36

(d-9-THCP) [30]. Both of these were identified initially with HPLC/ ESI-MS/MS with high resolution mass spectrometry followed up by extensive characterization and confirmation with synthesized standards. No GC/EI-MS analyses were done, which may have confirmed our tentative assignment.

The CBD-type cannabis contained each of the C1 thru C7 homologues of CBD. The retention times of the C2-, C4-, C6- and C7-homologues which were detected by the relationship of Fig. 9, all fall on the same line

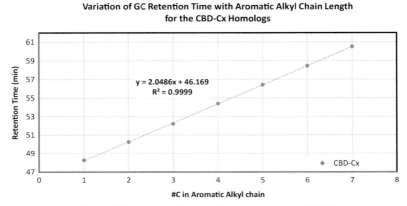

Fig. 10 Variation of the GC Retention Time with the Aromatic Alkyl Chain Length in the CBD-related Homologues.

very nicely in Fig. 10. This adds evidence to the tentative identification of these homologues.

The MW 316 cannabigerol (CBG-C5) and MW 310 cannabinol (CBN-C5) were readily detected and identified via good spectral matching with the NIST library (Fig. 11). CBN is usually the last eluting of the major phytocannabinoids. While CBG and CBN show good chromatographic separation in Fig. 11, in many of our analyses there is a lot of overlap of CBG and CBN. It's not clear at this point what causes this but it may be an indication that it's time to replace the column. As their major ions are relatively specific, this is not a major problem in distinguishing CBG and CBN. A search for the next most abundant C3-homologues for these two failed to detect either.

The neutral phytocannabinoid CBD-C5, CBC-C5, d-9-*trans*-THC-C5 and CBG-C5 and their aromatic alkyl homologues discussed above are the decarboxylated forms of their acidic parent compounds, which are enzymatically produced by cannabis [9]. These same reviewers also expressed a frequently-read opinion, and supported in some cases by data, that many of the other phytocannabinoids that are characterized in cannabis and cannabis-related products may be due to various degradation routes from the original acid forms. For example, it is well-known that CBN and CBNA are oxidative degradation products of d-9-*trans*-THC and d-9-*trans*-THCA [9].

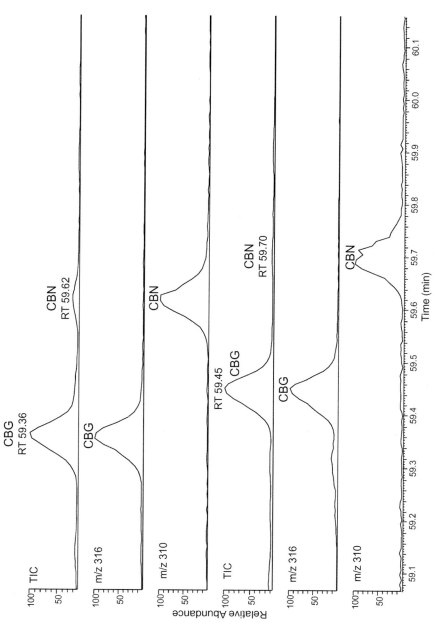

Fig. 11 GC/EI-MS chromatograms for CBG and CBN in the THC-type cannabis (top 3 traces) and in the CBD-type Cannabis (bottom three traces). The individual traces are (top) TIC, (2nd) m/z 316 and (3rd) m/z 310 for each cannabis type.

Table 7 GC/EI-MS characteristics of the MW 328 CBDM and CBCM.

Cannabinoid	RT (min)		%Total area		% Relative abundances				
	THC-type	CBD-type	THC-type	CBD-type	207	245	260	313	328
Cannabidiol monomethyl ether (CBDM)	54.36	54.47	88.72	89.50	10.29	100.00	19.97	1.30	3.25
Cannabichromene monomethyl ether (CBCM)	54.43	54.57	11.28	10.50	0.00	100.00	0.00	4.23	3.40

Cannabidiol monomethyl ether (CBDM) and cannabichromene mono-methyl ether (CBCM) were tentatively detected in both cannabis types (Table 7) with their GC/EI-MS mass chromatograms in the THC-type being shown in Fig. 12. Note that the CBDM had an m/z 260 ion, which was absent in the CBCM; the characteristic identification of CBD-type cannabinoids. The NIST EI mass spectra library did not contain spectra for these and the tentative identification are based upon the relative abundance of the ions in relation to CBD and CBC and Vree et al. [28] indicated CBDM eluted prior to CBD. Note that the ratio of the CBDM to CBCM was almost the same for both cannabis types. In the THC-type, the ratio of CBD to CBC was ca. 0.1 but the ratio of CBDM to CBCM was ca. 8.

Degradation of CBD results in the formation of a MW 330 cannabielsoin (CBE-C5) [9]. Two isomeric MW 330 compounds which produced nearly identical spectra are almost always observed eluting between the d-9-*cis*-THC-C5 and the d-9-*trans*-THC-C5-peaks in extracts of CBD-type cannabis (Fig. 13). The EI mass spectra of both of these MW 330 compounds match the NIST standard spectrum of 1,6–Dibenzofurandiol, 5a,6,7,8,9, 9a–hexahydro-6-methyl-9-(1-methylethenyl)-3-pentyl-, (5aα,6α,9α,9aα)-(CAS#: 54002-78-7 NIST#: 68058). This agrees with the structure shown for cannabielsoin (CBE-C5, 9). In a 1-year old methanol solution of 11 cannabinoid standards, these peaks went from not being detected when new to very obvious peaks after 1 year, even while the solution was kept in a −20 °C freezer when not in use.

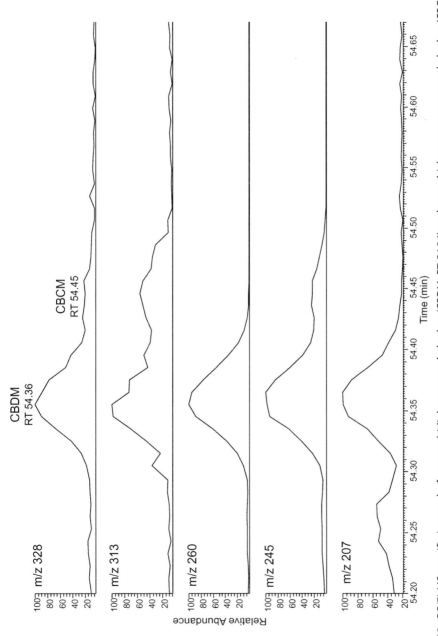

Fig. 12 GC/EI-MS specific ion-peaks for cannabidiol monomethyl ether (CBDM; RT 54.36) and cannabichromene monomethyl ether (CBCM; RT 54.45) in the THC-type cannabis extract. EI-MS chromatograms are (top to bottom): *m/z* 328, 313, 260, 245 and 207.

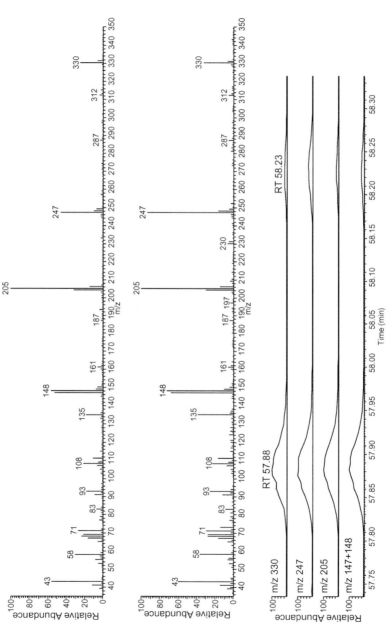

Fig. 13 EI mass spectra (top RT 57.88, middle RT 28.23) and GC/EI-MS mass chromatograms for two cannabielsoin (CBE) isomers (top to bottom): m/z 330, 247, 205, m/z 147 + 148.

It has been well-accepted that MW 310 CBN is an oxidation product of MW 314 d-9-*trans*-THC [9]. It seems there should be an intermediate MW 312 compound also. Three MW 312 isomers were detected in the extracts of both cannabis types. The chromatograms (Fig. 14) and the EI mass spectra (Fig. 15) of the MW 312 isomers are shown for the THC-type cannabis. Ross et al. reported the presence of a MW 312 dihydrocannabinol eluting just after CBN which produced *m/z* 312, 297, 269, 256, 231, 213, and 193 ions [31]. These characteristics match the RT 59.89 MW 312 isomer of Figs. 14 and 15. We commonly observe the RT 58.76 MW 312 isomer eluting in the tail of d-9-*trans*-THC-C5. The RT 59.45 isomer is often not observed but this may be due to the co-elution of CBG and CBN that we also commonly see, which would obscure this isomer.

Cannabispiran was detected at RT 55.38 in the extract of the CBD-type cannabis and its EI mass spectrum matched well with the cannabispiran spectrum in the NIST library (Fig. 16). In addition, Brenneisen et al. showed cannabispiran eluting between tetrahydrocannabivarin (THCV, i.e. d-9-*trans*-THC-C3) and CBD-C5 [32]. This matches the elution of the compound above. Cannabispiran was also detected in the extract of the THC-type cannabis but with a much lower abundance.

The possibility for the presence of deprenyl-cannabinoids (with a single isoprenyl unit) in cannabis was discussed but there has not been much evidence for them [9]. In both extracts, a MW 248 compound was detected, which produced ions consistent with a deprenyl-CBG-C5 structure (Fig. 17). A proposed structure and fragmentation scheme is shown as an inset of Fig. 17.

If the MW 248 deprenyl-CBG-C5 was present, it would also seem likely that a MW 246 deprenyl compound would exist, equivalent to the main MW 314 cannabinoids. As the MW 314 cannabinoid-C5s all elute before the MW 316 CBG-C5, an expected MW 246 deprenyl-cannabinoid-C5 should elute prior to the MW 248 deprentyl-CBG-C5. Indeed, a MW 246 compound was detected in both extracts at RT ca. 46.7 and produced ions consistent with a proposed structure of a MW 246 deprenyl-cannabinoid-C5 (Fig. 18).

Olivetolic acid is the precursor used to produce the phytocannabinoids [9]. Olivetolic acid was not detected likely due to its decarboxylation in the hot GC injection port as with the phytocannabinoid acids. The MW 180 olivetol (also known as 1,3-benzenediol, 5-pentyl) was detected in both

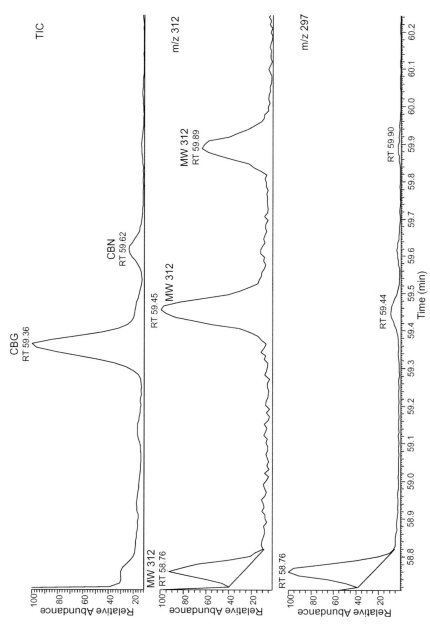

Fig. 14 MW 312 C5-cannabinoid isomers, which elute after d-9-*trans*-THC-C5 (eluting just to left of view). GC/EI-MS (top to bottom): TIC, *m/z* 312 and *m/z* 297. For reference, in the top TIC trace, RT 59.36 is CBG and RT 59.62 is CBN.

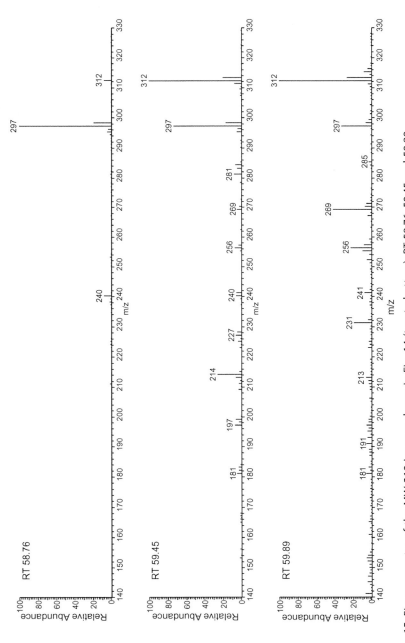

Fig. 15 EI mass spectra of the MW 312 isomers shown in Fig. 14 (top to bottom): RT 58.76, 59.45 and 59.89.

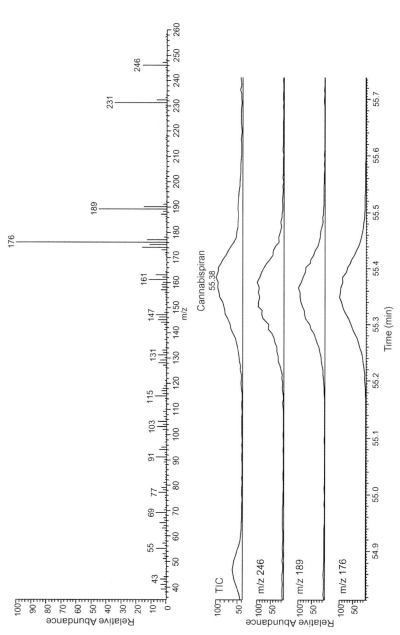

Fig. 16 MW 246 Cannabispiran eluted at RT 55.38 in the extract of the CBD-cannabis with the EI mass spectrum (top) and mass chromatograms on the bottom with traces for (top to bottom): TIC, m/z 246, 189 and 176.

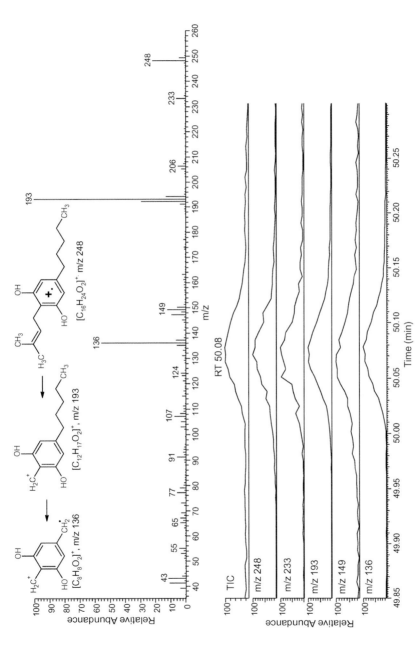

Fig. 17 MW 248 deprenyl-CBG-C5 in the extract of the CBD-type cannabis with the EI mass spectrum (top) and mass chromatograms on the bottom with traces for (top to bottom): TIC, *m/z* 248, 223, 193, 149 and 136. A tentative structure and dissociation pathway is shown in the inset.

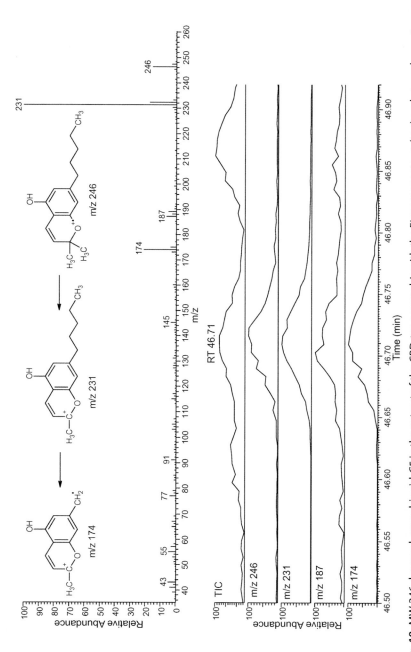

Fig. 18 MW 246 deprenyl-cannabinoid-C5 in the extract of the CBD-type cannabis with the EI mass spectrum (top) and mass chromatograms on the bottom with traces for (top to bottom): TIC, m/z 246, 231, 187, 149, and 174. A tentative structure and dissociation pathway is shown in the inset.

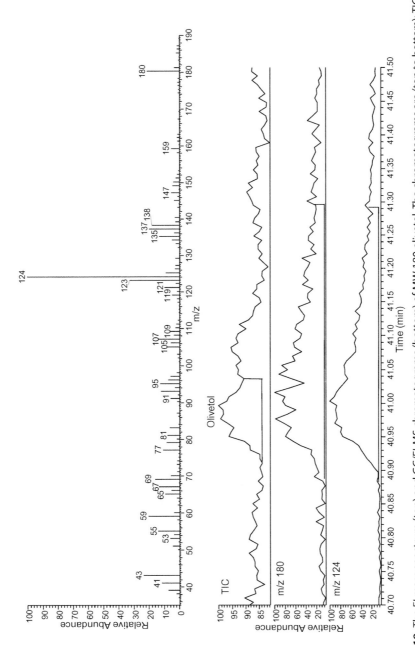

Fig. 19 The EI mass spectrum (top) and GC/EI-MS chromatograms (bottom) of MW 180 olivetol. The chromatograms are (top to bottom): TIC, m/z 180 and m/z 124.

extracts at ca. 41 min and was identified by a good match with the olivetol spectra in the NIST database (Fig. 19).

4. Concluding remarks

Since the advent of GC/EI-MS in the 1960s, it has been used in the characterization of cannabis compounds, both the volatile terpenes and sesquiterpenes and the semivolatile phytocannabinoids. GC/EI-MS offers high chromatographic resolution, large dynamic range and produces characteristic fingerprint spectra while being a robust, relatively inexpensive and easy-to-operate analytical instrument. These attributes will continue to see it as a powerful tool for characterizing the neutral phytocannabinoids and their degradation products in cannabis and cannabis-derived products. The major disadvantage of GC/EI-MS is the inability to detect the acidic forms of the phytocannabinoids without some type of derivatization to protect the acidic group from decarboxylation in the hot GC injection port. HPLC/MSn instruments are able to analyse the phytocannabinoids acids directly. While more expensive and somewhat more problematic from an operational standpoint, HPLC/MSn instruments with HRMS capabilities are making a dramatic impact on the characterization of phytocannabinoids and routine monitoring of cannabis.

References

[1] R. Mechoulam, Y. Gaoni, Recent advances in the chemistry of Hashish, Fortschr. Chem. Org. Naturst. 25 (1967) 175–213.

[2] R. Mechoulam, N.K. McCallum, S. Burstein, Recent advances in the chemistry and biochemistry of Cannabis, Chem. Rev. 76 (1976) 75–112.

[3] C.E. Turner, M.A. Elsohly, E.G. Boeren, Constituents of *Cannabis Sativa* L. XVII. A review of the natural constituents, J. Nat. Prod. 43 (1980) 169–234.

[4] Y. Gaoni, R. Mechoulam, Isolation, structure, and partial synthesis of an active constituent of hashish. J. Am. Chem. Soc. 86 (1964) 1646–1647. https://doi.org/10.1021/ja01062a046.

[5] https://pharmacy.olemiss.edu/marijuana/publications/.

[6] https://cannabinoids.huji.ac.il/people/raphael-mechoulam. A Google Scholar search of "Mechoulam cannabinoid" will result in hundreds of listings.

[7] J.A. Hartsel, *Cannabis sativa* and Hemp., Chapter 53. in: G. Ramesh (Ed.), Nutraceuticals: Efficacy, Safety and Toxicity, first ed., Elsevier, 2016, pp. 735–760. https://doi.org/10.1016/B978-0-12-802147-7.00053-X.

[8] M.A. ElSohly, W. Gul, Constituents of Cannabis sativa. in: R. Pertwee (Ed.), Handbook of Cannabis, University Press Scholarship, January, 2015, Online. https://doi.org/10.1093/acprof:oso/9780199662685.001.0001.

[9] L.O. Hanuš, S.M. Meyer, E. Muñoz, O. Taglialatela-Scafati, G. Appendino, Phytocannabinoids: a unified critical inventory, Nat. Prod. Rep. 33 (2016) 1357–1392.

[10] B.T. Borille, M. González, L. Steffens, R. Scorsatto Ortiz, R.P. Limberger, Cannabis sativa: a systematic review of plant analysis, Drug Anal. Res. 01 (2017) 1–23.

[11] L. Nahar, M. Guo, S.D. Sarker, Gas chromatographic analysis of naturally occurring cannabinoids: a review of literature published during the past decade. Phytochem. Anal. 31 (2020) 135–146. https://doi.org/10.1002/pca.2886.

[12] T. Béres, L. Černochová, S.Ć. Zeljković, S. Benická, T. Gucký, M. Berčák, P. Tarkowski, Intralaboratory comparison of analytical methods for quantification of major phytocannabinoids, Anal. Bioanal. Chem. 411 (2019) 3069–3079. https://doi.org/10.1007/s00216-019-01760-y.

[13] F.E. Dussy, C. Hamberg, M. Luginbühl, T. Schwerzmann, T.A. Briellmann, Isolation of Δ9-THC in cannabis products, Forensic Sci. Int. 149 (2005) 3–10.

[14] B. Fodor, I. Molnár-Perl, The role of derivatization techniques in the analysis of plant cannabinoids by gas chromatography mass spectrometry, Trends Anal. Chem. 95 (2017) 149–158. https://doi.org/10.1016/j.trac.2017.07.022.

[15] Jack Cochran 2015. blog.restek.com/cbdv-and-thcv-on-the-rxi-35sil-ms-gc-column-with-other-cannabinoids.

[16] M.W. Giese, M.A. Lewis, L. Giese, K.M. Smith, Development and validation of a reliable and robust method for the analysis of cannabinoids and terpenes in cannabis. J. AOAC Int. 98 (2015) 1503–1522. https://doi.org/10.5740/jaoacint.16-116.

[17] B. De Backer, B. Debrus, P. Lebrun, L. Theunis, N. Dubois, L. Decock, A. Verstraete, P. Hubert, C. Charlier, Innovative development and validation of an HPLC/DAD method for the qualitative and quantitative determination of major cannabinoids in cannabis plant material. J. Chromatogr. B 877 (2009) 4115–4124. https://doi.org/10.1016/j.jchromb.2009.11.004".

[18] J.V. Johnson, R.A. Yost, P.E. Kelly, D.C. Bradford, Tandem-in-space and tandem-in-time mass spectrometry: triple quadrupoles and quadrupole ion traps, Anal. Chem. 62 (1990) 2162–2172.

[19] M.M. Delgado-Povedano, C. Sánchez-Carnerero Callado, F. Priego-Capote, Untargeted characterization of extracts from *cannabis sativa* L. cultivars by gas and liquid chromatography coupled to mass spectrometry in high resolution mode. Talanta 208 (2020) 120384. https://doi.org/10.1016/j.talanta.2019.120384.

[20] R. Pavlovic, S. Panseri, L. Giupponi, V. Leoni, C. Citti, C. Cattaneo, M. Cavaletto, A. Giorgi, Phytochemical and ecological analysis of two varieties of hemp (Cannabis sativa L.) grown in a mountain environment of Italian Alps. Front. Plant Sci. 10 (2019) 1265. https://doi.org/10.3389/fpls.2019.01265.

[21] C. Citti, P. Linciano, S. Panseri, F. Vezzalini, F. Forni, M.A. Vandelli, G. Cannazza, Cannabinoid profiling of hemp seed oil by liquid chromatography coupled to high-resolution mass spectrometry. Front. Plant Sci. 10 (2019) 120, https://doi.org/10.3389/fpls.2019.00120.

[22] P. Berman, K. Futoran, G.M. Lewitus, D. Mukha, M. Benami, T. Shlomi, D. Meiri, A new ESI-LC/MS approach for comprehensive metabolic profiling of phytocannabinoids in Cannabis. Sci. Rep. 8 (2019) 14280, https://doi.org/10.1038/s41598-018-32651-4.

[23] H. Budzikiewicz, R.T. Alpin, D.A. Lightner, C. Djerassi, R. Mechoulam, Y. Gaoni, Massenspektroskopie und ihre Anwendung auf Strukturelle und Stereochemische Probleme-LXVIII, Tetrahedron 21 (1965) 1881–1888.

[24] A. Hazekamp, A. Peltenburg, R. Verpoorte, C. Giroud, Chromatographic and spectroscopic data of cannabinoids from *Cannabis sativa* L. J. Liq. Chromatogr. Relat. Technol. 28 (2005) 2361–2382. https://doi.org/10.1080/10826070500187558.

[25] NIST Mass Spectra Library. Chemdata.nist.gov/dokuwiki/doku.php

[26] Wiley Registry: Mass Spectra Library, 11th ed. www.wiley.com

[27] R.M. Smith, K.D. Kempfert, Δ1-3,4-*cis*-Tetrahydrocannabinol in Cannabis sativa, Phytochemistry 16 (1977) 1088–1089.

[28] T.B. Vree, D.D. Breimer, C.A.M. Van Ginneken, J.M. Van Rossum, Gas chromatography of cannabis constituents and their synthetic derivatives, J. Chromatogr. 74 (1972) 209–224.

[29] C. Citti, P. Linciano, F. Forni, M.A. Vandelli, G. Gigli, A. Laganà, G. Cannazza, Analysis of impurities of cannabidiol from hemp. Isolation, characterization and synthesis of cannabidibutol, the novel cannabidiol butyl analog, J. Pharm. Biomed. Anal. 175 (2019) 112752. https://doi.org/10.1016/j.jpba.2019.06.049.

[30] C. Citti, P. Linciano, F. Russo, L. Luongo, M. Iannotta, S. Maione, A. Laganà, A.L. Capriotti, F. Forni, M.A. Vandelli, G. Gigli, G. Cannazza, A novel phytocannabinoid isolated from *Cannabis sativa* L. with an in vivo cannabimimetic activity higher than Δ9-tetrahydrocannabinol: Δ9-tetrahydrocannabiphorol, Sci. Rep. 9 (2019) 20335. https://doi.org/10.1038/s41598-019-56785-1.

[31] S.A. Ross, M.A. ElSohly, G.N.N. Sultana, Z. Mehmedic, C.F. Hossain, S. Chandra, Flavonoid glycosides and cannabinoids from the pollen of *Cannabis sativa* L. Phytochem. Anal. 16 (2005) 45–48. https://doi.org/10.1002/pca.809.

[32] R. Brenneisen, M.A. ElSohly, Chromatographic and spectroscopic profiles of cannabis of different origins: part I, J. Forensic Sci. 33 (1988) 1385–1404.

PART 3

Applications of LC-MS techniques for cannabis analysis

CHAPTER NINE

The analysis of pesticides and cannabinoids in cannabis using LC-MS/MS

Paul Winkler*

SCIEX LLC, Framingham, MA, United States
*Corresponding author: e-mail address: paul.winkler@sciex.com

Contents

1. Introduction	277
2. LC-MS/MS instrumentation	279
3. Method development	286
3.1 Instrumental procedure	286
3.2 Sample preparation procedure	291
3.3 Method validation	292
4. Sample analysis	304
4.1 Quality assurance/quality control	304
4.2 Sample results	305
4.3 Cannabinoid analysis	308
5. Summary	313
References	313

1. Introduction

With the legalization of medical marijuana, first approved by the state of California in 1996, there has been the need to provide analyses of plant and edible samples to ensure the potency and safety of products that are used by consumers [1]. The need for a comprehensive set of analytical methods for plant and concentrate samples increased significantly with the legalization of adult use marijuana laws, first promulgated by the state of Colorado in 2012 [2]. The legalization of cannabis led to the need for states and various regulatory bodies to develop testing requirements to ensure the safety of cannabis products.

Perhaps the most complicated of all analysis required for cannabis samples is the determination of pesticides in plant and manufactured products. This may also be the most important analysis in terms of safety for consumption, especially for long term usage. Additionally, this is the analysis that may cause growers and manufactures the most problems because a failed pesticide test may result in the total destruction of the plant or product.

Given the potential impact of this analysis and the complexity of the data, it is important to provide a discussion of how the analysis is performed, how the quality of the data is determined and how to interpret the data from the laboratory.

There are many important considerations for sample analysis in cannabis samples. What concentration of pesticides for the calibration should be used? How is an acceptable calibration defined? How do I know that the system can detect a pesticide at the maximum residual limit (MRL)? How is the quality of the sample extraction procedure checked? How can I have confidence in the results?

This chapter is designed to aid the data end user in evaluating the results that have been reported on a Certificate of Analysis (COA) and to understand the analytical process enough to look at the results, including raw data, and to ask pertinent questions about the results. Often, the end user is at the complete mercy of the laboratory and cannot accurately evaluate a result or ask the correct questions to understand the result. The goal of this chapter is to give an overview of the entire process for the analysis of pesticides in cannabis samples in a way that a non-chemist can understand and to explain how to interpret the data when questions arise. This will aid those without a background in analytical chemistry to have a useful discussion with their laboratory.

The term pesticides is, in fact, a general term for synthetic compounds that may be applied to cannabis samples for one reason or another. Most compounds that are on a targeted analyte list in cannabis samples are for the extermination of pests, including mites, aphids and insects, but not all. As an example, Myclobutanil and Imazalil are actually fungicides and are used to prevent the growth of various fungi, which can spoil a product. Another example is daminozide which is used as a plant growth regulator. It should be noted that the use of daminozide on food crops has been banned since 1989 but it is still in use on ornamental plants. For clarity, the term pesticides will be used for any banned substance that is on a list of analytes required by a state or country regulator for cannabis testing.

Many pesticides are banned for use on cannabis but their presence has still been found. The possibility that these compounds may have a severe impact on the safety of the cannabis products has led most states in the United States to have some requirement for the analysis of pesticides on cannabis and related products such as concentrates and edibles. The actual target compound list, however, has been a difficult task for most states to develop. This is in part due to the lack of federal guidance and also to the lack of any specific toxicological data for these compounds in cannabis. As a result, each state has their own list of pesticides and action limits, a situation that has only further confused an already complex situation [3–6]. The analytical method of choice for the determination of most of the pesticides is Liquid Chromatography Tandem Mass Spectrometry (LC-MS/MS) as this technology provides the best overall performance for the compounds on the regulatory target analyte list. While Gas Chromatography Tandem Mass Spectrometry (GC-MS/MS) is also popularly used, this method cannot be used for all of the pesticides that are required for testing and the focus of this chapter is on the use of LC-MS/MS techniques, thus, GC-MS/MS methods are not discussed here.

Users of data from a laboratory are often at a disadvantage when attempting to understand the results. In certain cases, a pesticide that the grower is sure was not used may be found, but without a basic understanding of the method used to analyse the sample, or what criteria the laboratory used to report the result, the data end user cannot asses the quality of the data or the certainty of the results. The goal of this chapter is to familiarize the reader with the basics of pesticide analysis using LC-MS/MS and to discuss aspects of the method that need to be understood in order to make decisions related to laboratory results.

Regulations specify a maximum residual limit (MRL) for each pesticide and sample type, inhalable vs topical, for example, that the material can contain. In the case of cannabinoids, regulations typically specify how much tetrahydrocannabinol (THC) or cannabidiol (CBD) is contained in the sample to meet regulations for labelling. What is LC-MS/MS, how is it used to measure pesticides and how can the data user be comfortable with the results? The following discussion is intended to help answer these questions.

2. LC-MS/MS instrumentation

The goal of any analytical method is to provide accurate quantitation of a given compound in a sample. This is often a difficult task because of the chemical complexity of the sample, referred to as the sample matrix. Compounds from the matrix can interfere with target compounds by creating an analytical signal that mimics that of the target analyte, resulting in a reported analyte that is not actually in the sample. This is a false positive result. Chemicals in the matrix may also suppress the signal from a compound that is actually present in the sample making it appear as if there is no contamination in the sample. This is a false negative result. Given the extremely complex nature of the cannabis plant, it is necessary to have an analytical technique that can adequately separate the sample matrix from the target analytes of interest and then provide sufficient specificity for compound identification and quantitation if there is to be any confidence in the analytical results.

The technique of LC-MS/MS strives to separate the different chemicals in a matrix from the pesticides of interest using liquid chromatography and then detecting and quantitating those pesticides using mass spectrometry. The analysis of pesticides in foods and other agricultural products using LC-MS/MS has used been for several years and extraction methods and chromatographic methods have been fully developed. These methods provided an excellent starting point for working out a viable method for the analysis of pesticides in cannabis.

A typical LC-MS/MS experimental setup is shown in Fig. 1. The system requires a liquid pump (HPLC pump), an injector, a column and a mass spectrometer. The pump pushes liquid through the system at pressures from 3000 psi up to 18,000 psi. This liquid is the mobile phase and are called eluants. Usually, two different solvents are pumped through the system. One of the solvents is aqueous while the other is an organic solvent, typically methanol or acetonitrile. The solvent ratio between the aqueous and organic phase changes during the separation starting with a mostly aqueous solvent. Chemicals then separate based on their solubility. Compounds more soluble

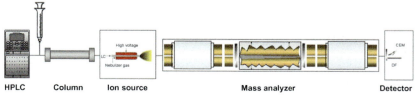

Fig. 1 Schematic diagram of a typical instrumental arrangement for an LC-MS/MS instrument.

in the aqueous phase move through the system faster than compounds more soluble in the organic phase. The time that is required for a compound to move through the system is called the retention time (RT).

To start an analysis, a portion of the extract, usually 0.5–4 µL, is injected into the system using the injector and is moved by the mobile phase into the column. A High Pressure Liquid Chromatography (HPLC) column is a metal tube (internal diameter 2–4.5 mm) filled with a powder that has been modified for the purpose of separating chemicals, this powder referred to as the stationary phase. Two common stationary phases for pesticide separations in cannabis are the C_{18} and biphenyl phases. Both column chemistries provide similar separations but have different selectivity [7]. A typical separation using a biphenyl column and a C_{18} column is shown in Fig. 2. Each peak that is seen in the figure is a unique compound and, as can be seen there, is a wide range in retention times and most of the compounds are separated from each other. Total separation is not required though, as the mass spectrometer is a highly specific instrument and can discriminate compounds due to their different molecular masses.

The experimental approach is to extract a sample using a known weight of sample and a known volume of solvent. The sample is shaken in the solvent to extract chemicals into the solvent. Extraction strategies have used either a solvent extraction of solid material using methanol or acetonitrile or samples have been extracted using the QuEChERS method [8,9]. The solvent extraction method has the advantage that all compounds are extracted with high efficiency while some of the pesticides can have low recoveries using the QuEChERS technique. Regardless of the extraction method used, the chromatographic separation that is used for the extract is the same.

Due to the large amount of matrix from a cannabis sample extraction, particularly if a solvent extraction is used, the best results are often obtained using a 4.6 mm diameter column. The larger column diameter has more mobile phase and therefore more capacity, thus larger injection volumes can be used without a degradation in the chromatographic separation due to overloading the column.

When the compound elutes off of the column, the liquid enters into the mass spectrometer. The eluant from the column enters into an ion source where the mobile phase is separated from the compounds of interest, the pesticides are ionized and then enter into the mass spectrometer. The process of removing the mobile phase and ionizing the compounds is mainly accomplished using a process known as electrospray (ESI) but another ionization technique, Atmospheric Chemical Pressure Ionization (APCI) is also used

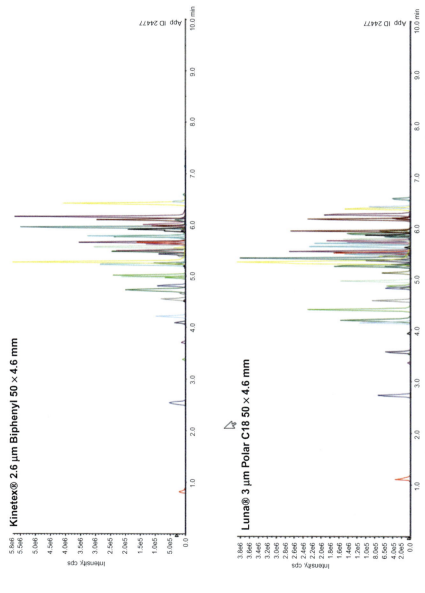

Fig. 2 Chromatogram of pesticides for the California list. Top chromatogram is a Biphenyl column and bottom chromatogram is a C_{18} column. Both columns provide adequate separation but different selectivities.

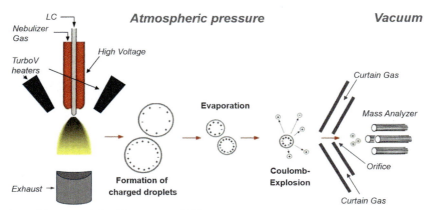

Fig. 3 Diagram of the Electrospray (ESI) process.

when ESI does not work well. This process is required for specific compounds such as chordane or pentachoronitrobenze (Quintozene) that do not form ions in the electrospray process if the entire list of pesticides is to be determined using LC-MS/MS.

The process of ESI is shown in Fig. 3. When using ESI, the eluant from the column flows through a metal tube, known as a capillary, that has a voltage applied to it and has a gas flowing around it to nebulize the liquid. The voltage that is applied to the capillary results in the droplets in the spray being charged. A heated gas may be focused on the spray from the capillary and this gas speeds up the process of evaporation of the spray. As the charged droplets evaporate, they become unstable. This results in ions being ejected from the droplet.

At this point, two objectives have been accomplished. The eluant has been removed and the compound of interest has been ionized and is now ready to be separated in the mass spectrometer. ESI is used for most of the pesticides that are analysed in cannabis samples because of the sensitivity and ease of use of the technique. There are, however, some issues related to electrospray that impact the quality of the results during cannabis analysis. The most common problem with electrospray is a phenomenon known as ionization suppression. When ions begin leaving the droplet (coulomb explosion, Fig. 3), if there are other compounds in the droplet because they eluted at the same time as the pesticide, these compounds may leave the droplet instead of the pesticide if they are easier to ionize. When this happens, the ionization of the pesticide out of the droplet is decreased and a loss of sensitivity is observed. This ionization suppression can and does result in problems with the quality of the data from the analysis. Situations often arise where the suppression is significant enough that it is not possible to detect a pesticide at the required maximum residual limit. In these cases, a false negative result may be obtained, which can cause problems later in the manufacturing process. For example, a pesticide may be undetected in the flower analysis but, after forming a concentrate, the pesticide is now at a high enough concentration to be observed, leading to disposal of the concentrate. Another problem that may occur with electrospray is that the pesticide simply does not form an ion in solution and therefore will not be ionized. In these situations, a pesticide will not be detected in a sample using ESI no matter how high the concentration. A good example of this is chlordane. For those compounds, electrospray is not an appropriate technique and the use of APCI will be required if the analysis is done only using LC-MS/MS.

The process of APCI is shown in Fig. 4. Unlike the ESI process, in APCI the spray is formed in a heated tube and the solvent and pesticides are vaporized. The solvent molecules are ionized by a corona discharge needle and then these ionized solvent molecules ionize the pesticide. This is different from ESI because ionization occurs in the gas phase and not the liquid phase. The APCI process is usually less sensitive than ESI but it is far less prone to ionization suppression and may offer advantages in certain situations. There are some compounds that are ionized more efficiently by this process than by the ESI process. Using a combination of ESI and APCI it is possible to analyse all of the pesticides that are currently regulated at this time using only LC-MS/MS.

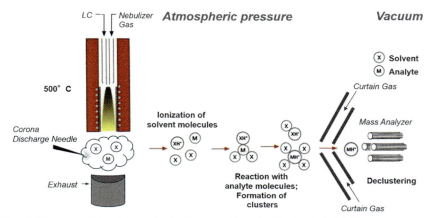

Fig. 4 Diagram of the Atmospheric Pressure Chemical Ionization (APCI) process.

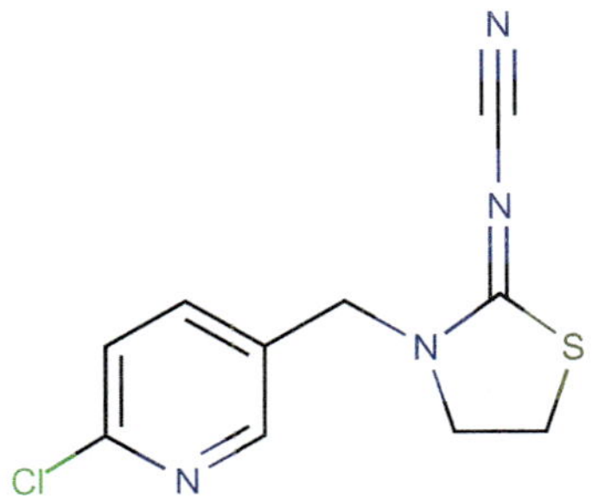

The analysis of pesticides and cannabinoids

Fig. 5 Thiacloprid molecule, molecular formula, $C_{10}H_9ClN_4S$, Molecular mass 252 amu.

In both ESI and APCI, the ion that is formed is a protonated molecular ion, $M + H^+$. Take as an example, the compound thiacloprid, a commonly used insecticide that is used for aphids. Thiacloprid is a compound with a molecular weight of 252 amu (amu). The structure of thiacloprid is shown in Fig. 5.

When ionized by ESI, an ion at mass 253 is observed. This ion then is focused into a triple quadrupole mass spectrometer. A diagram of a triple quadrupole mass spectrometer is shown in Fig. 6.

When using a triple quadrupole instrument, the protonated ion for the compound is selected to be transferred through Q1, this is an ion filtering step. All other ions are excluded from passing through Q1 during a particular scan. This ion is accelerated into Q2, where it is fragmented due to colliding with gas molecules that are in Q2. The ions formed during fragmentation are characteristic of the structure and are focused into Q3, which may be scanned to produce a MS/MS spectrum that is unique to the compound of interest or Q3 may set to only transmit one fragment mass for the purpose of quantitation. When the instrument is operated using Q1 to transmit the parent mass and Q3 to transmit the fragment mass, this scan type is called a Multiple Reaction Monitoring (MRM) scan. MRM scans are very sensitive and also form the basis for the specificity of LC–MS/MS methods. In the case of thiacloprid, the MS/MS spectrum is shown in Fig. 7.

The two most intense ions are at masses 126 and 186. These transitions, 253 to 126 and 253 to 186 are referred to as the quantifier ion and qualifier ion respectively. The quantifier ion is the more sensitive of the two transitions and is used to determine the concentration of the analyte and the qualifier ion is the less sensitive of the two transitions and is used to verify the identity of the compound. The ratio of the area of the quantifier ion to the area of the qualifier ion is used to confirm that the signal is due to the

General Setup of a Triple Quad Instrument

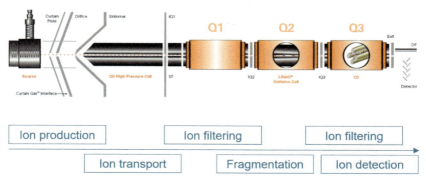

Fig. 6 Diagram of a triple quadrupole instrument.

compound being measured and not due to some interferent. In the case of thiacloprid, the transition from ion 253–126 is the most sensitive transition and the transition 253–186 is the confirmation transition. These two transitions are shown in Fig. 8.

The blue peak is the signal for the 253–126 transition and the pink trace is the signal for the 253–186 transition. The ratio between these two transitions is constant no matter what the concentration of the compound is in the sample. In practise, while performing a quantitative analysis, a series of different concentrations are analysed, as will be discussed later. The ratio of these two transitions is measured for a compound at each concentration of the calibration and the average ratio is set as a limit for sample analysis. If the ratio of the area for the two transitions is within the acceptance limits, then, along with the retention time, a positive result is reported. This procedure works well and minimizes the occurrence of false positive results.

When analysing a complex matrix such as cannabis, it is important to use all of the specificity of the method that is available in order to provide as accurate data as possible. When analysing cannabis samples, if a peak for a given pesticide is observed for the primary transition at the correct retention time, with the correct ratio of primary to secondary areas, a positive identification will be reported.

3. Method development
3.1 Instrumental procedure

The process of creating a method for the analysis of pesticides in cannabis first requires that each pesticide to be measured must have quantitation

The analysis of pesticides and cannabinoids

Fig. 7 MS/MS spectrum of thiacloprid showing main fragment at mass 126 and qualifier ion at mass 186.

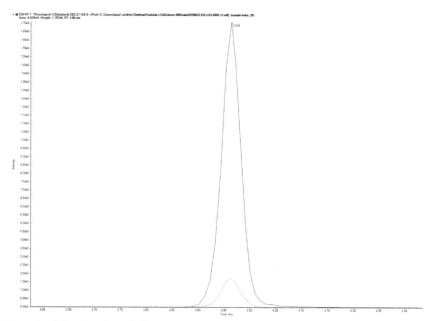

Fig. 8 Peak for thiacloprid showing both the quantitation ion in blue and the qualifier ion in pink. The quantitation transition is usually the most sensitive transition as seen in this figure.

and qualifier transitions determined. This requires that a pure standard of each compound be obtained and infused into the LC-MS/MS. The correct parameters for primary and secondary transitions are found and a table is created that the instrument uses to scan for each compound. This process may be time consuming depending on how many pesticides that method will be used for. In some states, this is as little as 13 compounds, in other states the list is over 60 compounds, but this process is the essential piece of information that is required for the analysis. Once the method has been created for the mass spectrometer, adequate liquid chromatography conditions must be found. This process is accomplished by making a solution that contains all of the pesticides for the method at a relatively large concentration. This standard is injected into the LC-MS/MS and the data are reviewed for adequate separation of the compounds. Typically, it is not necessary to have every compound in the standard separated fully from each other because they will have different molecular weights and fragment ions and therefore, the signal for these compounds do not interfere with each other. However, some separation is important because of the time that is required to measure each of the pesticides. If too many compounds elute at the same time, it will not be possible to measure a large list with enough precision to generate an accurate result. As a result, as seen in Fig. 2, a typical separation in a standard may be accomplished in as little as 7–10 min.

Once a base method has been created, it is imperative, especially for cannabis, to then test the method on an actual sample. This is important due to the severe matrix effects that are observed with cannabis. An example of how the matrix can result in interferences that are not observed in the standard is shown in Fig. 9. This data shows that, even with the specificity of MS/MS, there are still several peaks that are observed near the actual pesticide. In the top chromatogram, it can be seen that for both Pyrethrin I and II, there are many peaks that, without the secondary transition, could easily be confused with the real pyrethrin, which is shown in the bottom traces of Fig. 9.

In the absence of careful chromatographic method development, it could be possible to have the pyrethrins all elute at the same time which may result in a false positive result. Also note that if all of the extra peaks seen in Fig. 9 co-eluted, it is possible that even the ion ratios would be correct, resulting in the laboratory reporting a false positive result or an enhanced concentration result. This is an excellent example of the importance of testing a method in real matrices before using the method to report results.

One of the best practises during method development in matrix to find problems such as the co-elution shown in Fig. 9 is to perform matrix spiking studies. This is most usefully performed by extracting a representative cannabis sample and spiking aliquots of the extract with known concentrations of pesticides, starting at the maximum residual limit for the pesticide and increasing the spike amount until the extract concentration would be expected to be at a concentration near the highest concentration in the calibration curve. If a severe matrix interference is occurring, the spiked component does not appear because it is completely suppressed by the matrix. In these situations, the recoveries for the low concentration spikes are low or not observed. This continues until the spiked amount is high enough to produce a signal that is large enough to be observed over any matrix interference. In the case of the pyrethrins, the spiking study indicated that a 10 min analysis time resulted in unacceptable interferences and led to the development of a more complete separation. After chromatographic optimization, the method required 16 min to complete but was critical to a developing an accurate method in real matrices.

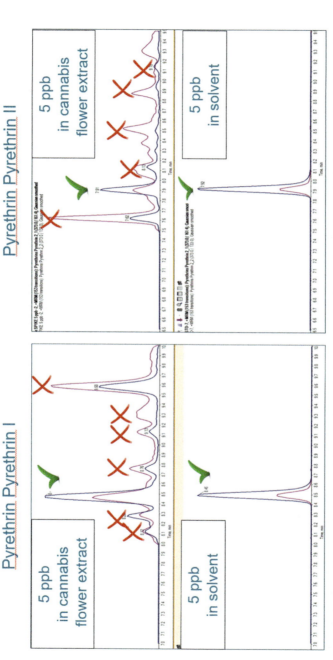

Fig. 9 Analysis of pyrethrin I and pyrethrin II in cannabis and solvent. The top chromatogram is for pyrethrin in cannabis matrix, the bottom chromatogram is pyrethrin in solvent. Interference from the matrix is shown in the top chromatogram.

3.2 Sample preparation procedure

In the previous section, the creation of an analytical method was discussed. To use the method, it is necessary to perform some type of sample preparation to obtain an extract that is suitable for analysis. Sample preparation in the world of cannabis is extremely complex due to the large number of different sample types. These range from plant material to creams and salves. Different matrices will require different preparation procedures but basically fall into two categories: solvent extraction and dilution.

Solvent extraction is a process where the sample is placed in a certain volume of solvent and time is allowed for the solvent to dissolve analytes and plant chemicals alike. Once the sample has been extracted with the solvent, it may be subjected to further processing in order to remove unwanted and interfering components that have been extracted along with the target analytes. These processing procedures are referred to as clean up steps. Depending on the type of instrumentation that the laboratory has, it may be possible to inject a portion of the raw extract into the system without further processing. In other cases, a cleanup step is mandatory to keep the LC–MS/MS system operational. In other cases, such as isolates or concentrates, the sample preparation procedure is a simple dilution with a solvent.

When a sample extract can be analysed without further cleanup the procedure is fairly simple. The solvent is transferred from the extraction tube and centrifuged for approximately 5 min. An aliquot of the extract is filtered and placed in a freezer for several hours. This "Winterization" of the extract is critical because this step removes lipids from the extract which can cause significant problems during the analysis. These problems range from fouling the column to causing ion suppression in the mass spectrometer. After winterization, the extract is filtered and is ready for chromatographic analysis. The advantage of this type of sample preparation is that it is simple, inexpensive, has no analyte losses and the extract may be used for pesticide and cannabinoid analysis.

A common cleanup procedure is the QuEChERS procedure [10]. With this procedure, a typical method is to take 1 g of dried sample and add 10 mL of deionized water and let the sample stand for 30 to 60 min to fully hydrate the sample. After the soaking time, an aliquot of acetonitrile is added to the sample and the sample is shaken. Then a salt is added to the solution so that the acetonitrile layer will separate from the aqueous layer. This separation is aided by centrifuging for a period of time, typically 5 min. An aliquot of the acetonitrile is removed and processed further with other cleanup materials such as primary secondary amine (PSA), C_{18} or graphitized carbon to further cleanup the extract. These different phases are used to remove different chemicals from the extract and allows a level of customization to the sample preparation procedure to match the requirements of the sample type. The QeEChERS cleanup is popular due to the excellent removal of interferences from the matrix but the procedure adds a significant cost to the sample preparation procedure in both time and materials and some of the target pesticides may be removed during the cleanup process. In addition, cannabinoids are lost to some extent during the cleanup. The portion of the extract that is used for the analysis of cannabinoids only requires a further dilution step after the extraction in order to be ready for analysis. The dilution is required for the analysis of the cannabinoids because of the relatively high concentrations of cannabinoids in the sample. This dilution step minimizes matrix interferences and dilutes the cannabinoids to a concentration that is in the range of what the instrument can accurately measure and into the concentration range of the calibration curve.

When evaluating data from the laboratory, it is essential to know how the laboratory prepared the samples, as the cleanup procedure directly relates to how to evaluate the quality of the data.

3.3 Method validation

After a candidate method has been created it must be validated to demonstrate that it is fit for purpose. The criteria used to demonstrate the effectiveness of a method are called Figures of Merit. The Figures of Merit that must be demonstrated before a laboratory can begin to use a method include accuracy, precision, linearity and limits of detection and quantitation.

A method is only a set of conditions that allow for an acceptable chromatographic separation of the extract and then measurement of the signal from target compounds in the extract. In order to accurately determine the quantitation of a target analyte, it is necessary to analyse a series of solutions with different concentrations so that when a sample extract is analysed, the signal from the sample may be related to a concentration. This is the process of calibrating the instrument. While this may seem a simple process, there are some critically important considerations that the laboratory must take into account in order to provide results of acceptable quality and a method that meets regulatory requirements.

3.3.1 Calibration

Proper calibration of the instrument is perhaps the most important criteria to be established before beginning sample analysis because a good calibration is critical for accurately measuring the concentration of an analyte in a sample and serves as a demonstration that the instrument is operating properly. Because the fundamental quality of the data produced during the analysis depends greatly on how the instrument was calibrated, regulations have very specific criteria to define how a calibration must be run and what is an acceptable calibration. Calibrations must have at least five calibration points, linearity must be demonstrated by having a correlation coefficient of greater than 0.99, the curve model should not have more than $1/x$ weighting and the curve must have a calibration point that is representative of the MRL for that compound. The MRL is often the lowest concentration or the second to lowest concentration used for the calibration.

A typical calibration curve is shown in Fig. 10. This is a calibration for the compound trifloxystrobin, a common fungicide. A series of solutions with increasing concentrations of trifloxystrobin were analysed and the peak areas were measured and plotted against concentration. Peaks from the trifloxystrobin calibration are shown in Fig. 11 to provide a picture of what the data from a calibration actually looks like. In this figure, peaks from both the primary and secondary transitions may be observed along with the acceptance lines for the peak ratios.

An examination of this calibration show that it meets all of the criteria that defines an acceptable calibration. A calibration curve is required to have at least five points on the curve. This is necessary to adequately model the instrument response to the analyte concentration. The example curve has nine calibration points, which is acceptable because the criteria specify at least five points on the curve.

A good calibration will appear as a straight line as seen in Fig. 10. This shows a linear relationship between concentration and signal. Regulations will require that calibrations meet a linearity requirement to prove that the method is capable of generating accurate quantitation results.

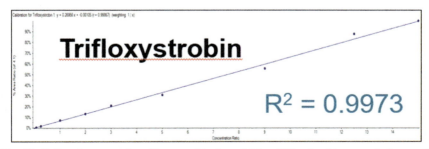

Fig. 10 Calibration curve for trifloxystrobin from 0.25 to 250 ppb with a correlation coefficient, r^2 of 0.9973, showing excellent linearity.

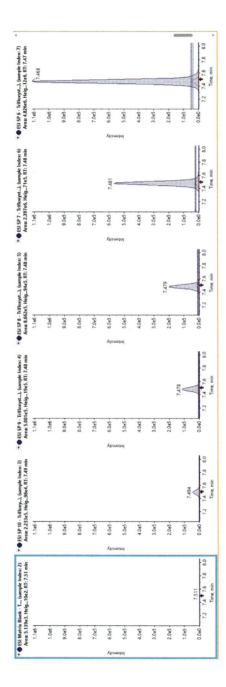

Fig. 11 Peaks from the trifloxystrobin calibration showing how the raw data should appear when reviewing data from a calibration.

As seen in Fig. 10, there is a number indicated as R^2. This number is the correlation coefficient and is a measure of how close each individual point of the calibration is to the calculated calibration line. The number is usually required to be greater than or equal to 0.99, see for example, the State of California requirements [11].

The calibration curve was calculated using a $1/x$ weighting, which is usually done as it improves the accuracy of quantitation for lower concentrations.

Finally, the MRL in California for trifloxystrobin is $0.1\,\mu g/g$ (100 ppb), which corresponds to the sixth point on the curve. The calibration concentrations used for this curve were 0.25, 0.5, 1, 2.5, 5, 10, 25, 50, 100 and 250 ng/mL (parts per billion). What is important to note is that the calibration must have a concentration point in the curve that corresponds to the MRL required for that compound. A typical sample preparation procedure would require the extraction of 1 g of dried plant material in 10 mL of solvent. For a plant that had trifloxystrobin present at 100 ppb ($0.1\,\mu g/g$), the extraction would result in a solution concentration of $0.1\,\mu g/10\,mL$ or $0.01\,\mu g/mL$, which is 10 ng/mL. The calibration must have at least a calibration concentration of 10 ng/mL to show that the instrument has the required sensitivity to meet the regulatory requirements. The calibration in Fig. 10 has several calibration points at concentrations that would be useful for quantitation well below the MRL. This allows the laboratory to perform some additional dilutions to minimize the amount of matrix injected onto the column and to reduce ionization suppression issues in the mass spectrometer. Using the trifloxystrobin example, an extract could be diluted by 10:1, which would result in a concentration of 1 ng/mL in the extract for a 10 ppb plant concentration. The third point in this calibration would correspond to the MRL if a 10:1 dilution were performed. The calibration curve in Fig. 10 demonstrates that the method is easily capable of measuring trifloxystrobin at the MRL. When reviewing data, it is good practise to inspect the peaks from the calibration curve at the concentration that is at the MRL for that compound to ensure that the data actually meets the regulatory requirements. Inspection of the raw data for matrix extracts spiked at the MRL is an important way for the data end user to know that the laboratory method actually is sensitive enough to accurately report data at the MRL.

When evaluating data from a laboratory, an inspection of the calibration curve is an important way to ascertain that the instrument was in good working order for the analysis and that the end user can have confidence in the results reported on the Certificate of Analysis (COA).

A common problem with a calibration is when several calibration points do not fit the line closely and may be an indication of poor instrument stability, poor standard stability, poor laboratory technique in making the calibration solutions or a compound that the method is simply not suitable for.

An example of a poor calibration is shown in Fig. 12, a calibration curve for cinerin I, one of the pyrethrin insecticides. Note that for this calibration, the calibration points at the low end of the curve do not fit the calculated calibration curve well. This is because the peaks at the lower concentrations are not observed or are of low quality. From left to right, the calibration points are 0.25, 0.5, 1, 2.5, 5.0 and 10.0 ng/mL respectively, the corresponding raw data is shown below the calibration.

It can be observed that until the concentration reaches 2.5 ng/mL, there is no peak. If the MRL required a calibration point at a lower concentration than 2.5 ng/mL, this calibration would not be acceptable. Often times, a laboratory will perform an additional extract dilution due to unusually severe matrix interferences and not check that the lower calibration point meets the requirements. It is important to know how the laboratory has prepared the sample and also how the laboratory has calibrated the instrument in order to know if a low enough calibration point was used. Also note that the r^2 value for this curve is greater the 0.99! This value would appear to meet the calibration criteria, even though there are no peaks for the first three calibration concentrations. This demonstrates the need for the data end user to ask to see data when evaluating results from a laboratory.

Another issue related to calibration is the practise of matrix matching. Matrix matching is where instead of using neat solvents to make calibration solutions, the calibrations are made in a sample extract. The intent of matrix matching is to account for whatever interferences the matrix may cause during the calibration. If the matrix caused a certain compound to be suppressed, then the suppression would be the same for a sample extract or a calibration point. This is an excellent means of correcting for complex matrix interferences but must be used with extreme caution. One consideration is that the matrix extract used to construct the calibration curve must be clean, that is that the extract cannot have any of the target compounds present in the sample or this will interfere with the calibration. Typically, when such a problem exists, it is usually apparent from the calibration curve, because there will be problems achieving an acceptable correlation coefficient. The second and more significant problem with matrix matching in cannabis analysis is that the sample used for the calibration must be exactly

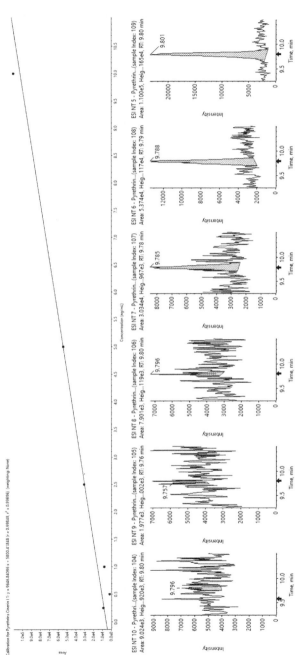

Fig. 12 Calibration curve for cinerin I showing poor fit for second and third calibration point for the top pane in the figure. The bottom pane are the actual peaks for the calibration and show that a peak is not actually observed until the fourth calibration point.

the same as the one being analysed. The interference caused from one strain to another can be significantly different, thus rendering the matrix matching invalid. If matrix matching is used, it is imperative that a clean matrix blank be generated and that only the sample strain or type that is being analysed is used for the calibration. Most laboratories do not analyse just one strain or one sample type and so matrix matching is not commonly used, but the data end user should know how the calibration was created in order to assess any data quality issues when they arise.

3.3.2 Limit of quantitation and detection

When analysing samples, it is important to know what concentration of pesticides could be accurately quantitated and what concentration of pesticides could be detected. This information must be generated to show that a method is Fit for Purpose and will generate data that will meet the requirements of the regulation. The Limit of Quantitation (LOQ) is a concentration value where the result that is generated can be expected to be reported accurately, with known precision. In the measurement of pesticides using LC-MS/MS, even in the absence of the pesticide in the sample, there may be some signal that is measured that is not due to the target compound. This is usually caused by something in the matrix that interferes with the pesticide signal. The LOQ is a concentration where the signal strength from the pesticide is large enough that any error in the measurement due to extraneous signals from the matrix is too small to affect the quantitated value.

While there are a number of methods to find an LOQ, this value is often determined by extracting seven samples of a clean matrix such as a flower sample and then spiking them at a concentration that is at a concentration that produces a signal from the pesticide, where the signal is approximately 10 times that of the signal from a blank matrix. This can be stated as spiking a sample at a concentration that produces an analyte response with a signal to noise (S/N) ratio of approximately 10. The response for the seven spiked samples is tabulated and the standard deviation of the seven replicates is calculated and then multiplied by 10. This value is then divided by the slope of the calibration line for that analyte. This produces a calculated value for the LOQ for that analyte. It is important to then spike a blank matrix at the calculated LOQ concentration to verify that the LOQ is real in matrix. There are situations where a calculated LOQ cannot be measured when spiked into a real matrix and this should be checked carefully because this shows that the laboratory method is fundamentally capable of meeting the regulatory criteria.

The Limit of Detection (LOD) is a value where the laboratory can determine with confidence that the analyte is present in the sample but cannot measure that concentration with acceptable accuracy or precision. This concentration is lower than the concentration for the LOQ. The LOD is determined in the same manner as the LOQ but it is spiked at a lower concentration, one where the S/N is closer to 3 than to 10. Again, it should be emphasized that the actual S/N of the response is not important for determining the LOD or LOQ but it is a useful means for determining the correct spiking concentration for the LOQ or LOD study.

Calculated LOQ and LOD values (ppb in plant) for the California list is shown in Table 1. Example data from a compound from the study is shown in Fig. 13.

This data is for the pesticide acequinocyl and was generated using ESI ionization and spiked into an extract that was made by extracting 1 g of dried flower into 10 mL of acetonitrile. The top row of data are the peaks for the seven samples spiked with acequinocyl at a concentration of 2.5 ppb (ng/g in plant) for the LOD study. The calculated LOD for this compound is 2.9 ppb.

The RSD for the response is 7.95. The data show that the calculated value is an accurate representation of the LOD and, furthermore, that it is possible to observe a peak for acequinocyl when it is present in an extract at a concentration of 2.5 ppb and verifies the calculated LOD.

The bottom row is the data for seven samples spiked at a concentration of 10 ppb (ng/g in plant). It can be seen that the S/N for the LOQ data is much better than that for the LOD. The calculated LOQ is 8.8 ppb. The RSD for the response is 5.14. This data verifies the calculated MRL in matrix.

The MRL for acequinocyl for the state of California is 100 ppb and in this example it is shown that the method is easily capable of meeting the regulatory requirements. This data serve as an example for how the LOD and LOQ are used but the regulatory requirements can be very different depending on the regulating body. The acequinocyl LOQ for Canada is 50 ppb and the method is still capable of meeting the regulatory requirements but it is important to verify the method performance to ensure that regulatory requirements can be met. It is also highly important to verify by spiking, that the calculated LOD or LOQ value can be measured as expected in a representative matrix extract. As a data end user, these data should be checked carefully when evaluating a laboratories method. This will ensure that the data generated by the laboratory are of high quality and will meet the regulatory requirements.

Table 1 Calculated limits of detection (LOD) and limits of quantitation (LOQ) for the California list of pesticides.

Analyte	Spiked concentration (ng/g)	LOQ value (ng/g)	LOD value (ng/g)
Acephate	0.5	4.9	1.6
Acequinocyl	2.5	88.1	29.1
Acetamiprid	0.5	1.3	0.4
Aldicarb	0.25	1.9	0.6
Avermectin B1a	0.25	2.1	0.7
Avermectin B1b	5	34.2	11.3
Azoxystrobin	0.25	3.8	1.3
Bifenazate	0.5	8.9	2.9
Bifenthrin	0.25	2.6	0.9
Boscalid	0.25	9.5	3.1
Captan	10	282.7	93.3
Carbaryl	0.25	2.2	0.7
Carbofuran	0.25	2.9	1.0
Chlorantraniliprole	0.25	2.6	0.9
Chlorefenapyr	5	72.0	23.8
Chlorpyrifos	1	5.8	1.9
Clofentezine	0.25	4.3	1.4
Coumaphos	0.5	8.8	2.9
Cyfluthrin	5	39.1	12.9
Cypermethrin	0.5	23.4	7.7
Daminozide	0.5	5.4	1.8
Diazinon	0.25	4.1	1.4
Dichlorvos	0.25	2.7	0.9
Dimethoate	0.25	0.9	0.3
Dimethomorph I	0.5	4.0	1.3
Dimethomorph II	0.25	5.5	1.8

The analysis of pesticides and cannabinoids

Table 1 Calculated limits of detection (LOD) and limits of quantitation (LOQ) for the California list of pesticides.—cont'd

Analyte	Spiked concentration (ng/g)	LOQ value (ng/g)	LOD value (ng/g)
Ethoprophos	0.5	6.8	2.2
Etofenprox	2.5	60.7	20.0
Etoxazole	2.5	7.5	2.5
Fenhexamid	0.5	7.0	2.3
Fenoxycarb	0.25	3.5	1.2
Fenpyroximate	2.5	20.8	6.9
Flonicamid	0.25	1.5	0.5
Fludioxonil	0.25	4.4	1.5
Hexythiazox	0.25	5.4	1.8
Imazalil	1	17.5	5.8
Imidacloprid	1	6.6	2.2
Kresoxim–methyl	0.25	4.4	1.5
Malathion A	0.25	4.2	1.4
Metalaxyl	0.25	3.2	1.1
Methiocarb	0.25	3.3	1.1
Methomyl	0.25	0.9	0.3
Methyl Parathion	2.5	72.6	24.0
Mevinphos I	0.25	3.4	1.1
Mevinphos II	1	8.6	2.8
MGK 264 I	0.25	5.8	1.9
MGK 264 II	1	26.1	8.6
Myclobutanil	0.5	10.9	3.6
Naled	0.25	2.7	0.9
Oxamyl	0.25	0.8	0.3
Paclobutrazol	0.5	6.8	2.2
Permethrin trans	2.5	22.2	7.3

Continued

Table 1 Calculated limits of detection (LOD) and limits of quantitation (LOQ) for the California list of pesticides.—cont'd

Analyte	Spiked concentration (ng/g)	LOQ value (ng/g)	LOD value (ng/g)
Permethrin cis	0.25	3.2	1.1
Phosmet	0.25	4.2	1.4
Piperonyl butoxide	0.25	7.1	2.3
Prallethrin	0.5	34.6	11.4
Propiconazole	5	78.4	25.9
Propoxure	0.25	1.8	0.6
Pyrethrins Cinerin I	5	146.6	48.4
Pyrethrins Jasmolin I	25	464.3	153.2
Pyrethrins Pyrethrin I	0.5	6.9	2.3
Pyridaben	0.25	2.6	0.9
Spinetoram	2.5	96.5	31.8
Spinosyn A	2.5	61.6	20.3
Spinosyn D	2.5	75.3	24.8
Spiromesifen	5	33.0	10.9
Spirotetramat	0.5	8.9	2.9
Spiroxamine	0.5	7.0	2.3
Tebuconazole	0.25	15.3	5.0
Thiacloprid	0.25	1.0	0.3
Thiamethoxam	0.25	1.3	0.4
Trifloxystrobin	0.25	1.3	0.4
Fipronil	0.25	4.3	1.4
Aflatoxin B1	1	10.1	3.4
Aflatoxin B2	10	183.3	60.5
Aflatoxin G1	5	64.5	21.3
Aflatoxin G2	25	145.0	47.9
Ochratoxin A	5	39.2	12.9

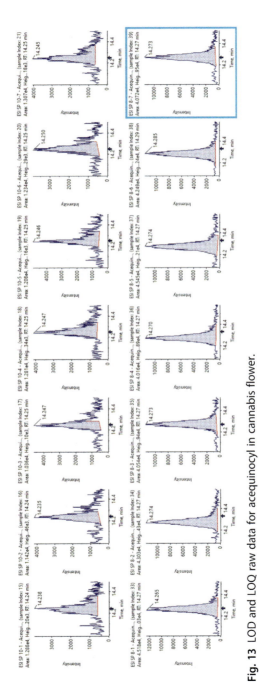

Fig. 13 LOD and LOQ raw data for acequinocyl in cannabis flower.

4. Sample analysis

When a method has been successfully validated, the laboratory may now use this method for the analysis of samples. While the validation process demonstrates that the method can work, it does not assure that the method is working or that the samples were properly prepared. There are several requirements that must be met during samples analysis that assure that the quality of the data are acceptable. When reviewing data from the laboratory all of these requirements, which are described below, should be carefully reviewed to make certain that the analysis was done in accordance with the quality requirements of the method and/or the regulation.

4.1 Quality assurance/quality control

An important component of the analytical procedure is the Quality Assurance/Quality Control procedures that are used to ensure data quality and acceptability. Quality assurance (QA) is a policy that describes what experiments must be done to monitor the entire analytical process and what the acceptable limits for the data are. Quality Control (QC) is the execution of those experiments and the evaluation of the data from those experiments. Every time samples are analysed, there are certain QC samples that are prepared and analysed along with the samples. These QC data are evaluated to demonstrate that the entire analytical procedure is working properly and in control.

The QC that is used to evaluate the quality of the data from a set of samples includes:

- Method blank (MB): The method blank is a sample that is prepared that does not contain any sample at all. This sample is used to prove that there are no pesticides or interferents in any of the reagents or materials used during the analysis.
- Laboratory control sample (LCS): Often referred to as a blank spike sample. The LCS is a sample that is either used with a clean matrix of the type being analysed or often is simply a method blank that is spiked with a known concentration of the analytes. This sample is important for demonstrating that in the absence of matrix, the analysis results in an accurate quantitation and shows that the method is operating correctly. If a MB or LCS fails to meet the method requirements the batch may need to be prepared again. The exception to this are situations where the recovery of one or more of the pesticides in the LCS are higher than allowed and

there are no detections for the samples in that batch. In this case a high bias does not impact the quality or usability of the reported results.

- Matrix spike (MS) and matrix spike duplicate (MSD): Two sample aliquots of the same sample are measured out and spiked with a standard mix prior to extracting the samples. The samples are then extracted and analysed. The MS and MSD provide information about the accuracy of the extraction in the matrix that is being tested and also provides information about the precision of the analysis. In addition to calculating the percent recovery of the pesticides that were spiked into the matrix, the relative percent difference (RPD) is calculated for each pesticide that was spiked.
- Sample Duplicates (SD): In every batch of analysed samples, a sample duplicate is prepared. The SD is when two portions from the same sample are prepared separately. The results from these two analyses are compared and if there are any target analytes observed, the two results should agree, typically within 20% of each other. The SD provides valuable information about the homogeneity of the sample being analysed.
- Continuing Calibration Verification (CCV): The CCV is when one of the standards used for the calibration curve are analysed during the analysis. Typically, a CCV is analysed every 10 samples and the standard used for each CCV varies throughout the sequence to demonstrate the validity of the calibration across the concentration range. The accuracy of each of the pesticides must be within the accuracy specified in the regulation.

4.2 Sample results

A lot of work is involved to demonstrate that the method is fit for purpose and that during sample analysis the instrument is operating correctly. If all of the QC requirements of the method have been met for a batch of samples, then, finally, comes the analysis and evaluation of actual sample data. A typical sample results table is shown in Fig. 14. This example is intended to demonstrate how a laboratory reviews and evaluates the data from their sample analysis. The data in Fig. 14 is for the insecticide Imidacloprid. In the column, Sample Name, are the various samples that have been analysed in the sequence. The sample types are either Standard or Unknown. The unknown samples are cannabis flower that have been analysed. In the column, Calculated Concentration, is the *extract* concentration in parts per billion. The sample preparation procedure for this

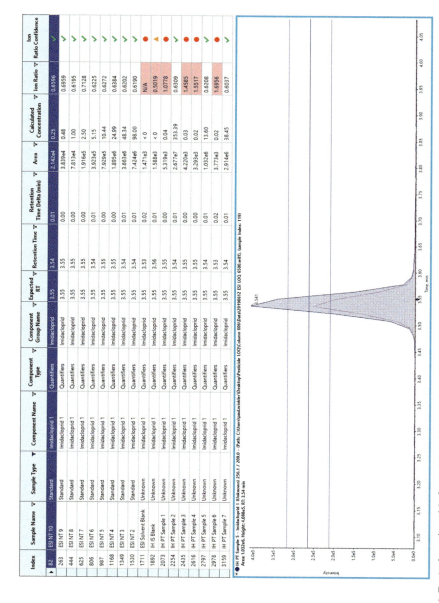

Fig. 14 Typical results table for an LC-MS/MS analysis showing ion ratio and retention time fields in the top pane and an imidacloprid peak for one of the samples in the bottom pane. The peak shows acceptable ion ratios.

example requires that the extract concentration be multiplied by 10 to calculate the actual sample concentration in the plant. Several of the samples have a quantitated result for Imidacloprid while others do not. The results for IH PT Sample 1, for example, has a concentration of 0.04 ppb but the calculated ion ratio shown in the Ion Ratio column is shaded pink, indicating that while a signal was observed, the quantitation and qualifier ion ratios were outside of the acceptance range. The Ion Ratio Confidence column has a red dot in the column also indicating that the ratios are not acceptable.

Other results, such as for sample IH PT Sample 5 and IH PT Sample 2, have a green check mark in the Ion Ratio Confidence column and the Ion Ratio result is not shaded. This is an indication that the ion ratio for the two ions was acceptable. This data alone is not enough to positively identify the signal as Imidacloprid. The compound must also have the correct retention time. In the column Expected RT, it can be seen that Imidacloprid is expected to have a RT of 3.55 min. The Retention Time column shows the actual RT for Imidacloprid for that sample. The Retention Time Delta column next to the RT column shows the time difference between the expected RT and the observed RT. Typically, this value should be 0.06 or less. As can be seen, the observed RT's are all within 0.06 min. Results that have an acceptable ion ratio and the correct RT may be reported on the COA, if the quantitated value is at or above the action limit that is required. The laboratory will also look at the peak from the results table to ensure that the correct peak has been integrated and that the peak has been integrated correctly. The peak in Fig. 14 is for IH PT Sample 5.

The calculated concentration for this sample is 136 ppb in the flower sample. The peak for the quantitation and qualification ion are integrated correctly and the ratio is within the acceptance limits. This sample can be confidently considered to contain imidacloprid, but does the result need to be reported on the COA?

The decision to report a compound on the COA is dependent on the action limit concentration requirements of the particular regulation that the sample is being tested against and can vary widely. Imidacloprid has an action level of 5000 ppb for inhalable samples for the State of California. If the samples in Fig. 14 were being tested for the state of California, the results for all of the samples would be listed as less than the action limit (<5) and would pass the testing criteria even though Imidacloprid is present in the sample. If the samples were being tested for Canada, the results would be above the action limit of 10 ppb, the result would be reported on the COA and the samples would fail the testing criteria.

When evaluating a result on the COA, the raw data should be reviewed carefully. This review should include an inspection of the peak to ensure that ion ratios are correct, that the retention time is correct and the actual peak itself should be inspected to make sure that it is a real peak and not noise, and that it has been integrated correctly. If the criteria above are satisfied and if the quality control results indicate that the sample preparation and analysis are in control, then the data user can have confidence that the reported result is real and is correct.

4.3 Cannabinoid analysis

All samples must be analysed for the content of cannabinoids. The most common requirements are to analyse for the presence and concentration of the main five cannabinoids in the sample. These are Δ9-Tetrahydrocannabinol (THC), Δ9-Tetrahydrocannabinolic acid (THCA), Cannabidiol (CBD), Cannabidiolic acid (CBDA) and Cannabinol (CBN). These analytes are typically determined using liquid chromatography with ultraviolet detection (LC/UV) due to their relatively high sample concentrations and the ease of use of the technique.

This class of compounds are also easily done using an LC-MS/MS or an LC-UV/MS/MS instrumental configuration. To the analytical chemist, cannabinoids are just another group of compounds to be analysed and method development is the same as it is for the pesticides. The same method validation procedures, analytical sequence QC and data review procedures apply to the analysis of cannabinoids as to the pesticides. Unlike pesticides, however, that are expected to either not be present in the sample or present at low concentrations, the analysis of cannabinoids is more complicated because of the wide difference in concentration between the major and minor cannabinoids. When performing this analysis using LC/UV, two injections at two different dilutions are required as a more concentrated extract is required for the minor cannabinoids. Even then, there are uncertainties in the data because LC/UV is not a highly specific technique and identification is accomplished solely on the basis of retention time.

The challenge of analysing cannabinoids with LC-MS/MS is related more to the very high concentrations of the main five cannabinoids found in the samples. This high concentration causes the mass spectrometer to saturate quickly and therefore it is difficult to analyse a wide range of concentrations. By comparison, UV has a much higher range of concentrations before saturation, but it is far less selective than mass spectrometry.

A typical chromatogram from the LC-MS/MS analysis of cannabinoids using LC-MS/MS is shown in Fig. 15. The analysis of the cannabinoids is usually performed using negative ion polarity for the acidic compounds, as they are negative ions in solution and are much more sensitive in this mode. Therefore, the peaks in Fig. 15 show both positive and negative ion data.

It is also important for full chromatographic separation of some of the cannabinoids, such as THC and CBD or THCA and CBDA, because they have the same quantifier and qualifier transitions and would therefore interfere with each other. Fortunately, the separation of these cannabinoids is not difficult.

When the analysis of a full range of cannabinoids in the same injection is desirable, a useful means to achieve accurate quantitation is to use LC-UV/MS/MS, where there is an ultraviolet detector between the column and the mass spectrometer. The UV signal is used for the quantitation of the major cannabinoids and the mass spectrometer is used for the quantitation of the minor cannabinoids. An example of how this is accomplished is shown in Fig. 16.

In this figure, two different calibrations ranges are shown for CBD. The concentrations for the mass spectrometer is 0.1, 0.25, 0.5, 2.5, 5 and 10 ppb while the calibration curve for the UV is essentially 100 times more concentrated. This allows for using either the UV detector or the mass spectrometer for quantitation. As example, if the sample is a marijuana sample, a low CBD concentration may be expected and if the sample is hemp, a high CBD concentration is expected. Using LC-UV/MS/MS, it is possible to successfully analyse both types of samples in the same sequence and dilution. The data system can be programmed to determine which detector to use to report the result.

The advantage of using the LC-UV/MS/MS for the analysis of minor cannabinoids is shown in Fig. 17. This example is for the analysis of cannabidivarin, a compound thought to have anticonvulsant properties. On the left side of the figure, the light blue trace is a wavy line but there does appear to be a possible peak at the retention time for CBDV, but it would be difficult to have confidence that this was actually CBDV and it would also be difficult to accurately quantitate this peak. On the right side of the figure is the signal for the mass spectrometer. There is clearly a large peak that would be satisfactory for quantitation. This is a demonstration of the importance of specificity when analysing samples from a complex matrix.

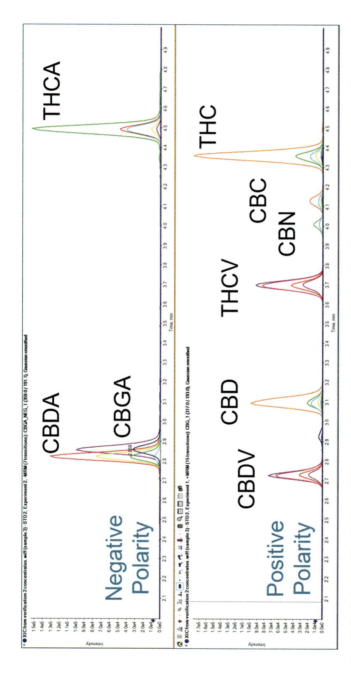

Fig. 15 Typical cannabinoid chromatogram using LC-MS/MS. The use of both negative and positive ions is needed for this analysis.

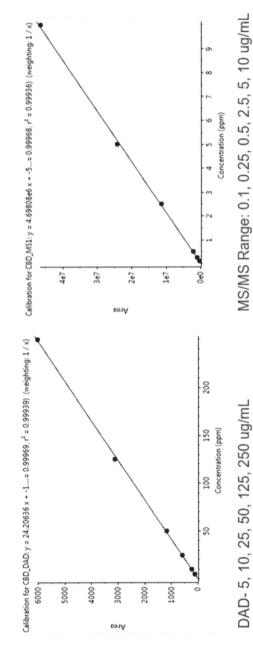

Fig. 16 Two CBD calibration curves, one for the mass spectrometer and one for the UV detector. This approach shows linearity for two different concentration ranges.

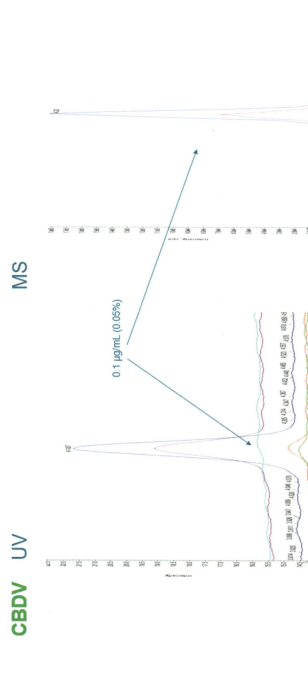

Fig. 17 LC/UV and LC-MS/MS data for CBDV in cannabis flower matrix. The left pane shows the data for the UV detector and the right pane shows the data for the mass spectrometer. The light blue trace is the wavelength for CBDV. A large difference in signal is observed between the UV and mass spectrometer.

5. Summary

The intention of this chapter was to provide a simple, yet useful, explanation of the analysis of pesticides and cannabinoids to a non-technical reader. The information that has been discussed will hopefully provide the reader with a basic understanding of the technology that is used for the analysis. This is important information to have when discussing results with a laboratory.

The concepts of method validation and quality control processes were discussed to provide a basic knowledge of how and why a laboratory decides a result is valid and how they determine to report data. With this knowledge, the data end user can have an informed discussion with a laboratory about the results that have been reported.

Finally, it is important to have the ability to ask the laboratory to provide the opportunity to show raw data and not simple numbers on a form. The discussions in this chapter will provide the end user with some understanding of how that data appears and will hopefully remove some of the mystery of this complex analysis.

References

[1] http://www.ncsl.org/research/health/state-medical-marijuana-laws.aspx.
[2] https://www.colorado.gov/pacific/sites/default/files/Section%2016%20-%20%20Retail.pdf.
[3] http://www.oregon.gov/oha/ph/PreventionWellness/marijuana/Documents/oha-8964-technical-report-marijuana-contaminant-testing.pdf.
[4] https://bcc.ca.gov/law_regs/cannabis_order_of_adoption.pdf.
[5] https://www.canada.ca/en/public-health/services/publications/drugs-health-products/cannabis-testing-pesticide-list-limits.html.
[6] https://www.sos.state.co.us/CCR/GenerateRulePdf.do?ruleVersionId=8439&fileName=1%20CCR%20212-3.
[7] S. Krepich, J. Layne, Determination of Pesticide Residues in Cannabis by LC-MS/MS, Phenomenex Application Note TN-1224 (2017). https://az621941.vo.msecnd.net/documents/7c79d055-9844-4369-b7f0-c9c621f97c02.pdf.
[8] M. Anastassiades, S.J. Lehotay, D. Stajnbaher, F.J. Schenck, Fast and easy multiresidue method employing acetonitrile extraction/partitioning and "dispersive solid-phase extraction" for the determination of pesticide residues in produce, J. AOAC Int. 86 (2) (2003) 412–431.
[9] J. Kowalski, J. Dahl, A. Rigdon, J. Cochran, D. Laine, G. Fagras, Evaluation of modified QuEChERS for pesticide analysis in cannabis, LCGC North Am. 35 (2017) 8–22.
[10] X. Wang, D. Mackowsky, J. Searfoss, M. Telepchak, Determination of cannabinoid content and pesticide residues in cannabis edibles and beverages, LCGC 34 (2016) 20–27.
[11] Bureau of Cannabis Control Text of Regulations, https://bcc.ca.gov/law_regs/cannabis_order_of_adoption.pdf.

CHAPTER TEN

Cannabis and hemp analyzers for improved cannabinoid potency accuracy and reproducibility

Masayuki Nishimura[a], Tairo Ogura[a], Yohei Arao[a], Taka Iriki[a], Craig Young[a], Andy Sasaki[a], Bob Clifford[a], A.J. Harmon-Glaus[a], Raz Volz[a], Niloufar Pezeshk[a], Jeff Dahl[a], Will Bankert[a], Paul Winkler[a], Max Wang[a], Jordan Frost[a], Sandy Mangan[b], John Easterling[c,d], Scott Kuzdzal[a,*]

[a]Shimadzu Scientific Instruments, Columbia, MD, United States
[b]SPEX SamplePrep LLC, Metuchen, NJ, United States
[c]Happy Tree Microbes, Los Angeles, CA, United States
[d]Laughing Dog Farms, Portland, OR, United States
*Corresponding author: e-mail address: sakuzdzal@SHIMADZU.com

Contents

1. A new era of medical cannabis and cannabis quality control testing emerges	316
2. The importance of sample homogenization	319
3. Potency testing instrumentation considerations	320
4. The development of cannabis and hemp analyzers	320
5. Development of easy to use, fit for purpose overlay software	323
6. Ability to operate in 21 CFR 11 compliance mode	324
7. Sample preparation and performance data	325
8. Cannabis flower sample preparation	325
9. Important notes regarding cannabis flower sample filtration	325
10. Proper handling of cannabinoid standards	326
11. Methods section	326
11.1 Unknown cannabis sample preparation	327
12. Calibration of the HPLC system by use of a standard solution	328
13. Standard curves	329
14. Total THC potency formula	330
15. Application of the cannabis analyzer for the quantitative determination of cannabinoids in cannabis flower and edible products	331
16. The need for a hemp analyzer	333
17. Future directions of cannabis and hemp analyzer development	333
18. Expansion to a 15-cannabinoid standards analysis	334

Comprehensive Analytical Chemistry, Volume 90
ISSN 0166-526X
https://doi.org/10.1016/bs.coac.2020.04.010

© 2020 Elsevier B.V.
All rights reserved.

315

19. 'Full Spectrum' cannabis and automated analysis of tinctures/oils	334
20. Summary	336
Disclaimers	336
References	337

1. A new era of medical cannabis and cannabis quality control testing emerges

The San Francisco Cannabis Buyers Club, which opened in 1992 after the passing of Proposition P, was the first public marijuana dispensary in the United States. Later the passage of California's Proposition 215 in 1996 was another significant victory for medical marijuana, ushering in a new era with an increased need for quality, and lab tested cannabis products. In 2008, 12 years after Proposition 215 passed, the medical marijuana advocacy group 'Americans for Safe Access' estimated that California had more than 200,000 doctor-qualified medical cannabis users. By July 2019, the Marijuana Policy Project estimated the total number of all state-legal U.S. medical cannabis patients to be 3,099,934. This rapid growth in medical cannabis patients created the demand for cannabis quality-control (QC) testing laboratories dedicated to performing analytical testing on cannabis products to ensure that state requirements were being met.

Cannabis QC testing involves a variety of different analytical tests, and the requirements for each test vary from state to state. As of March 2020, cannabis remains federally illegal, which forces each state to set their own requirements for cannabis testing. Cannabis QC analytical testing ranges from visual inspection for contaminants and chromatographic determination of cannabinoid potency to mass spectrometric determination of low level contaminants such as pesticides, residual solvents and heavy metals (see Table 1 for a partial listing of key cannabis QC analytical testing types). The dynamic range comparisons of various cannabis analytical tests are illustrated in Fig. 1. Since the more abundant cannabinoids are typically in the mg/g range, HPLC or UHPLC with UV or PDA detection provides sufficient cannabinoid detection sensitivity. Instrument manufacturers are also demonstrating the utility of liquid chromatograph mass spectrometry (LC-MS) for ultra low-level cannabinoid detection [1]. Other technologies, such as handheld mid-IR spectroscopy instruments, have shown utility for cannabinoid measurements [2]. While these technologies have benefits such

Table 1 A partial listing of some of the key analytical test types offered by cannabis QC testing labs. The last four tests represent screens for potentially dangerous contaminants.

Cannabis QC analytical testing

Potency testing
Terpene profiling
Moisture content
Genetics (cultivar typing)
Pesticide screening
Residual solvents
Mycotoxins/aflotoxins
Heavy metals

as portability, ease of use, reduced cost and less sample preparation, they generally lack the accuracy and reproducibility of HPLC platforms and have been relegated for use as an internal quality control for growers.

'Potency testing' refers to the accurate quantitation of cannabinoids in cannabis products. Cannabis plants do not produce THC, they produce THCA, the carboxylic acid form of THC. THCA is converted to THC through a process known as decarboxylation. THC was first discovered in 1964 at the Weizmann Institute of Science in Rehovot, Israel, by Dr. Raphael Mechoulam and his colleagues, Drs. Yehiel Gaoni and Haviv Edery. While determination of THC is a primary potency test, it is important to note that there are hundreds of cannabinoids present in cannabis. The total number of cannabinoids being both discovered by researchers (as well as those being reported by cannabis QC testing labs) have been steadily increasing since 1996.

Cannabis was widely utilized in the United States as a patent medicine during the 19th and early 20th centuries, but the first written accounts of the medicinal value of cannabis date back to Shen-Nung (c. 2700 B.C.), who is recognized as the Father of Chinese Medicine. Medicinal uses of cannabis were described in the *United States Pharmacopoeia* in 1850. Federal restriction of cannabis use and cannabis sale first occurred in 1937, and cannabis was removed from the *United States Pharmacopoeia* in 1942.

The total number of cannabinoids in cannabis is unknown and various literature sources provide different numbers. Most manuscripts report

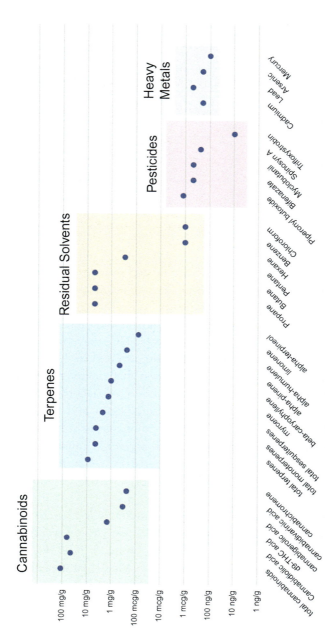

Fig. 1 Cannabis component dynamic range comparisons. Cannabinoid and terpene data extracted from Polyploidization for the Genetic Improvement of *Cannabis Sativa*. Frontiers in Plant Science, published 30 April 2019. Article 476. Residual solvents, pesticides and heavy metal data extracted from California State regulation limits.

between 113 and 150 different cannabinoids [3,4]. While the physiological importance of cannabinoids is actively under investigation, it is generally believed that phytocannabinoids (i.e., naturally occurring cannabinoids in nature) mimic the endocannabinoid anandamide, which is produced endogenously within our bodies. Anandamide (or N-arachidonoylethanolamine) binds to receptors in our endocannabinoid system, acting as a neurotransmitter and mood-enhancer. Anandamide interacts with cannabinoid receptors in the brain and it plays key roles in memory, thought processes, movement control, as well as management of appetite, pain and fertility.

2. The importance of sample homogenization

It is important to note that many pre-analytical variables directly impact potency values, from the genome of the cannabis plant and various cultivation variables such as soil, water, nutrients, lighting, temperature and humidity to the conditions in harvesting and drying. In order to achieve accurate sampling, cannabis samples should be properly homogenized to ensure accurate test results. In the modern cannabis analytical QC lab homogenization is achieved through mechanical milling.

Good sample preparation is a crucial step and can be a main factor in the outcome of your final cannabis test results. Consistency, reproducibility and the elimination of cross contamination are key in any industry. Sample homogenization is critical, and mechanical mills are an ideal solution for tough or temperature sensitive materials. A freezer mill like the SPEX Freezer/Mill® cools cannabis samples to cryogenic temperatures, then pulverizes them by magnetically shuttling a steel impactor back and forth against two stationary end plugs. This allows for a complete sample to be ground for true homogenized representative sampling. These mills work for multiple forms of cannabis products; even in the toughest samples, like gummy bears or cookies and granola bars, which may include raisins and nuts. The milling takes only 1–2 min and results in a fine-powered sample with an enlarged surface area, making it much easier to extract the THC for more accurate potency results.

Another powerful homogenizing tool commonly employed prior to potency testing is the Geno/Grinder®, a high throughput mechanical homogenizer platform which can accommodate multiple sample formats, from microtiter plates to 80 g batch samples. Cannabis samples are typically ground within 1 min. Minimal heat is generated in this time frame, so there is no loss of volatiles compounds nor decarboxylation of cannabinoids.

Sample preparation should not be the primary bottleneck in the lab. Using mechanical disruption methods helps streamline the whole sample preparation step. It is not only the very first step, but it is often considered the most important step.

3. Potency testing instrumentation considerations

Cannabis is a complex composite containing hundreds of different types of compounds. Also, cannabis products can range from flower (or dry product) to oils/tinctures and a large variety of edibles, beverages, topicals, suppositories, etc.

High Performance Liquid Chromatography (HPLC) has emerged as the gold standard for potency determinations because separation and detection of the cannabinoids is completed without causing any decomposition of the naturally abundant THCA. This acid form undergoes decarboxylation to THC under the influence of heat and light. HPLC is preferred over gas chromatography (GC) because the heated source in GC can decarboxylate THCA and other cannabinoids. It is for this reason that GC typically reports 'total THC' and not individual values for THCA and THC.

HPLC can utilize many forms of detectors, from ultraviolet spectroscopy (UV) and photodiode array (PDA) to various types of mass spectrometers for increased sensitivityand selectivity. HPLC with UV detection is sufficient for potency determinations of most cannabis products in the range illustrated in Fig. 1. LC–MS or LC–MS/MS will play an important role in higher sensitivity applications, such as the determination of cannabinoids and their metabolites in biological fluids.

4. The development of cannabis and hemp analyzers

The term 'analyzer' has been used to describe a platform that is comprised of analytical hardware, a simplified, easy-to-use interface and the consumables/reagents necessary for analysis. Clinical chemistry analyzers, also referred to as biochemistry analyzers, utilize various measurement technologies in a streamlined, built-for purpose (often referred to as 'push–button' or 'turn–key') approach to simplify the analysis of clinical samples such as whole blood, sera, plasma, and urine. Chemistry analyzers are utilized in many types of laboratories, from small point-of-care clinics to high-throughput clinical labs in hospitals. While there are many types of analyzers employing an even wider variety of analytical technologies, they all have similar attributes. Analyzers reduce the amount of instrument parameters and enable technicians to essentially load samples and initiate analysis faster. By streamlining analysis, they frequently reduce analytical variables and thereby increase intra- and inter-lab reproducibility.

Today's HPLC instruments offer great flexibility in their operation and method parameters such as column selection, mobile phase reagents, separation gradients, etc. These parameters are typically optimized for many different sample types from small molecules to proteins and polymers. As illustrated in Fig. 2, method development for a set of analytes such as cannabinoids can take three or more months. In the early days of cannabis QC laboratories, most labs purchased an HPLC and developed their own, proprietary methods for cannabis potency testing. Because each lab was using different methods and conditions, different laboratories were obtaining different results for the same samples [5].

A significant advantage of analyzers is that they can greatly reduce the time it takes to get up and running. A new cannabis QC testing laboratory can employ HPLC instruments using their own method development and be operational in three to 6 months. Alternatively, they can be trained on a turn-key analyzer in under 1 day and achieve exceptional reproducibility.

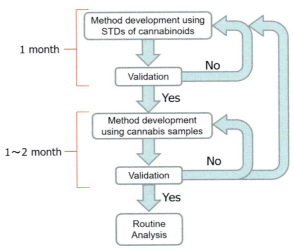

Fig. 2 HPLC method development includes development on standards, followed by validation and application to real world samples, often exceeding 3 months of laborious work. Implementation of a turn-key analyzer approach can enable greater cannabinoid potency reproducibility and accuracy while eliminating the need for method development. Operational training on a cannabis analyzer can be achieved in under 1 day while maintaining exceptional accuracy and reproducibility.

Another major benefit here is that manufacturer support for a customer with an analyzer is greatly simplified over that for an HPLC with proprietary methods. Because the analyzer sample prep methods, analytical conditions and consumables/reagents have been pre-optimized and field-tested, technical support is greatly simplified.

A multi-year project was undertaken by the Shimadzu Scientific Instruments USA Innovation Center under the guidance of Dr. Masayuki Nishimura, with the goal of developing cannabis and hemp analyzers for potency. Shimadzu Innovations Center scientists (shown in Fig. 3) optimized separation conditions for the 11 cannabinoids listed in Table 2.

One dozen cannabis QC testing laboratories were surveyed in 2015 and this target list (see Table 2) was designed to satisfy the testing needs of the most demanding QC labs. This 11-cannabinoid mix is provided along with the analyzers for both convenience and enhanced reproducibility. This team of engineers also specified the hardware and software requirements necessary and worked with a team of software engineers to create an overlay software that was easy to use. In addition, the three packages with unique modes of analysis below were desired to suit the needs of different laboratories, from QC testing labs to research labs.

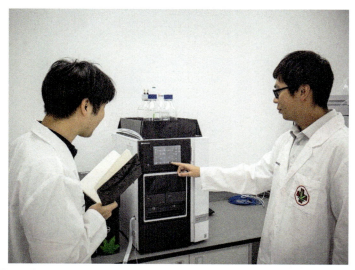

Fig. 3 Shimadzu Innovation Center engineers Tairo Ogura (left) and Yohei Arao developing the Cannabis Analyzer for Potency.

Table 2 The original 11 cannabinoid target compound list for the Cannabis Analyzer.

Target compound list

Tetrahydrocannabivarin (THCV)
d8-Tetrahydrocannabinoid (d8-THC)
d9-Tetrahydrocannabinoid (d9-THC)
d9-Tetrahydrocannabinolic acid (THCA)
Cannabidiol (CBD)
Cannabidiolic acid (CBDA)
Cannabidivarin (CBDV)
Cannabinol (CBN)
Cannabigerol (CBG)
Cannabigerolic acid (CBGA)
Cannabichromene (CBC)

1. *High throughput method*

   Designed for analysis of the 10 most commonly requested cannabinoids in under 8 min. This is the original method developed in collaboration with industry laboratories. (Does not include THCV.)

2. *High sensitivity method*

   Adds THCV to the target analyte list, with an instrument cycle time of under 10 min. The short analysis time produces the sharpest chromatographic peaks for the best overall sensitivity.

3. *High resolution method*

   Presents full baseline resolution for 11 cannabinoids and an analysis time under 30 min. This method is preferred for research purposes, or when additional compounds must be added to the analysis in response to new state regulatory requirements.

5. Development of easy to use, fit for purpose overlay software

Modern HPLC software allows the user to access and control large numbers of instrument and method parameters. However, in an analyzer, these parameters are well defined and optimized for any particular set of

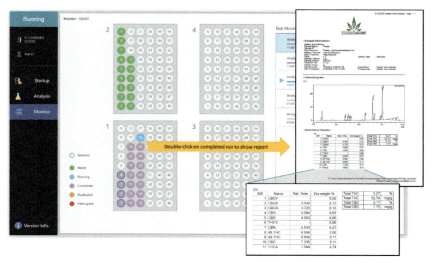

Fig. 4 Screenshot of Cannabis Analyzer for Potency user interface with insets of potency report and cannabinoid quantitation table (showing total THC at 5.27%).

analytes. Additionally, an analyzer interface should be much simpler to operate so that training is minimized and running of samples is enhanced.

Development of 'overlay' software, or software that sits over a higher functionality instrument control software platform is a powerful approach to streamlining analyzer operation. A team of Shimadzu Innovation Center engineers developed an easy to use overlay software with graphical screens for startup, analysis and monitoring modes. This enables users to load samples into the analyzer, easily enter sample information, as well as expedite calibration and analysis. Another critical component of an analyzer is the ability to generate meaningful sample reports in an automated manner.

The monitoring mode of the user interface of the Cannabis Analyzer for Potency software is shown in Fig. 4. Colour coded wells indicate the status of samples (ready, running, completed, etc.). Users can double-click on any completed run to view the corresponding sample report.

6. Ability to operate in 21 CFR 11 compliance mode

Title 21 CFR Part 11 is the part of Title 21 of the Code of Federal Regulations that establishes the United States Food and Drug Administration (FDA) regulations on electronic records and electronic signatures. Part 11, as it is commonly called, defines the criteria under which electronic records and electronic signatures are considered trustworthy, reliable, and equivalent to paper records (Title 21 CFR Part 11 section 11.1 (a)) [6]. While cannabis QC testing labs do not currently need 21CFR11 compliance, it was determined that this would be essential for any government work, or if cannabis becomes federally legal in the coming years.

The Cannabis Analyzer for Potency is 21CFR11 compliant in LabSolutions mode. This mode is widely utilized in environmental and pharmaceutical laboratories.

7. Sample preparation and performance data

The sections below provide details regarding sample preparation and performance criteria including standard curves.

8. Cannabis flower sample preparation

Weigh 200 mg of flowers or cuttings into a 50 mL centrifuge tube. Add two 9.5 mm steel balls into the tube. Shake at 1000 rpm for 1 min with the SPEX 2010 Geno/Grinder®. Add 20 mL of methanol to the tube. Shake at 1000 rpm for 1 min.

Wait for 15 min. Mix using a vortex mixer for 1 min. Transfer 1 mL of the mixture into a 1.5 mL micro-tube and centrifuge at 3000 rpm for 5 min.

Transfer 100 μL of supernatant to a new 1.5 mL micro-tube. Add 900 μL of methanol. Filter the mixture through a 0.45 μm syringe filter and transfer to a 1.5 mL sample vial.

9. Important notes regarding cannabis flower sample filtration

Sample filtration is a critical step in preparing cannabis and hemp samples for HPLC potency analysis. Fine particles must be removed to make the sample suitable for HPLC injection, following extraction into a suitable solvent. Syringe filters, while effective for particulate removal, can sometimes be problematic in terms of analyte adsorption, resulting in some loss of target recovery.

A filtration efficiency study was conducted by analysing 10 separate replicates per syringe filter containing a 10 ppm spike of 11 phytocannabinoids. The results show that the Nylon, Polypropylene, and PTFE syringe filters were ideal candidates, as they presented minimal hold-up of the phytocannabinoids and stable recoveries among 10 replicates (Nylon, Polypropylene, and PTFE showed a %RSD of 1.30, 1.16, and 1.27, respectively). The CA, PES and both hydrophilic- and hydrophobic-PVDF syringe filters showed a %RSD of 1.86, 2.64, 4.90 and 1.55, respectively. Notably, a clear correlation between the hydrophilicity/hydrophobicity of the syringe filter's material properties and the concentration of the cannabis recovery was observed. This implies that hydrophilicity does impact filtration in a statistically significant manner. For a detailed discussion on filter types including additional data on recoveries, see reference [7].

10. Proper handling of cannabinoid standards

When preparing standards for use in a calibration curve, the common practice is to use Certified Reference Material (CRM) from an accredited ISO-17034 manufacturer. Labs rely on these standards so that they can accurately quantitate samples. These standards typically come from multiple vendors and this practice has been shown to be acceptable when ISO-17034 guidelines are followed by the standard manufacturer. It has been demonstrated that adhering to these guidelines enables reciprocation among reference material manufacturers, allowing the end user to obtain valid and reliable results [8].

Cannabis standards are available from multiple vendors including Shimadzu, Cayman Chemicals, LGC, and other suppliers in both single component and multi-component mixtures. At this time, the largest ISO-17034 CRM mixture is a 11-component cannabinoid mixture offered by Shimadzu. Testing has verified that a pre-made CRM mixture, when properly maintained and handled, can be used as accurately as a comparative set of single CRM components. Additionally, it has been demonstrated that premade multicomponent CRM mixtures improve quantitation accuracy by avoiding some of the preparation errors that can occur while preparing a stock standard using single-component CRMs [9–11]. A pre-made mixture offers a simpler approach, which can save time and effort, while giving the user confidence in their analytical data.

11. Methods section

A Cannabis Analyzer for Potency (Shimadzu) was used to determine the cannabinoid content of flowers, leaf and trim, concentrates, edibles and

Accuracy and reproducibility   327

other samples. Eleven cannabinoids, Δ9-THCA, Δ9-THC, CBDA, CBD, CBN, CBC, CBGA, CBG, CBDV, THCV and Δ8-THC, can be analysed in a single chromatographic run with a cycle time of approximately 8 min.

Pre-modified mobile phase specific to the following methods is available for purchase directly from Shimadzu. The 'high sensitivity' method of the Cannabis Analyzer was utilized. Only HPLC or UHPLC grade water and acetonitrile should be used for mobile phase. Standards Mixture Certified Reference Material (CRM) standards may be purchased directly from Shimadzu.

A calibration curve was generated using six standards in the calibration range: 0.5–100 mg/L. First the 100.0 mg/L standard mixture (11 components) was made by transferring 400 μL of the standard mixture to an HPLC 1.5 mL vial and adding 600 μL of methanol. This solution was vortexed for 30 s. For the 50.0 mg/L standards mixture, 500 μL of 100.0 mg/L standard mixture was transferred to an HPLC vial and 500 μL of methanol was added. This solution was vortexed for 30 s. For the 10.0 mg/L standard mixture, 200 μL of 50.0 mg/L standard mixture was transferred to an HPLC vial and 800 μL of methanol was added. This solution was vortexed for 30 s. The 5.0 mg/L standard mixture was made by transferring 500 μL of 10.0 mg/L standard mixture to an HPLC vial and adding 500 μL of methanol. This solution was vortexed for 30 s. Next the 1.0 mg/L standard mixture was made by transferring 200 μL of 5.0 mg/L standard mixture to an HPLC vial and adding 800 μL of methanol. This solution was vortexed for 30 s. Lastly, the 0.5 mg/L standard mixture was made by transferring 500 μL of 1.0 mg/L standard mixture to an HPLC vial and adding 500 μL of methanol. This solution was vortexed for 30 s.

11.1 Unknown cannabis sample preparation

Cannabis flowers (dry product) were prepared using a 2010 Geno/Grinder (SPEX SamplePrep, LLC) and shaker (to powder the samples) using the following procedure:

1. Weigh 200 mg into a 50 mL centrifuge tube.
2. Transfer two 9.5 mm O.D. steel balls into the tube.
3. Shake at 1000 rpm for 1 min using a 2010 Geno/Grinder.
4. Add 20 mL of methanol to the tube. (Extraction Vol. 20 mL).

5. Shake at 1000 rpm for 1 min using a 2010 Geno/Grinder.
6. Wait 15 min.
7. Mix using a vortex mixer for 1 min.
8. Transfer 1 mL of solvent into a 1.5 mL microtube, and centrifuge at 3000 rpm for 5 min.
9. Transfer 50 µL of supernatant into a new 1.5 mL microtube.
10. Add 950 µL of methanol and vortex for 30 s. (20 times dilution, Dilution Factor: 20).
11. Filter using a 0.45 µm syringe filter, and transfer into a 1.5 mL HPLC vial.
12. The sample is now ready for analysis using the Analyzer.

12. Calibration of the HPLC system by use of a standard solution

Rapid, automated and accurate calibration of an analyzer is essential for achieving high accuracy potency values. Fig. 5 shows the chromatogram of 10.0 mg/L standard mixture. Gradient elution conditions with acid modified water and acetonitrile were employed with a C18 column chemistry to achieve the separation in under 10 min.

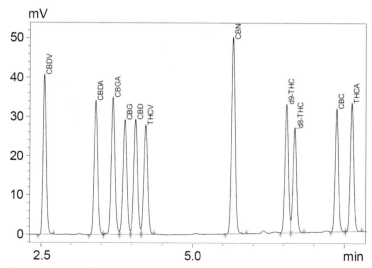

Fig. 5 Chromatogram of a 10.0 mg/L standard mixture.

13. Standard curves

Standard curves (Fig. 6) were prepared for each target analyte with a minimum acceptable correlation coefficient (R2) of 0.999 over 6 standard levels. A linear dynamic range was established at 0.5–100 mg/L (0.05%–10%) in each analyte except THCA and CBDA. In many cases, the abundance of THCA and CBDA in plant material is exceedingly high, therefore the linear dynamic range for those analytes was established from 0.5 to 250 mg/L (0.05%–25%).

The definition of Accuracy% appears below.
- Accuracy% = Cr/Cc × 100.
- Cr: Concentration value from the quantitative calculation.
- Cc: Standard concentration value of the corresponding level in Compound Table.

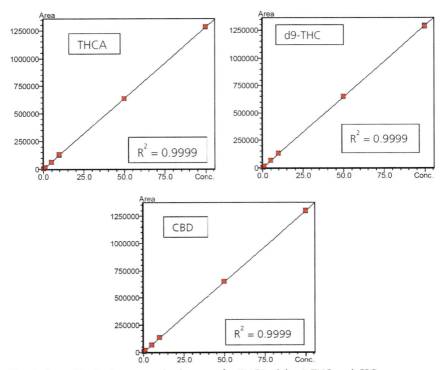

Fig. 6 Cannabis Analyzer standard curves for THCA, delta-9 THC and CBD.

Table 3 Excellent accuracy % of cannabinoids in a 70 ppm standard mixture.

70 ppm	Conc. (mg/L)	
#	Name	Avg. accuracy (%)
1	CBDV	100.03
2	CBDA	99.47
3	CBGA	99.39
4	CBG	100.09
5	CBD	99.96
6	THCV	100.09
7	CBN	99.73
8	d9-THC	100.21
9	d98-THC	99.82
10	CBC	99.89
11	THCA	99.17

From the results of Table 3, the quantitation accuracy of all compounds are within $\pm 3\%$ for low, mid and high quantitation points. Thus, it was confirmed that active ingredients in cannabis flowers could be quantified accurately by using the corresponding calibration curves.

14. Total THC potency formula

The maximum amount of THC present in any cannabis sample can be calculated by a specific formula with great accuracy, *but* the calculation assumes 100% conversion of THCA into THC, which will only occur under ideal conditions. The total THC potency is defined as the molar sum of delta-9 THC and delta-9 tetrahydrocannabinolic acid (THCA).

The potency calculation below is necessary to determine the total THC potency. For a more detailed discussion of these calculations, see reference [8].

$$\%THCA = [THCA] \times (DIL) \times (VOL/MG) \times 100$$
$$Potency : (\%THCA \times 0.877) + \%\Delta 9 - THC$$

[THCA]: Concentration of THCA,
DIL = Dilution Factor, VOL = External Volume,
MG = dry sample weight (mg),
0.877 = molecular weight ratio of cannabinoids to cannabinoid acids.

15. Application of the cannabis analyzer for the quantitative determination of cannabinoids in cannabis flower and edible products

This new analyzer was tested in state approved cannabis QC laboratories using a variety of cannabis products. Fig. 7 shows the chromatogram of an extract from a cannabis flower sample (with overlay of cannabinoid standards). The analyzer provides exceptional accuracy and run to run reproducibility. Edible samples, including gummy bears, were homogenized using the SPEX Freezer/Mill® (Fig. 8) and prepared using the sample preparation methods described in the Cannabis Analyzer for Potency manual. Exceptional recoveries for 11 cannabinoids were obtained (see Fig. 9).

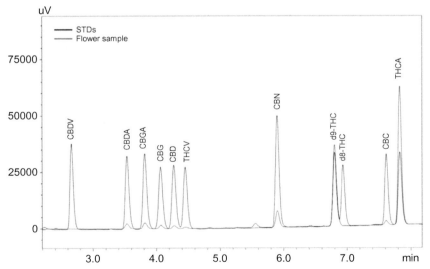

Fig. 7 Cannabis Analyzer for Potency chromatograms for flower sample (purple) with overlay of the 11-cannabinoids standards mix (black).

Fig. 8 Before (left) and after (right) freezer mill homogenization of infused gummy bears (using a SPEX Freezer/Mill).

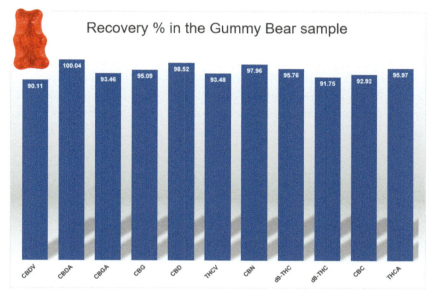

Fig. 9 Recoveries of cannabinoids in an infused gummy bear sample ranged from 91.75% to 100.04%.

After homogenization, the powder needs to be dissolved in a solvent. We propose testing hot water and several other compounds as solvent. The time required for the complete dissolving of the homogenized substance will also require testing. We need to control the temperature of the heating, to ensure we do not convert products to CBN, which generally occurs above 180 °C. The question of the variance between Gummy Bears in a single batch also arises, and we propose testing multiple Gummy Bears to see variation within batches and compare it to that between batches.

16. The need for a hemp analyzer

With the passage of the 2018 Farm Bill, we have entered our first decade of commercial hemp production in the United States since the 1920s. Hemp production varies from medical cannabis cultivation because hemp production is typically associated with many applications beyond nutraceutical/medicinal value. Hemp is used for food, fuel, fibre and much, much more [9]. Cannabidiol in hemp has been coveted for its potential medicinal value in a wide variety of human disorders and diseases [10]. The hemp industry is projected to be over a $90B industry by 2030 [11].

The 2018 Farm Bill defined hemp as 'the plant species *Cannabis Sativa* L. and any part of that plant ... with a delta-9 tetrahydrocannabinol concentration of not more than 0.3 percent on a dry weight basis'. The USDA adopted a total THC testing requirement as described in the aforementioned formula (see Section 14). This 0.3% THC cutoff is believed to have originated in the work of Dr. Ernest Small, a botanist who began studying and writing about cannabis in the 1970s. Dr. Small and his colleague, Arthur Cronquist published a paper in 1976 [12], suggesting a cutoff between hemp and marijuana at 0.3% THC for the explicit purpose of establishing a biological taxonomy. (Note: it was never intended as a legal threshold.)

This ≤0.3% THC threshold value discriminating hemp from cannabis has become the default value for legal hemp. According to the 2018 Farm Bill, growing industrial hemp in the United States is legal if the sample contains not more than 0.3 dry weight percent (wt%) total tetrahydrocannabinol (THC) [13,14]. The federally illegal status of cannabis prohibits transfer of cannabis over state lines. As a result, hemp industry professionals, departments of agriculture and crime labs need a reliable, rapid, reproducible and accurate means of quantification of THC. A hemp analyzer, in large part built upon the Cannabis Analyzer for Potency but with important considerations including hemp analysis optimization and reporting, was developed.

17. Future directions of cannabis and hemp analyzer development

Recent efforts have been underway to expand the number of cannabinoids analysed, as well as add important functionality to the Cannabis

Analyzer for Potency overlay software. We have partnered with a number of cannabis QC labs, academic centers and private industries to evaluate and improve our analyzers. A brief summary of such examples is provided below.

18. Expansion to a 15-cannabinoid standards analysis

Since our initial efforts in cannabis analyzer development, medical cannabis users and researchers have increased interest in a number of additional cannabinoids, including Cannabidivarinic acid (CBDVA), cannabicyclol (CBL), Tetrahydrocannabivarin acid (THCVA) and cannabichromenic Acid (CBCA). Automated quantitation capabilities of these cannabinoids, along with CBCA and CBDVA were added to the analyzers (Fig. 10).

19. 'Full Spectrum' cannabis and automated analysis of tinctures/oils

A cannabis concentrate that preserves the full cannabinoid (and terpene) content of the raw plant rather than simply increase THC content is referred to as a 'full spectrum extract'. The goal of full spectrum extracts is to maintain the natural complexity of the cannabis plant with minimal

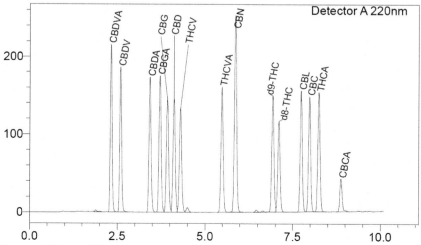

Fig. 10 Addition of four cannabinoids (CBDVA, CBL, THCVA and CBCA) to the standard mixture.

alteration via decarboxylation or oxidation. Full Spectrum Cannabis Oil (FSCO) and Full Extract Cannabis Oil (FECO) are commonly used terms to describe this whole plant approach to extract production.

An emerging trend in the industry has seen researchers and cannabis oil formulators working to create more 'full spectrum' cannabis tinctures with greater concentrations of more cannabinoids. In essence, these tinctures aim to preserve all of the natural, therapeutic compounds of the raw cannabis cultivars and deliver them to medical cannabis patients in a more convenient format. Developers of these tinctures work closely with cultivators to harvest a large number of cannabis cultivars that are grown for specific cannabinoid content. We have seen up to 26 cultivars grown for the expressed purpose of developing a tincture with high cannabinoid (and terpene) content.

We have extended the capabilities of the Cannabis Analyzer for Potency to rapidly determine the mg/ml concentrations of cannabinoids in tinctures and oils. The software interface was modified so that the user can check a box if the sample is a tincture or oil (Fig. 11). If this tincture/oil check box is selected, cannabinoid results are reported in mg/ml. This new functionality enables flower products and tinctures/oils to be batched and analysed together. An example chromatogram of a cannabis tincture sample, along with quantitation results is shown in Fig. 12.

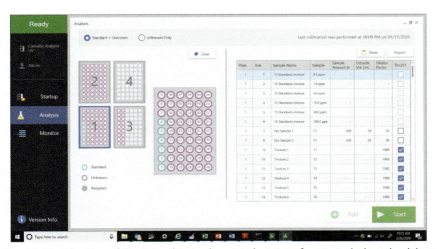

Fig. 11 Modification of the Cannabis Analyzer analysis interface to include a check box (far right) for tincture or oil samples.

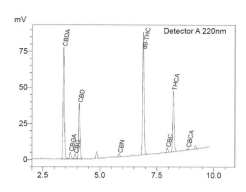

Fig. 12 Automated analysis of Happy Tree Microbe's 'Tangie Bliss XP' tincture with total THC at 40.78 mg/mL and total CBD at 33.01 mg/mL.

20. Summary

The development of automated, easy to use analyzers for the rapid, accurate and reproducible analysis of cannabinoids in cannabis and hemp represents a giant step forward for the routine quantitation of cannabinoids worldwide. Because the interface is simple to use, technicians can be trained in less than 1 day. In spite of this increased ease-of-use, there is no sacrifice in analytical performance. The methods are built for purpose and are robust enough even for the most demanding, high-throughput laboratories. The requisite sample prep and analysis methods are included, along with standards, reagents, columns and 3 years of instrument warranty. These uniform sample preparation, methods and consumables greatly reduce intra- and inter-laboratory variability.

Disclaimers

Shimadzu does not support or promote the use of its products or services in connection with illegal use, cultivation or trade of cannabis products.

Shimadzu products are intended to be used for research use only purposes or state approved medical research.

Shimadzu Scientific Instruments is not condoning the use of recreational nor medical marijuana, we are merely providing a market summary of the cannabis testing industry.

References

[1] Shimadzu Application News, Quantitative Analysis of Cannabinoids using the LCMS-2020 Single Quad MS, Shimadzu Application News SSI-LCMS-045, 2019.

[2] B. Smith, Quantitation of Cannabinoids in Dried Ground Hemp by Mid-Infrared Spectroscopy, vol. 2, Cannabis Science and Technology, 2016, 6.

[3] O. Aizpurua-Olaizola, U. Soydaner, E. Öztürk, D. Schibano, Y. Simsir, P. Navarro, N. Etxebarria, A. Usobiaga, Evolution of the cannabinoid and terpene content during the growth of Cannabis sativa plants from different chemotypes, J. Nat. Prod. 79 (2) (2016) 324–331.

[4] L.O. Hanuš, S.M. Meyer, E. Muñoz, O. Taglialatela-Scafati, G. Appendino, Phytocannabinoids: a unified critical inventory, Nat. Prod. Rep. **33** (2016) 1357–1392. https://doi.org/10.1039/C6NP00074F.

[5] B. Smith, Inter-lab variation in the cannabis industry, in: Part I: Problem and Causes, vol. 2, Cannabis Science & Technology, 2019. Issue 2.

[6] U.S. Food & Drug Administration, CFR—Code of Federal Regulations Title 21, U.S. Food & Drug Administration, 2016. Retrieved 15 September.

[7] J. Edwardsen, Sample Preparation of Cannabis Products by Syringe Filtration using the Shimadzu Cannabis Analyzer, Shimadzu Application News, 2019. No. CONS-001.

[8] Comparison of CRM Concentrations among Four Different Reference Material Producers, Cayman Chemicals and Shimadzu Application Note, 2020.

[9] International Standard ISO/IEC 17025, General Requirements for the Competence of Testing and Calibration Laboratories, 2017.

[10] R. Franckowski, Precision testing, Grow Opportunity 3 (4) (2019) 18.

[11] CRM Mixtures Improve Quantitation Accuracy, Cayman Chemicals Application Note, 2019.

[12] E. Small, A. Cronquist, A practical and natural taxonomy for Cannabis, Taxon 25 (4) (1976) 405–435.

[13] P. Bouloc, Hemp: industrial production and uses, CABI, 2013.

[14] C. Conrad, Hemp for Health: The Medicinal and Nutritional Uses of Cannabis Sativa, Simon and Schuster, 1997. ISBN 162055027X, 9781620550274.

CHAPTER ELEVEN

Marihuana safety: Potency of cannabinoids, pesticide residues, and mycotoxin in one analysis by LC/MS/MS

Jerry Zweigenbaum[a,*], Agustin Pierri[b]

[a]Agilent Technologies, Wilmington, DE, United States
[b]Weck Laboratories, City of Industry, CA, United States
[*]Corresponding author: e-mail address: j_zweigenbaum@agilent.com

Contents

1. Introduction	340
1.1 History	340
1.2 Taxonomy	340
1.3 Medical use	341
1.4 Recreational use today	342
1.5 Regulations	343
2. Method	343
2.1 Definition	343
2.2 Rationale	344
2.3 Materials, equipment and reagents	344
2.4 Protocol	345
2.5 Calibration, quality assurance and quality control	353
2.6 Safety considerations and standards	354
2.7 Data analysis and statistics	354
2.8 Alternative methods/procedures	355
3. Results and discussion	355
3.1 Sample preparation	355
3.2 Pesticides	355
3.3 Mycotoxins	357
3.4 Cannabinoids	357
4. Summary	363
Acknowledgements	363
References	363

Comprehensive Analytical Chemistry, Volume 90
ISSN 0166-526X
https://doi.org/10.1016/bs.coac.2020.04.008

© 2020 Elsevier B.V.
All rights reserved.

1. Introduction

Marihuana has throughout history found use as both a medicinal and psychoactive plant. Its scientific or botanical name is *Cannabis sativa* L. (the L indicating it was first classified by Carl Linnaeus). Recently marihuana and hemp have been distinguished legally by the content of the psychoactive ingredient Δ^9-tetrahydrobannabinol (THC), whereas hemp has less than 0.3% THC and marihuana has more (typically between 5% and 35%). The long history of the plant includes both varieties.

1.1 History

Cannabis use may very well date back before recorded history [1] and hemp cultivated for fibre may have dated to 10,000 years ago [2]. Evidence of its human use can be traced to pollen deposits in China [3] dating back to 2500 BC and to the Roman Empire about 120 BC [2]. It was noted by Wills that there was little evidence of its use by the Romans and Greeks for its intoxicating effects and Wills gives both a chronological and geographic description of its use [4]. The psychoactive effects were either unknown or not appealing to those societies. Although it is difficult to find cannabis in the literature for recreational or spiritual use, its pharmaceutical use is well documented. Russo provides a detailed chronological account of both the timeline and its uses including some ritualistic and medical applications [5]. Likewise, Bonini et al. provide a historical account of its use dating back to Asia citing Taiwan having several uses as long as 12,000 years ago [6]. In this historical account medical use was identified 5000 years ago with emperor Chen Nung. From Rome and Greece, the use of cannabis was spread through Europe and was introduced to the Americas by the European colonists [6]. However, a geographic representation of its domesticated spread by Warf shows the plant originating in central Asia and eventually coming from South Africa to South America and then to North America [7]. Again, the primary use was as a source of fibre.

1.2 Taxonomy

Both cannabis and hops are in the family of Cannabaceae although cannabis is much more complex than hops. *Cannabis sativa* L. was classified by Carl Linnaeus but it is believed its classification was made much before Linnaeus by Leonhart Fuchs in 1543 [8]. The plant has three recognized varieties, sativa, Indica and ruderalis [8,9]. There has been debate whether sativa and Indica are in fact different varieties and it is not clear whether that debate has been resolved [1,10]. Regardless, both sativa and Indica contain the cannabinoids responsible for the effects of cannabis and both produce tetrahydrocannabinolic acid (THCA). The most psychoactive ingredient is Δ^9-tetrahydrobannabinol formed from decarboxylation of THCA either by drying or heat. Indica is characterized by a shorter bush with broader leaves while sativa is a much taller bush with narrow leaves [11]. Ruderous is a variety that has much lower cannabinoid content. One recommendation is that all be recognized as a single species, one covar with narcotic effects and one non-narcotic [11]. Whether or not Indica is a separate variety than sativa, both have the psychoactive qualities of containing significant THCA/THC.

The variation in the cannabis plant for production of cannabinoids ranges from high THCA content to high CBDA content. What is now considered hemp contains low THC content and high CBD content. The hemp plant has been known through time for its fibre content and the oil obtained from the seeds. The plant is also known for its high terpene content that gives its aromatic characteristics. The cannabinoids, terpenes and essential oils obtained from the plant are responsible for its many qualities.

Whether sativa or Indica, the origin of the plant is central Asia and India. The plant is known for flourishing on the side of roads in wet soil areas and thus has been designated as a weed. The three characteristic compositions with their Indian names are bhang, containing most parts of the plant including the seeds; ganja (sinsemilla), containing the flowers only; and charas (or hashish), containing the resin from the trichomes of the plant. Middle Eastern varieties were known to contain similar portions of THC and CBD and known to be more relaxing than generating a "high", but the narcotic variety of cannabis has been bred to contain only THC in both North America and Europe [1].

1.3 Medical use

The medical use of cannabis is documented by reviews in both a historical view and by application, ranging from treatment of headaches, to appetite loss, and inflammation among many others [3,5,12–16]. The plant is rich in cannabinoids with over 120 identified and with at least as many terpenes found [17]. The physiological and psychoactive effects of the cannabinoids have been related to CB1 and CB2 receptors [18], with the most actively binding cannabinoids identified as THC and CBD [19].

A major advance for the medical application of Cannabis has most recently been the treatment of epilepsy. Its use for convulsions and seizures has been traced back to ancient civilization by Freidmand and Sirven [20]. A recent Australian survey showed positive effects of using cannabis for its treatment including the use of Epidiolex (CBD) in epileptic children [21]. This preparation of pure CBD was the first cannabis product approved for medicinal use by the US FDA [22–24].

Because of the legal status of cannabis throughout the world, scientific studies have been limited. However, comprehensive and valid studies have been undertaken. Alsherbiny and Li provided a list of studies on treating nausea from chemotherapy and cancer pain and included drug interactions [25]. In a single-arm cohort study, the use of CBD in conjunction with opioid use reduction was examined [26], leading to a hypothesis that it may help addicts [27]. The effects of CBD on blood pressure and heart rate was examined and showed no effect under normal conditions but a positive effect under stress [28]. In one more example of a study to examine the effects of CBD, rats were used to test the extinction of fear [29]. Statistically there was a positive effect implying that CBD could have application in post-traumatic stress disorder and phobias. A review by White gives extensive information on scientific studies evaluating treatments by CBD [30].

A recent study on the use of cannabis and its effects on the different stages of sleep was conducted concluding only the resting phase was affected and that more investigation was needed [31]. Not all scientific studies on marihuana are based on therapeutic use. An epidemiological study on the daily use of marihuana and its correlation to psychosis was conducted [32]. Di Forti et al. found a positive effect with dose dependence, that is higher potency marihuana lead to an earlier first contact with psychiatric services. Another cohort study examined the correlation of marihuana use and schizophrenia and found that cannabis users had a poorer prognosis [33]. These examples, whether positive or negative for the use of cannabis, demonstrate that as the legal restrictions are relaxed more good scientific studies on this complex plant will ensue.

1.4 Recreational use today

From the accounts of cannabis use, it appears historic societal acceptance for recreation has oscillated between positive and negative views [34]. Today in a number of countries that has changed, with Uruguay being the first and now Canada. It should be noted that the Netherlands has not legalized recreational use but has implemented a "tolerance" policy towards it. Although illegal federally in the United States, specific states have legalized it for recreational use. As of now in 2020, they include California, Oregon, Washington State, Nevada, Alaska, Colorado, Illinois, the District of Columbia, Vermont, Maine, Michigan, and Massachusetts. With each country and state legalizing its use recreationally, assuring its safety becomes a challenging task. This includes the concerns around the general populations' use, frequency and dose, along with the various possible forms of productisation. Whereas, medicinal use implies treatment with a dose response relationship. For example vaping THC has been related to lung injury [35]. Thus, regulation around crop production, distribution, and sale is needed.

1.5 Regulations

Legalization requires regulations for crop production, processing, distribution, and retail sales. For any crop consumed by the public, regulation of pesticide use is needed to assure the safety of the crop. This is complicated in the USA because the setting of limits for pesticide residues is a function of the US EPA and because it is still federally illegal, that agency cannot set limits [36]. This remains an issue for states legalizing recreational use, but should be resolved for the general growing and use of hemp (defined as dried cannabis containing less than 0.3% THC) with the 2018 Farm Bill [37]. Although now legal to grow and possess hemp as defined, the total implications and regulations are still not clear.

Countries and states that have legalized marihuana use have put in place regulations for pesticide residue limits, mycotoxins, mould, bacteria, and heavy metals. They also include potency, typically selecting 6–10 of the most prevalent cannabinoids for quantitative analysis. In this chapter, we will use the current California regulations as an example [38] and limit the methodology to only those substances that respond well to liquid chromatography/tandem mass spectrometry (LC/MS/MS) with a triple quadrupole mass spectrometer.

2. Method
2.1 Definition

Detection of the four aflatoxins and ochratoxin at very low levels, pesticides at intermediate concentrations and cannabinoids at percent levels is demonstrated. The use of the very sensitive and selective electrospray LC/MS/MS with ultra-high-pressure liquid chromatography and a triple quadrupole mass spectrometer is presented.

2.2 Rationale

Quantitative analysis by most techniques have a limited dynamic range where concentrations of analyte can be accurately measured. Although triple quadrupole mass spectrometers can often achieve three to four orders of magnitude dynamic range, it is still limited. However, the use of the native carbon-13 stable isotopes of a molecule, readily separated with unit resolution can be employed to extend the concentrations that are measured. This is done in the analysis of cannabinoids while at the same time analysing the native monoisotopic carbon-12 transitions for pesticides and mycotoxins.

2.3 Materials, equipment and reagents

Note that materials, equipment and reagents specified in the following can be substituted for equivalents. This method utilizes 50 mL centrifuge tubes and ceramic homogenizers such as those obtained from Agilent, part number 5982-9313. A Spex 2010 Geno/Grinder or equivalent can be used. An analytical balance that can measure to 0.0001 g, e.g. VWR part number 75802-858, is needed for both standards and sample preparation. Pipets that can accurately deliver from 100 µL to 1.0 mL such as VWR 89079-974 with 1.0 mL pipet tips and a solvent dispenser that can deliver accurately 5–15 mL, e.g., VWR 10015-344, are also needed. Finally, the method requires solid phase extraction cartridges that can hold 6 mL with 500 mg C18 packing material (Agilent SampliQ C18 EC 6 mL 500 mg SPE).

The methodology includes a liquid chromatograph capable of high-quality separations such as the Agilent 1260 Infinity II flexible pump (pressure to 800 bar), with the 1260 Infinity II Multisampler and Multicolumn compartment, and a triple quadrupole mass spectrometer, such as the Agilent Ultivo. A C_{18} reversed-phase column, Agilent InfinityLab Poroshell 120 EC-C18, 2.1×50 mm, 1.9 µm (p/n 699675-902) is used for chromatographic separation.

A water supply that provides ultra-high purity water can be obtained from a system such as the Milli-Q Integral 10 or IQ 7010 (Millipore Sigma). Acetonitrile that is HPLC grade (VWR p/n K981-4L) is used for sample preparation. Acetonitrile that is LC/MS grade is used for the LC/MS/MS mobile phase (Agilent InfinityLab Ultra-Pure LC/MS acetonitrile, p/n 5191-4496). Isopropanol that is LC/MS grade (VWR Hipersolv Chromanorm for LC-MS, p/n 84881.290P) is used for the LC/MS/MS injector rinsing. LC/MS grade acetic acid can be obtained from Millipore Sigma (Lichropur LC-MS grade).

Standards of pesticides can be obtained from Agilent, mycotoxins can be obtained from Millipore Sigma, and cannabinoids from Cerilliant.

2.4 Protocol

The protocol described here was developed to meet the California Code of Regulations Title 16, Division 42, Bureau of Cannabis Control. To meet other regulations, this protocol can be modified to add or delete analytes. Any protocol used or modified to meet regulations must be validated by appropriate procedures.

2.4.1 Sample preparation

For harvested batches of cannabis flower, the number of sample increments per batch are dependent on size. For up to a 10-pound batch, 8 increments per sample are required. For 10–20 pounds, 16 increments are needed per sample, and 20–30, 23 increments, 30–40, 29 increments and 40–50, 34 increments. Harvested batches greater than 50 pounds are not permitted. The increments are to be taken at random locations at different heights and widths of the batch and should be equal in weight and the sample size should be at least 0.35% of the total batch weight. For a 10-pound batch the total weight of the 8 increments should be at least 15 g. The 15 g sample of flowers is homogenized by placing in a 50 mL centrifuge tube with the ceramic bars and either shaken manually or with the Geno/Grinder. A 1 g aliquot of the homogenized sample is weighed into another 50 mL tube and 15 mL HPLC grade acetonitrile is added. The tube is then shaken for 120 s. The liquid is decanted into an unconditioned SPE cartridge, allowed to elute by gravity and collected in another container. The tube containing the 1 g aliquot of sample is washed with 5 mL of acetonitrile and this is decanted into the SPE cartridge and collected. This is repeated one more time with 5 mL acetonitrile. The total volume, less than 25 mL is brought to 25 mL. Finally, 100 µL of this extract is pipetted into a 2 mL autosampler vial and diluted with 900 µL of acetonitrile for a final concentration of 4 mg/mL of cannabis flower.

The 1 mL sample is then analysed for the LC/MS amenable pesticides, mycotoxins and cannabinoids specified in the California regulation. The same sample can also be analysed by GC/MS/MS for those pesticides in the California regulation amenable to GC/MS (not covered in this method). These methods can be applied to other regulations provided the analytes are amenable to LC/MS/MS and the validation of the method demonstrates the regulatory limits can be met.

2.4.2 LC/MS/MS conditions

The chromatographic conditions to achieve appropriate separation of the analytes employs the reversed-phase C_{18} column with a mobile phase consisting of (A) 0.1% acetic acid and 5 mM ammonium acetate in ultra-high purity water and (B) LC/MS grade acetonitrile. The column is maintained at a constant temperature of 25 °C. The sample is injected by drawing 10 μL water, 2 μL of sample, and then 10 μL of water using an injection program. The injection program includes drawing 30 μL of isopropanol into the injection loop near the end of the run as a final rinse. A gradient starting at 5% B (acetonitrile) mobile phase holding for 0.5 min and then ramping linearly to 76% B at 4 min, then to 83% at 5.5 min and finally to 100% at 7.5 min, with a hold until 8 min (end of run) is used. The mobile phase flow is constant at 0.65 mL/min.

The mass spectrometers source is operated with positive/negative mode switching and the Agilent Jetstream electrospray ionization sheath gas temperature is set at 250 °C with its flow at 12 L/min. The drying gas is set to 250 °C with a flow of 9 L/min. The electrospray nebulizer pressure is set to 35 psi, and the capillary voltage is 4 KV in positive and 3.5 KV in negative modes. Both modes have 0 V on the nozzle.

Table 1 provides the MS/MS precursor and product transitions (multi-reaction monitoring or MRM), collision energies and fragmentor voltages for the pesticides in the method. These conditions were optimized for the instrument used and it is recommended that optimization of any specific instrument for the MS/MS should be undertaken before running the method.

Table 2 shows the optimized conditions for the mycotoxins analysed in this method.

Finally, the MRM transitions and their optimized conditions for the cannabinoids are given in Table 3. The cannabinoids required for potency in the California regulations are tetrahydrocannabinol (THC, no distinction is made between Δ^8 and Δ^9), tetrahydrocannabinolic acid (THCA), cannabidiol (CBD), cannabidiolic acid (CBDA), cannabinol (CBN), and cannabigerol (CBG). Other regulatory bodies may include other predominant cannabinoids.

Table 1 Pesticide transitions and their MRM conditions.

Compound name	Precursor ion	Product ions	Retention time	Fragmentor	Collision energy	Polarity
Abamectin B1a	890.5	567 305	5.63	130	20 0	Positive
Abamectin B1b	876.5	553 291	5.49	100	12 0	Positive
Acephate	184	143 49.2	0.39	75	0 12	Positive
Aldicarb	116	89 70	2.81	75	5 5	Positive
Azoxystrobin	404	372 344	4.02	110	4 20	Positive
Bifenazate	301.2	198 170	4.17	95	0 12	Positive
Boscalid	343	307 271	4.07	145	12 28	Positive
Carbaryl	202.1	145.1 127	3.33	106	0 24	Positive
Chlorantraniliprole	484	453 286	4.39	105	8 4	Positive
Carbofuran	222.1	165.1 123	3.23	83	4 16	Positive

Continued

Table 1 Pesticide transitions and their MRM conditions.—cont'd

Compound name	Precursor ion	Product ions	Retention time	Fragmentor	Collision energy	Polarity
Clofentezine	303	138 102	4.61	97	4 36	Positive
Coumaphos	363	307 227	4.6	111	28 40	Positive
Daminozide	161.1	143.1 43.9	0.21	83	0 20	Positive
Diazinon	305.1	169 153	4.92	115	16 16	Positive
Dimethoate	230	199 124.9	2.41	74	0 16	Positive
Dimethomorph	388.1	301 165	3.85	120	12 28	Positive
Ethoprop (ethoprophos)	243.1	130.9 96.9	4.05	100	15 35	Positive
Etofenprox	394.5	177 107	5.88	111	4 44	Positive
Etoxazole	360.2	141 113	5.22	109	28 55	Positive
Fenhexamid	302.1	97.2 55.1	4.07	100	35 60	Positive

Fenoxycarb	302.1	116 / 88	4.26	115	0 / 12	Positive
Fenpyroximate	422.2	366 / 138	5.17	118	8 / 28	Positive
Fipronil	435	330 / 250	4.44	120	22 / 40	Negative
Flonicamid	230.1	203 / 174	1.54	112	12 / 12	Positive
Fludioxonil	247	126	3.89	141	28	Negative
Fludioxonil	229	185	3.89	120	8	Positive
Hexythiazox	353.1	228 / 168	5.09	112	4 / 20	Positive
Imazalil	297.1	200.9 / 159	3.41	124	20 / 16	Positive
Imidacloprid	256.1	209 / 175	2.4	89	8 / 12	Positive
Jasmolin I	331.2	126.9 / 98.9	4.42	97	4 / 20	Positive
Malathion	331.1	126.9 / 98.9	3.55	103	4 / 20	Positive

Continued

Table 1 Pesticide transitions and their MRM conditions.—cont'd

Compound name	Precursor ion	Product ions	Retention time	Fragmentor	Collision energy	Polarity
Metalaxyl	280	220 192	3.49	120	4 8	Positive
Methiocarb	226.1	169 121	3.82	86	0 12	Positive
Methomyl	163.1	106 88	1.15	75	0 0	Positive
Mevinphos	225	193.1 127.1	2.68	74	0 8	Positive
Myclobutanil	289.1	125 70	4	112	16 16	Positive
Oxamyl	237	90 72	1.1	75	0 10	Positive
Paclobutrazol	294.1	125 70	3.81	112	36 12	Positive
Phosmet	318	160 133	4.09	75	8 36	Positive
Piperonyl Butoxide	356	177 119	5.39	90	0 32	Positive
Propiconazole	342	159 69	4.3	115	32 12	Positive

Propoxur	210.1	110.9 92.9	3.19	83	4 20	Positive
Pyridaben	365.2	309 147	5.43	80	0 20	Positive
Spinetoram	760.5	142 98	4.55	150	20 48	Positive
Spinosyn	732.5	142 98	4.96	80	24 50	Positive
Spiromesifen	371.2	355.1 255.1	6	97	0 16	Positive
Spirotetramat	374.2	330 302	3.95	120	4 8	Positive
Spiroxamine	298.3	143.9 100	3.66	115	12 28	Positive
Tebuconazole	308.2	125 70	4.14	100	40 12	Positive
Thiacloprid	253	125.9 98.9	2.81	100	16 40	Positive
Thiamethoxam	292	211 181	1.91	77	4 16	Positive
Trifloxystrobin	409.1	206 186	4.82	120	4 8	Positive

Table 2 Mycotoxin transitions and MRM conditions.

Compound Name	Precursor Ion	Product Ions	Fragmentor voltage (V)	Collision Energy (V)	Retention time (min)	Polarity
Aflatoxin B1	313	285.1	159	16	3.13	Positive
		241.1		36		
Aflatoxin B2	315	287.1	167	20	3.01	Positive
		259		24		
Aflatoxin G1	329	243	153	24	3.00	Positive
		200		40		
Aflatoxin G2	331	313.1	153	20	2.88	Positive
		189.1		40		
Ochratoxin A	404	239	115	16	3.62	Positive
		221		32		

© Agilent Technologies, Inc. 2019
Reproduced with Permission, Courtesy of Agilent Technologies, Inc.

Table 3 C^{13} precursor transitions for cannabinoids and their dynamic MRM settings.

Compound name	Precursor ion	Product ions	Fragmentor voltage (V)	Collision energy (V)	Retention time (min)	Polarity
THC	317	193	100	20	5.88	Positive
		123		36		
THCA	359	245	128	36	5.51	Negative
		313		24		
CBD	316	193	128	24	5.31	Positive
		123		32		
CBDA	359	245	134	32	4.97	Negative
		339		20		
CBN	311	279	162	32	5.63	Negative
		222		50		
CBG	318	123	118	32	5.28	Positive
		193		12		

© Agilent Technologies, Inc. 2019
Reproduced with Permission, Courtesy of Agilent Technologies, Inc.

2.5 Calibration, quality assurance and quality control

The mass spectrometer should be tuned and calibrated according to the laboratory's standard operating procedure. The instrument used here is very stable and requires only a checktune once a week and tune and calibration when specifications are not met, or when major maintenance or acquisition software/firmware changes have been made. The LC should be checked routinely by running a selected reference standard and comparing retention times and peak widths to assure proper operation.

Calibration of reference standards for quantitative analysis should be done after any instrument tune or calibration. Continuous calibration should be performed per the standard operating procedure (SOP) of the laboratory. Calibration for pesticides and mycotoxins should be done by preparing standard concentrations of each analyte in a zero–cannabis matrix. A zero matrix is one that has been analysed and found not to contain the analytes at or below 0.1 times the action level set by the operating regulations. Given the biological variation of cannabis described in the introduction, the matrix should be matched by at least the three categories of marihuana (*cannabis sativa*), hemp (*cannabis sativa*) and *cannabis Indica*. The 250-fold dilution in this method mitigates most of the variability within a category. For the cannabinoids a zero matrix is not possible, so hops are used (also of the Cannabaceae family- Humulus) as the matrix. Some laboratories require calibration at the beginning and end of each batch. At a minimum, a check standard prepared in matrix at the lower level of the calibration curve should be run at regular intervals of sample analysis. The calibration range should be at a minimum of 0.5 times the action level to 2 orders of magnitude above. In addition, zero–matrix samples (not spiked) and solvent blanks should be run at regular intervals with samples. These should comprise at a minimum those required by the SOPs of the laboratory to maintain quality control of these analyses.

The method must be validated in each laboratory by analysing zero–matrix samples spiked both before and after sample preparation at the action level and above, and quantified with standard curves prepared in solvent. This will demonstrate method efficiency. In addition, calibration curve linearity must be demonstrated along with accuracy and precision of the method. In day repeatability with a single instrument and a single analyst should be determined along with multi–day multi–analyst and multi–instrument (if applicable) reproducibility. Statistical analysis should demonstrate capability of the laboratory to make accurate and precise determinations of the analytes. Documentation of the validation, training and experience of the analysts and reviewers should all be maintained as record of the quality assurance of the laboratory. A quality officer of the lab should be independent of both the analysts and reviewer and assure procedures are followed and documented.

2.6 Safety considerations and standards

Laboratories performing this method should adhere to good laboratory practices and regulatory standards (such as USA OSHA) for handling toxic materials. All standards, reagents, solvents and samples should be properly labelled and handled with personal protective equipment (PPE). PPE should include nitrile gloves, safety glasses, and laboratory coats. All reference standards, (pesticides, mycotoxins, cannabinoids) should be obtained as solutions (e.g.) 1 mg/mL or less to avoid handling the solids. Aflatoxins solids should be handled with the utmost care using protective equipment and isolation from skin and inhalation contact. Fume hoods should be used for handling all solutions and solvents.

Because this method is a trace analysis for mycotoxins and pesticides and the inclusion of cannabinoids with reduced response, utmost cleanliness in the laboratory is required. Instrumentation, laboratory equipment, reagents, solvents, standards and samples should be handled using standard operating procedures that minimize contamination.

2.7 Data analysis and statistics

Data analysis is carried out with MassHunter Quantitative Analysis software (Agilent Technologies). The quantitative transition of each analyte is integrated using the Agile2 parameter less integrator, and the qualitative/quantitative ratios are determined. Calibration standards are used to prepare curves that are either a linear or a power function fit. For the pesticides and mycotoxins, a linear fit is used and for the cannabinoids a quadratic fit. All employ a $1/x$ weighting to provide more influence for the lower level of detection. Mean, standard deviation and relative standard deviations are used to determine accuracy and precision of the method. Where large discrepancies exist in solvent calibration and pre and post spikes, a nested ANOVA can be used. Reproducibility is determined with pooled standard deviations of the various day-to-day, analyst-to-analyst experiments.

2.8 Alternative methods/procedures

Chlordane, chlorfenapyr, coumaphos, ethoprop, methyl parathion, acequinocyl, captan, cyfluthrin, cypermethrin, kresoxim-methyl, naled, PCNB, and permethrin are all pesticides included in the California regulation. They are not included in this method because they are best analysed by GC/MS/MS. Trying to analyse them in the LC/MS/MS method would degrade the quality of the results.

The use of native isotopes to extend the dynamic range of mass spectrometric measurement is an advanced technique that should only be performed by analysts highly trained and experienced with mass spectrometry. The use of HPLC with a UV detector to determine cannabinoid potency is the method of choice for most regulations and laboratories.

3. Results and discussion

3.1 Sample preparation

Cannabis is a very complex matrix of plant cell material, terpenes, and cannabinoids. Fig. 1 provides a visual of this complexity, even after extraction and clean-up.

Fig. 2 shows the effect of further dilution on the sample. A clean extract is obtained with a 250-fold dilution and still there is ample instrument sensitivity to meet even the mycotoxin regulatory limits for California. In addition, as described in the method section, hops is used as a surrogate matrix to determine the accuracy of the potency measurement. Even though hops are in the same botanical family as cannabis, the extracted matrix of hops has nowhere near the complexity of cannabis. Regardless of this, because the dilution mitigates much of the matrix effects, hops is a reasonable substitution given that it is not possible to obtain a "zero" matrix of cannabis for all the cannabinoids.

3.2 Pesticides

Pesticides regulated by the state of California represent a diverse list of compounds from acephate with a Log P of −0.85, to an intermediate compound such as imazalil with a Log P 3.82, to a relatively large compound such as Spinosyn with a Log P of 4.43 [39]. For this work each of the pesticides was spiked into hops at the action level in inhalable products [38]. The pesticide with the lowest signal-to-noise at the action level was fenhexamid, s/n

Fig. 1 Picture on left is 1 g cannabis sample in 15 mL acetonitrile. Vial on right is 25 mL dilution after SPE. (Photograph by author).

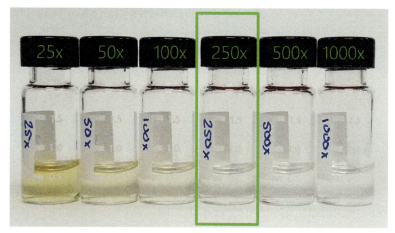

Fig. 2 Dilution effect on the cannabis extract. (Photograph by author).

130 and action level 100 parts per billion (ppb). The highest was azoxystrobin with a s/n 15,000 at 100 ppb. All showed more than sufficient signal to be able to determine the pesticides at least at 1/10th the action level. Acephate is barely retained but still demonstrates good linearity and recovery. The recovery of each pesticide in a marihuana flower sample containing no pesticides was between 70% and 105%. The only pesticide detected in the marihuana flower samples analysed was imidacloprid. The measured concentration was 1.5 µg/g and the California limit is 5 ppm. Imidacloprid is a neonicotinoid pesticide effective in controlling aphids and whiteflies [40] common to cannabis plants. It does not control spider mites, also a problem for cannabis plants, and may even cause adverse effects [41].

3.3 Mycotoxins

The four aflatoxins B1, B2, G1 and G2 have a total action level of 20 ppb. If all four were present in a sample at 5 ppb, the total would be at the action level and this sets the limit of quantification (LOQ) required for each of them. In contrast, the action level of ochratoxin is also 20 ppb but only for that compound. The signal-to-noise of B1 was the lowest observed and was 210 at 5 ppb. For B2, G1 and G2 the s/n was 2300, 3900, and 1000 respectively. All had more than enough response to exceed the needed LOQ. Ochratoxin at 20 ppb spike produced a s/n of 1100 also more than sufficient. The recoveries of these naturally produced toxins, spiked at three times the action level into a cannabis sample was 78%, 79%, 77%, and 76% for B1, B2, G1, and G2 and 60% for ochratoxin. For the three cannabis flower samples tested, no mycotoxins were detected.

3.4 Cannabinoids

The analysis of both pesticides and mycotoxins in food and beverage samples has been applied and is not unique [42–45]. What makes this method unique is the addition of the cannabinoids at a much higher concentration. This demonstrates that the dynamic range of an LC/MS/MS method can be extended tremendously using native isotopes of the analytes as transitions. However, we were not the first to examine the ability to perform this type of analysis using a triple quadrupole mass spectrometer [46], and the technique was extended to the analysis of amphetamines [47]. The use of native isotopes has found interest in time-of-flight mass spectrometers [48] with associated errors evaluated [49]. Most recently an algorithmic approach to address IMS-QTOF saturation in metabolomic studies was reported [50].

The abundance of the ^{13}C isotopes can be calculated using the binomial probability equation:

$$A(C) = \frac{n!}{C!(n-C)!} \pi^C (1-\pi)^{n-C}$$

where:

A(C): abundance of ^{13}C

C: number of ^{13}C atoms in molecule

n: number of total carbons in molecular

π: natural abundance of ^{13}C

For THC with 21 carbon atoms, the abundance of the $^{13}C_1$ isotope would be:

$$A(C) = \frac{21!}{1!(21-1)!} 0.011^1 (1-0.011)^{21-1}$$
$$A(C) = 0.185 \ or \ 18.5\%$$

The abundance of $^{13}C_2$ in THC would be:

$$A(C) = \frac{21!}{2!(21-2)!} 0.011^2 (1-0.011)^{21-2}$$
$$A(C) = .02037 \ or \ 2.037\%$$

The typical setting for transitions on a triple quadrupole mass spectrometer is designated as "unit resolution". Note this is the commonly used term albeit the terms for resolution defined by IUPAC (R05318) and resolving power (R05321) have added confusion to their usage [51,52]. The final recommendation by Murray et al. supports the IUPAC definitions [53]. However, many mass spectroscopists have always used resolution and resolving power in the opposite terminology [54]. Specifically, according to IUPAC, resolution is the $m/z \div \Delta m/z$ where $\Delta m/z$ is the separation of two adjacent m/z by 10% valley (double focusing mass spectrometers) and $\Delta m/z$ is the full width at half maximum of the peak height, (FWHM used for quadrupole, time-of-flight and FT mass spectrometers). For quadrupole mass spectrometers, unit resolution is typically set at 0.7 FWHM across the entire m/z range. For THC the separation of the $^{12}C_{21}$ isotope, $^{13}C_1$ and $^{13}C_2$ isotopes at m/z 315, 316 and 317 (nominal mass) is shown in Fig. 3 with unit resolution. Note that the abundance of the m/z 317 is closer to 2.35% because of the contribution of the $^{18}O_2$ isotope.

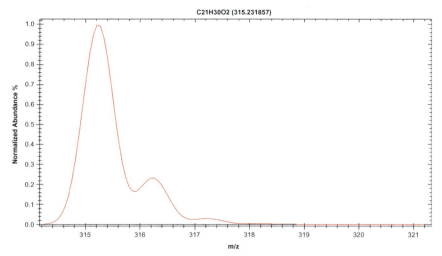

Fig. 3 Separation of isotopes of the THC [M+H]$^+$ ion at unit resolution.

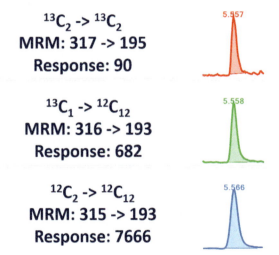

Fig. 4 Response of THC for the native isotopes of THC (© Agilent Technologies, Inc. 2019. *Reproduced with permission, courtesy of Agilent Technologies, Inc.*).

We have evaluated the response of cannabinoids using their monoisotopic ^{12}C M+H or M−H ion as the precursor for their MRM transitions, the ^{13}C$_1$ isomer, and the ^{13}C$_2$ isomer. As an example, Fig. 4 shows the difference in response for each of the isomer transitions of THC. For the ^{13}C$_2$ precursor we show the product ion also having ^{13}C$_2$. Using this transition, the response is dramatically reduced. This demonstrates that the signal can be reduced by using transitions with a product ion having at a maximum ^{13}C as that selected in the precursor ion or less, or a product with all ^{12}C. In this way the response can be modulated within the range of the expected concentration of the analyte. This example does not represent the actual transitions used (see Table 3). The transitions selected were those that gave the desired response in the calibration range needed. The transitions selected were from either 1 or 2 ^{13}C precursors with product ions that were all ^{12}C fragments.

Because the native isotopes are of the same molecule, there should be no difference in the ionization and response of the transitions selected over the normally used more abundant all ^{12}C transitions. However, there is a concern with the possibility of ion suppression. A common and simplified theory of suppression with electrospray ionization is the presence of less volatile compounds that change droplet formation and/or droplet evaporation [55–57] and reduce maximum ionization efficiency due to a co-eluting high concentration interference [58]. One way to evaluate ion suppression or enhancement is to compare the slope of a calibration curve [59,60]. The use of isotope dilution for internal calibration is the preferred way to address matrix effects, but for multi-residue analysis the cost can be very prohibitive. Reduction of matrix effects for suppression or enhancement is often associated with sample cleanup and the use of matrix matched standards [56,61,62]. However, dilution has found a major role in reduction or elimination of ion suppression or enhancement [58,63].

Given that dilution is a recognized solution for matrix effects, two calibration curves were constructed in solvent for determining the concentrations of the cannabinoids. One, shown in Fig. 5 gives the results for the use of the native ^{13}C isotopes for THC with reduced signal at percent levels with the 250-fold dilution prescribed in the method. Another calibration curve was constructed using the ^{12}C isotope transitions of THC at a 10,000-fold dilution and is shown in Fig. 5. To determine the accuracy of using the ^{13}C isotope transition with the method's SPE and 250-fold dilution, hops extracts were spiked with the equivalent of 2.5% of each of the cannabinoids. An aliquot of the spikes was then diluted 40 times to match the 10,000-fold dilution of the ^{12}C transition calibration curves. Both were then quantified and compared with the results in Table 4. The data shows that the results are comparable and with the very high dilution all matrix effects are minimized. Recovery of hops spikes at 5% each cannabinoid were between 85% and 105%.

Three real marihuana flower samples were analysed by the single run method (SPE with 250-fold dilution) and then 10,000-fold dilution.

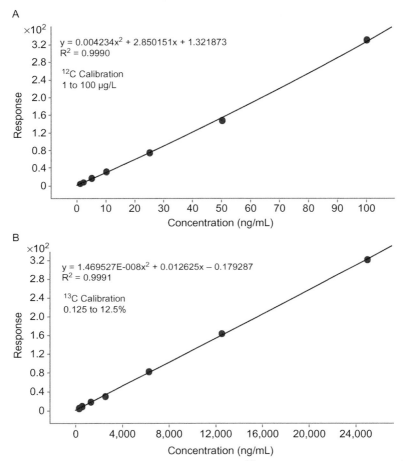

Fig. 5 Calibration curves for THC with A) the native ^{12}C transition and B) the native ^{13}C transition. © Agilent Technologies, Inc. 2019. *Reproduced with Permission, Courtesy of Agilent Technologies, Inc.*

Table 4 Hops spiked with 2.5% each cannabinoid. Diluted is the 10,000-fold dilution determined with ^{12}C transitions and the ^{13}C is the 250-fold dilution dtermined with ^{13}C transitions.

	Spiked	Sample 1 Diluted	Sample 1 ^{13}C	Sample 2 Diluted	Sample 2 ^{13}C
THC	2.5	1.85	2.39	2.31	3.1
THCA	2.5	2.65	1.97	1.7	1.98
CBD	2.5	3.03	3.3	1.85	2.06
CBDA	2.5	2.12	2	2.1	2.41
CBG	2.5	2.71	2.34	3.33	2.91
CBN	2.5	2.78	2.56	2.53	2.18

© Agilent Technologies, Inc. 2019
Reproduced with Permission, Courtesy of Agilent Technologies, Inc.

Table 5 Results of 3 real marihuana flower samples determined the same way the results in Table 4.

	Sample 1		Sample 2		Sample 3	
	Diluted	^{13}C	Diluted	^{13}C	Diluted	^{13}C
THC	7.03	6.88	0.5	0.86	0.95	0.66
THCA	9.29	9.8	7.17	7.92	4.63	4.42
CBD	ND	ND	ND	ND	ND	ND
CBDA	ND	ND	ND	ND	ND	ND
CBG	0.11	0.15	ND	ND	ND	ND
CBN	0.09	0.12	ND	ND	ND	ND

© Agilent Technologies, Inc. 2019
Reproduced with Permission, Courtesy of Agilent Technologies, Inc.

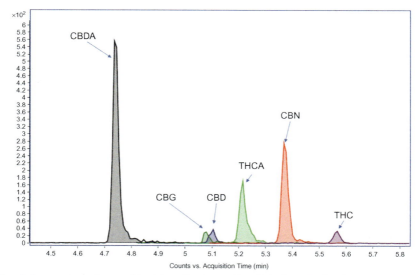

Fig. 6 Separation of cannabinoids spiked at 0.12% © Agilent Technologies, Inc. 2019. *Reproduced with Permission, Courtesy of Agilent Technologies, Inc.*

These results are given in Table 5. Again, the results labelled "Diluted" using the ^{12}C transitions (10,000-fold dilution) are comparable and show that the use of the ^{13}C isotope transitions with the single method provides accurate results. It is interesting that the flowers of these marihuana samples contain mostly THC and THCA with very little to none of the other cannabinoids. Fig. 6 shows the separation achieved using the Eclipse Plus C_{18} column for the 6 cannabinoids listed in the California regulation at the regulatory limit for potency.

4. Summary

Assuring the absence of harmful residues and measuring the potency of legalized cannabis is a safety issue. This work has shown the capability of LC/MS/MS to meet regulations of the flowers of cannabis for pesticide residues, mycotoxins and potency in one analysis for safe cannabis use. The use of native carbon 13 transitions demonstrated the ability to extend the dynamic range of the MS detection from $1\,\mu g/L$ to $24,000\,\mu g/L$. This capability can be more broadly applied. With the legalization of marihuana, not only will the need to assure safety continue but scientific studies will increase dramatically. One area of research will be further exploration of the entourage effect [17,64] and whether there is a synergistic effect between the over 120 cannabinoids in the plant. This will require quantitative measurement of those cannabinoids present and the range of concentrations may benefit from using mass spectrometry with the native isotopes of the compounds. For this type of measurement, even a single quadrupole mass spectrometer or other mass analyser (e.g. time-of-flight) will have the selectivity and sensitivity to perform the needed measurements in one analysis.

Acknowledgements

Both authors would like to thank Andre Santos and Tarun Anumol from Agilent Technologies for their support of this project.

References

[1] E.B. Russo, History of cannabis and its preparations in saga, science, and sobriquet, Chem. Biodivers. 4 (8) (2007) 1614–1648.

[2] A.M. Mercuri, C.A. Accorsi, M.B. Mazzanti, The long history of Cannabis and its cultivation by the romans in Central Italy, shown by pollen records from Lago Albano and Lago di Nemi, Veg. Hist. Archaeobotany 11 (4) (2002) 263–276.

[3] A. Hand, A. Blake, P. Kerrigan, P. Samuel, J. Friedberg, History of medical cannabis, J. Pain Manag. 9 (4) (2016) 387.

[4] S. Wills, Cannabis use and abuse by man: an historical perspective, in: D.T. Brown (Ed.), Cannabis, CRC Press, 1998, pp. 16–46.

[5] E.B. Russo, The pharmacological history of Cannabis, in: R.G. Pertwee (Ed.), Handbook of Cannabis, Oxford University Press, United States, 2014, pp. 23–43.

[6] S.A. Bonini, et al., Cannabis sativa: a comprehensive ethnopharmacological review of a medicinal plant with a long history, J. Ethnopharmacol. 227 (2018) 300–315.

[7] B. Warf, High points: an historical geography of cannabis, Geogr. Rev. 104 (4) (2014) 414–438.

[8] G. Piluzza, G. Delogu, A. Cabras, S. Marceddu, S. Bullitta, Differentiation between fiber and drug types of hemp (*Cannabis sativa* L.) from a collection of wild and domesticated accessions, Genet. Resour. Crop. Evol. 60 (8) (2013) 2331–2342.

[9] J.A. Beutler, A.H. Marderosian, Chemotaxonomy of Cannabis I. Crossbreeding between *Cannabis sativa* and *C. ruderalis*, with analysis of cannabinoid content, Econ. Bot. 32 (4) (1978) 387.

[10] D. Piomelli, E.B. Russo, The Cannabis sativa versus Cannabis indica debate: an interview with Ethan Russo, MD, Cannabis Cannabinoid Res. 1 (1) (2016) 44–46.

[11] E. Small, Evolution and classification of *Cannabis sativa* (Marijuana, Hemp) in relation to human utilization, Bot. Rev. 81 (3) (2015) 189–294 2015/09/01.

[12] A.W. Zuardi, History of cannabis as a medicine: a review, Braz. J. Psychiatry 28 (2) (2006) 153–157.

[13] H. Peters, G.G. Nahas, A brief history of four Millennia (BC 2000–AD 1974), in: N. Pace, H. Frick, K. Sutin, W. Manger, G. Hyman, G. Nahas (Eds.), Marihuana and Medicine, Humana Press Inc., Totowa, NJ, 1999, pp. 3–7.

[14] E.B. Russo, History of cannabis as medicine: nineteenth century irish physicians and correlations of their observations to modern research, in: *Cannabis sativa* L.-Botany and Biotechnology, Springer, 2017, pp. 63–78.

[15] E. Russo, Hemp for headache: an in-depth historical and scientific review of cannabis in migraine treatment, J. Cannabis Ther. 1 (2) (2001) 21–92.

[16] M. Kuddus, I.A. Ginawi, A. Al-Hazimi, Cannabis sativa: an ancient wild edible plant of India, Emir. J. Food Agr. 25 (10) (2013) 736–745.

[17] M.O. Bonn-Miller, M.A. ElSohly, M.J. Loflin, S. Chandra, R. Vandrey, Cannabis and cannabinoid drug development: evaluating botanical versus single molecule approaches, Int. Rev. Psychiatry 30 (3) (2018) 277–284.

[18] V. Di Marzo, A brief history of cannabinoid and endocannabinoid pharmacology as inspired by the work of British scientists, Trends Pharmacol. Sci. 27 (3) (2006) 134–140. 2006/03/01.

[19] P. Morales, P.H. Reggio, N. Jagerovic, "An overview on medicinal chemistry of synthetic and natural derivatives of Cannabidiol," (in English), Front. Pharmacol. 8 (2017) 422. 2017/06/28.

[20] D. Friedman, J.I. Sirven, Historical perspective on the medical use of cannabis for epilepsy: ancient times to the 1980s, Epilepsy Behav. 70 (2017) 298–301.

[21] A.S. Suraev, et al., An Australian nationwide survey on medicinal cannabis use for epilepsy: history of antiepileptic drug treatment predicts medicinal cannabis use, Epilepsy Behav. 70 (Pt. B) (2017) 334–340.

[22] K. Sekar, A. Pack, Epidiolex as adjunct therapy for treatment of refractory epilepsy: a comprehensive review with a focus on adverse effects. F1000Res. 8 (2019) pii: F1000 Faculty Rev-234. https://doi.org/10.12688/f1000research.16515.1.

[23] Clinical Policy: Cannabidiol (Epidiolex), Reference Number: CP.PMN.164, Effective Date: 07.17.18, Last Review Date: 08.19.

[24] Y.T. Yang, J.P. Szaflarski, The US Food and Drug Administration's authorization of the first Cannabis-derived pharmaceutical: are we out of the haze? JAMA Neurol. 76 (2) (2019) 135–136.

[25] M.A. Alsherbiny, C.G. Li, Medicinal cannabis—potential drug interactions, Medicines 6 (1) (2019) 3.

[26] A. Capano, R. Weaver, E. Burkman, Evaluation of the effects of CBD hemp extract on opioid use and quality of life indicators in chronic pain patients: a prospective cohort study, Postgrad. Med. 132 (1) (2020) 56–61. 2020/01/02.

[27] A. Knopf, CBD may help prevent relapse in abstinent heroin addicts, Alcohol. Drug Abuse Weekly 31 (22) (2019) 3–4.

[28] S.R. Sultan, S.A. Millar, T.J. England, S.E. O'Sullivan, "A systematic review and meta-analysis of the haemodynamic effects of Cannabidiol," (in English), Front. Pharmacol. 8 (2017) 81. Original Research 2017-February-24.

[29] C. Song, C.W. Stevenson, F.S. Guimaraes, J.L.C. Lee, "Bidirectional effects of Cannabidiol on contextual fear memory extinction," (in English), Front. Pharmacol. 7 (2016) 493. Original Research 2016-December-16.

[30] C.M. White, A review of human studies assessing cannabidiol's (CBD) therapeutic actions and potential, J. Clin. Pharmacol. 59 (7) (2019) 923–934.

[31] A. Mondino, et al., Acute effect of vaporized Cannabis on sleep and electrocortical activity, Pharmacol. Biochem. Behav. 179 (2019) 113–123.

[32] M. Di Forti, et al., Daily use, especially of high-potency cannabis, drives the earlier onset of psychosis in cannabis users, Schizophr. Bull. 40 (6) (2014) 1509–1517.

[33] E. Manrique-Garcia, S. Zammit, C. Dalman, T. Hemmingsson, S. Andreasson, P. Allebeck, Prognosis of schizophrenia in persons with and without a history of cannabis use, Psychol. Med. 44 (12) (2014) 2513–2521.

[34] R.M. Murray, P.D. Morrison, C. Henquet, M. Di Forti, Cannabis, the mind and society: the hash realities, Nat. Rev. Neurosci. 8 (11) (2007) 885–895.

[35] I. Ghinai, et al., E-cigarette product use, or vaping, among persons with associated lung injury—Illinois and Wisconsin, April–September 2019, Morb. Mortal. Wkly Rep. 68 (39) (2019) 865.

[36] L.N. Sandler, J.L. Beckerman, F. Whitford, K.A. Gibson, Cannabis as conundrum, Crop Prot. 117 (2019) 37–44. 2019/03/01.

[37] Public Law 115–334—Agriculture Improvement Act of 2018, 2018.

[38] Bureau of Cannabis Control Text of Regulations, D. B. O. C. Control, 2018.

[39] A.J. Williams, et al., The CompTox chemistry dashboard: a community data resource for environmental chemistry, J. Cheminf. 9 (1) (2017) 61. 2017/11/28.

[40] J. Mullins, Imidacloprid: A New Nitroguanidine Insecticide, ACS Publications, 1993.

[41] A. Szczepaniec, M.J. Raupp, R.D. Parker, D. Kerns, M.D. Eubanks, Neonicotinoid insecticides alter induced defenses and increase susceptibility to spider mites in distantly related crop plants. PLoS One 8 (5) (2013) e62620. https://doi.org/10.1371/journal.pone.0062620.

[42] J.-M. Zhang, Y.-L. Wu, Y.-B. Lu, Simultaneous determination of carbamate insecticides and mycotoxins in cereals by reversed phase liquid chromatography tandem mass spectrometry using a quick, easy, cheap, effective, rugged and safe extraction procedure, J. Chromatogr. B 915 (2013) 13–20.

[43] J. Xie, et al., Multi-residue analysis of veterinary drugs, pesticides and mycotoxins in dairy products by liquid chromatography–tandem mass spectrometry using low-temperature cleanup and solid phase extraction, J. Chromatogr. B 1002 (2015) 19–29.

[44] H.G.J. Mol, P. Plaza-Bolaños, P. Zomer, T.C. de Rijk, A.A.M. Stolker, P.P.J. Mulder, Toward a generic extraction method for simultaneous determination of pesticides, mycotoxins, plant toxins, and veterinary drugs in feed and food matrixes, Anal. Chem. 80 (24) (2008) 9450–9459. 2008/12/15.

[45] C.F. Amate, H. Unterluggauer, R. Fischer, A. Fernández-Alba, S. Masselter, Development and validation of a LC–MS/MS method for the simultaneous determination of aflatoxins, dyes and pesticides in spices, Anal. Bioanal. Chem. 397 (1) (2010) 93–107.

[46] H. Liu, L. Lam, P.K. Dasgupta, Expanding the linear dynamic range for multiple reaction monitoring in quantitative liquid chromatography–tandem mass spectrometry utilizing natural isotopologue transitions, Talanta 87 (2011) 307–310. 2011/12/15.

[47] A.M. Miller, M.M. Goggin, A. Nguyen, S.D. Gozum, G.C. Janis, Profiting from probability; combining low and high probability isotopes as a tool extending the dynamic range of an assay measuring AMPHETAMINE and methamphetamine in urine, J. Anal. Toxicol. 41 (5) (2017) 355–359.

[48] H. Liu, L. Lam, L. Yan, B. Chi, P.K. Dasgupta, Expanding the linear dynamic range for quantitative liquid chromatography-high resolution mass spectrometry utilizing natural isotopologue signals, Anal. Chim. Acta 850 (2014) 65–70. 2014/11/19.

[49] S. Trobbiani, P. Stockham, T. Scott, Increasing the linear dynamic range in LC–MS: is it valid to use a less abundant isotopologue? Drug Test. Anal. 9 (10) (2017) 1630–1636.

[50] A. Bilbao, et al., An algorithm to correct saturated mass spectrometry ion abundances for enhanced quantitation and mass accuracy in omic studies, Int. J. Mass Spectrom. 427 (2018) 91–99. 2018/04/01.

[51] K.K. Murray, et al., IUPAC standard definitions of terms relating to mass spectrometry, in: IUPAC MS Terms and Definitions, First Public Draft, 2005.

[52] J. Urban, N.K. Afseth, D. Štys, Fundamental definitions and confusions in mass spectrometry about mass assignment, centroiding and resolution, TrAC Trends Anal. Chem. 53 (2014) 126–136.

[53] K.K. Murray, R.K. Boyd, M.N. Eberlin, G.J. Langley, L. Li, Y. Naito, Definitions of terms relating to mass spectrometry (IUPAC recommendations 2013), Pure Appl. Chem. 85 (7) (2013) 1515–1609.

[54] A.G. Marshall, C.L. Hendrickson, S.D.-H. Shi (Eds.), Peer Reviewed: Scaling Ms Plateaus with High-Resolution FT-ICRMS, ACS Publications, 2002.

[55] T.M. Annesley, Ion suppression in mass spectrometry, Clin. Chem. 49 (7) (2003) 1041–1044.

[56] H. Faccin, C. Viana, P.C. do Nascimento, D. Bohrer, L.M. de Carvalho, Study of ion suppression for phenolic compounds in medicinal plant extracts using liquid chromatography-electrospray tandem mass spectrometry, J. Chromatogr. A 1427 (2016) 111–124. Jan 4.

[57] A. Furey, M. Moriarty, V. Bane, B. Kinsella, M. Lehane, Ion suppression; a critical review on causes, evaluation, prevention and applications, Talanta 115 (Oct 15 2013) 104–122.

[58] J.-P. Antignac, K. de Wasch, F. Monteau, H. De Brabander, F. Andre, B. Le Bizec, The ion suppression phenomenon in liquid chromatography–mass spectrometry and its consequences in the field of residue analysis, Anal. Chim. Acta 529 (1–2) (2005) 129–136.

[59] B.K. Matuszewski, Standard line slopes as a measure of a relative matrix effect in quantitative HPLC-MS bioanalysis, J. Chromatogr. B Analyt. Technol. Biomed. Life Sci. 830 (2) (2006) 293–300. Jan 18.

[60] F. Gosetti, E. Mazzucco, D. Zampieri, M.C. Gennaro, Signal suppression/enhancement in high-performance liquid chromatography tandem mass spectrometry, J. Chromatogr. A 1217 (25) (2010) 3929–3937.

[61] A. Kruve, A. Künnapas, K. Herodes, I. Leito, Matrix effects in pesticide multi-residue analysis by liquid chromatography–mass spectrometry, J. Chromatogr. A 1187 (1) (2008) 58–66. 2008/04/11.

[62] J. Kilcoyne, E. Fux, Strategies for the elimination of matrix effects in the liquid chromatography tandem mass spectrometry analysis of the lipophilic toxins okadaic acid and azaspiracid-1 in molluscan shellfish, J. Chromatogr. A 1217 (45) (2010) 7123–7130. Nov 5.

[63] C. Ferrer, A. Lozano, A. Aguera, A.J. Giron, A.R. Fernandez-Alba, Overcoming matrix effects using the dilution approach in multiresidue methods for fruits and vegetables, J. Chromatogr. A 1218 (42) (2011) 7634–7639. Oct 21.

[64] E.B. Russo, The case for the entourage effect and conventional breeding of clinical cannabis: no "strain," no gain, Front. Plant Sci. 9 (2019) 1969.

CHAPTER TWELVE

Using sesame seed oil to preserve and concentrate cannabinoids for paper spray mass spectrometry[☆]

Brandon J. Bills*, Nicholas E. Manicke
Department of Chemistry and Chemical Biology, Indiana University Purdue University at Indianapolis, Indianapolis, IN, United States
*Corresponding author: e-mail address: bjbills@iupui.edu

Contents

1. Introduction	367
2. Methods	371
2.1 Sesame oil preservation	371
2.2 Evaluating preconcentration by paper strip extraction	374
2.3 Measuring improvements to detection limits: Preconcentration and preservation	376
2.4 Automated analysis	377
3. Results and discussion	379
3.1 Sesame oil preservation	379
3.2 Evaluating preconcentration using paper strip extraction	381
3.3 Measuring improvements to detection limits: Preservation and preconcentration	383
3.4 Automated analysis	390
4. Suitability for cannabinoid testing	391
5. Conclusion	392
Acknowledgements	393
References	393

1. Introduction

Laws regarding marijuana and synthetic cannabinoids (synthetic compounds that mimic the active ingredient in marijuana, tetrahydrocannabinol or THC) in the United States are variable and in a state of flux.

[☆] This chapter was adapted with permission from [1] B. Bills, N. Manicke, Using sesame seed oil to preserve and preconcentrate cannabinoids for paper spray mass spectrometry, J. Am. Soc. Mass Spectrom. (2020), Ahead of Print. Copyright 2020 American Chemical Society.

Comprehensive Analytical Chemistry, Volume 90
ISSN 0166-526X
https://doi.org/10.1016/bs.coac.2020.04.003
© 2020 Elsevier B.V.
All rights reserved.

At the national level marijuana is still a schedule 1 narcotic (the most restricted category) [2]. However, at the state level 26 states have decriminalized marijuana allowing people to possess small amounts without fear of jailtime, and 11 states have even legalized recreational marijuana use for adults [2]. The number of states that have decriminalized or legalized recreational use of marijuana can shift on a regular basis, and laws regarding driving while under the influence of THC can vary extensively. With the legality of marijuana shifting, it may be necessary to reevaluate what is considered an acceptable amount of THC in the system while driving. Even if THC is legal in a state, it is still a psychoactive substance that has been shown to impair driving [3,4]. Currently, as shown in Fig. 1, 29 states have no specific law regarding THC [5]. This means that drug impairment for THC is treated as all other scheduled drugs.

The remaining 21 states have a mix of zero tolerance laws (any amount of THC or its metabolites in the system are illegal while driving) and *per se* or reasonable inference laws, a driver is considered intoxicated or impaired respectively if a concentration above the cut-off level is detected in their system.

Complicating matters is the prevalence of synthetic cannabinoids. These compounds were developed during pharmaceutical research into THC and are now being marketed as incense to skirt law enforcement [6,7]. Some of these substances end up being scheduled and are treated accordingly with regards to driving while impaired [8]. However, there are over 100 possible synthetic cannabinoids (see Fig. 2 for examples of synthetic cannabinoids as well as THC and its metabolites).

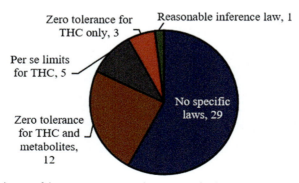

Fig. 1 Breakdown of how many states have specific laws regarding driving under the influence of marijuana. Zero tolerance indicates it is illegal to drive with measurable quantities of THC or its metabolites. A *per se* limit is a cap on how much THC can be present in the system before the driver is considered intoxicate. Reasonable inference indicates that over a certain quantity of THC in the system it is permissible to assume that the driver is under the influence.

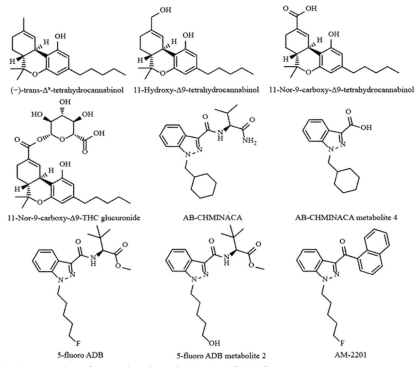

Fig. 2 Examples of natural and synthetic cannabinoids.

Beyond the issue of legality, these synthetic cannabinoids can have variable potency and unknown side effects [9–12]. As a result, it is more likely that someone using these drugs will end up in the emergency room with little to no idea of what substance they consumed.

These factors signal a need for adequate screening that is capable of detecting both natural and synthetic cannabinoids simultaneously. Traditionally, screening for THC involved an immunoassay, a test involving an antibody and a colour changing reagent, followed up with gas chromatography–mass spectrometry for confirmation [13,14]. This type of test was viable when the number of possible targets was limited and the concentrations of THC or its metabolites were high enough. However, with the range of potencies and high number of potential targets possible with synthetic cannabinoids, immunoassays are no longer adequate for the current challenge [15]. Mass spectrometry overcomes many of the challenges associated with synthetic cannabinoids. Steps like sample cleanup and preconcentration coupled with the increasing sensitivity of modern mass spectrometers are capable of detecting lower concentrations [16].

Simultaneously, coupling liquid or gas chromatography with tandem mass spectrometry offers a means to separate out components of the sample and enough selectivity to differentiate a large number of similar analytes [17,18]. For these reasons, chromatography paired with mass spectrometry is a useful tool with regards to confirmatory testing. However, many of the aspects that improve the sensitivity and selectivity of these methods also increase analysis time and cost, making them impractical as a rapid screening technique.

Ambient ionization techniques, such as direct analysis in real time (DART) or Desorption Electrospray Ionization (DESI), allow for much more rapid screening by directly ionizing samples with minimal preparation under atmospheric pressure [19–21]. These techniques often employ streams of charged solvent or gas to dissolve, dislodge or otherwise get analytes charged and in the gas phase directly out of the sample in order to obtain results as quickly as possible. Paper spray mass spectrometry is particularly useful for screening cannabinoids due to its ability to directly sample biofluids such as urine or oral fluid. First described in 2010, paper spray involves spotting a drop of biofluid on a piece of paper that comes to a point [22,23]. By spotting solvent behind the biofluid spot, analytes are eluted as the solvent wicks to the tip of the paper then ionized when the paper is charged with electricity (see Fig. 3).

Pairing the alacrity of paper spray with the selectivity of mass spectrometry could offer a powerful option for rapid screening of cannabinoids. However, several issues hinder the use of paper spray for this application. Due to the lack of a cleanup or chromatography, the entirety of the biofluid sample is analysed simultaneously. This leads to decreased sensitivity due to matrix effects [24,25]. In addition, natural cannabinoids pose a problem because they are often labile with a tendency to volatilize or otherwise degrade when in a dried aqueous matrix [26].

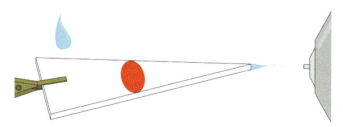

Fig. 3 Paper spray mass spectrometry set up.

Several techniques have been investigated to mitigate matrix effects with paper spray. Some of these techniques have included: building a cartridge to perform solid phase extraction and paper spray from the same device, using hydrophobic treated paper to perform a liquid-liquid extraction coupled with paper spray, and studying how the physical properties of the paper impact various matrix effects [27–29]. However, these techniques do not address how labile natural cannabinoids are. Pharmaceuticals that use THC as the active ingredient, like Dronabinol, utilize sesame oil inside tablets to store the cannabinoid. Research into the shelf life of these tablets found that at room temperature over several months the concentration of THC did not change significantly [30].

The concept that THC can be stable for prolonged periods of time in sesame oil prompted a series of experiments designed to determine whether this phenomenon could be used to facilitate simultaneous analysis of natural and synthetic cannabinoids by paper spray mass spectrometry. Initial experiments investigated whether sesame oil could preserve natural cannabinoids in dried spots by spotting spiked urine samples onto paper containing sesame oil and comparing results to untreated paper, as well as paper treated with alternate oils. A similar experiment was performed to verify whether the antioxidants within sesame oil were responsible for the preservation of the cannabinoids by comparing a synthetic sesame oil with different antioxidants. This work also investigated whether cannabinoids, which are exceptionally hydrophobic, can be preconcentrated by passing urine or oral fluid through an oil spot on a piece of paper [31]. Finally, samples were prepared with and without sesame oil to determine whether samples could be preserved over 27 days.

2. Methods
2.1 Sesame oil preservation
2.1.1 Rationale
Sesame oil is used to preserve THC within the pharmaceutical Dronabinol. This protocol was designed to determine whether sesame oil could be used to preserve THC in a urine sample stored on paper. Also tested are oleic acid (a major component of sesame oil), mineral oil (a nonfatty acid based oil), and a synthetic version of sesame oil with different antioxidants added (to evaluate whether the antioxidants present in sesame oil are the major preserving factor).

2.1.2 Materials and equipment

Fentanyl, THC, and metabolites of THC were obtained along with their stable isotopic labels from Cerilliant (Round Rock, TX, USA). AB-CHMINACA and AM-2201 and their labels were obtained from Cayman Chemical (Ann Arbor, MI, USA). HPLC grade methanol, acetonitrile, palmitic acid, linoleic acid and oleic acid came from Fisher Scientific (Waltham, MA, USA). Concentrated sulphuric acid came from EMD Millipore (Burlington, MA, USA). Whatman grade 31 ET chromatography paper came from GE Healthcare Life Sciences (Pittsburgh, PA, USA). Sesame oil, sesamolin, mineral oil and butylated hydroxytoluene came from Sigma-Aldrich (St. Louis, MO, USA). Sesamin came from ApexBio (Taiwan). Urine was provided by a single volunteer. Analysis was carried out on a Q-exactive focus from Thermo Fisher Scientific (San Jose, CA, USA).

2.1.3 Sample preparation and analysis

1. Urine was spiked at $1\,\mu g/mL$ for $(-)$-*trans*-Δ^9-tetrahydrocannabinol (THC), 11-Hydroxy-Δ^9-tetrahydrocannabinol (11-OH-THC), 11-Nor-9-carboxy-Δ^9-tetrahydrocannabinol (THC-COOH), 11-Nor-9-carboxy-Δ^9-THC glucuronide (THC glucuronide), AB-CHMINACA, AM-2201 and fentanyl (a noncannabinoid for comparison) in a glass vial. The structures for the cannabinoids are shown in Fig. 2.

2. A synthetic sesame oil was prepared by mixing oleic, linoleic and palmitic acid together in a 48:37:8 ratio and then adding either 0.6% sesamin, 0.5% sesamolin or both sesamin and sesamolin (same %) [32]. The food preservative butylated hydroxytoluene (BHT) was added as a synthetic alternative to another aliquot of synthetic sesame oil.

3. The various oils were spotted onto $5\,mm \times 5\,mm$ squares of 31ET chromatography paper in $5\,\mu L$ aliquots.

4. Urine samples were spotted in $5\,\mu L$ aliquots using a glass capillary onto the prepared squares of chromatography paper, and allowed to dry for either 1 h or 1 day at room temperature on the counter.

5. Internal standards were spotted and allowed to dry immediately prior to analysis to minimize the amount of degradation of the labile labelled compounds. Sample squares were spotted with $2.5\,\mu L$ of methanol containing $500\,ng/mL$ of the stable isotopic labels (SILs) THC D3, THC-COOH D3, AB-CHMINACA D4, AM-2201 D5 and fentanyl D5 in methanol. THC glucuronide and 11-OH-THC used THC D3 for an internal standard.

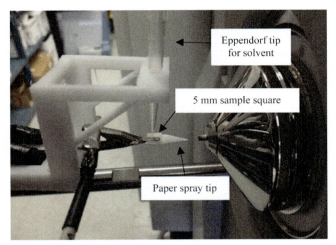

Fig. 4 Paper spray setup. A 3D printed support holding an alligator clip and Eppendorf tip to act as sample holder and solvent well respectively.

6. Analysis was carried out using a holder to minimize contact between plastic and paper as shown in Fig. 4.

   The holder supports a plastic pipette tip that holds solvent and an alligator clip that holds the sample square and paper tip. The paper tip was positioned in front of the inlet of a Thermo Fisher Scientific Q-Exactive Focus orbitrap mass spectrometer. 60 μL of spray solvent, 80:20 acetonitrile:methanol with 25 mM sulphuric acid, was added to the pipette tip. After the solvent flowed through the sample square to the paper spray tip clipped beneath it, 4.5 kV of voltage was applied via a secondary alligator clip for 1 min. The instrument was run in parallel reaction monitoring (PRM) mode, and the area under the curve (AUC) for the most abundant fragment for each compound was integrated. THC, 11-OH-THC, and COOH-THC all used the same fragment of 193.1221 (within 5 ppm precision).

2.1.4 Data analysis

The AUC while the voltage was on was determined for each analyte and SIL. The ratio between the analyte (which either had 24 h or minimal time to dry and degrade) and the SIL (which only had minimal time to dry and degrade) was determined for each analyte for each oil type. If the ratio decreased significantly in the 24 h sample then it was reasoned that the analyte of interest had degraded or volatilized, and the oil used was not a useful preservative for that analyte.

2.1.5 Summary

This method uses the change in ratios between an analyte and its internal standard to measure the preserving capabilities of different oils. If an analyte is well preserved by an oil, then the ratio between the analyte (which is spiked either 24 h or immediately prior to analysis) and the stable isotopic label (which is always spiked immediately prior to analysis) should be consistent regardless of how long the analyte sat before analysis. If the ratio drops, then the analyte volatilized or degraded in those 24 h, and the oil is shown to be a poor preservative.

2.2 Evaluating preconcentration by paper strip extraction

2.2.1 Rationale

Natural and synthetic cannabinoids are often exceptionally hydrophobic. This protocol was designed to determine whether passing a urine sample containing a mixture of cannabinoids and other pharmaceuticals through a spot of sesame oil on a strip of paper would lead to the hydrophobic compounds preconcentrating in the oil spot.

2.2.2 Materials

In addition to the materials listed in previous sections, analytes (atenolol, alprazolam, carbamazepine, diazepam, and gabapentin) and their stable isotopic labels (SIL) were obtained from Cerilliant (Round Rock, TX, USA) with the exception of atenolol D7 which was obtained from CDN isotopes (Pointe-Claire, QC, Canada) as a powder ($\geq$98% purity). 3MM chromatography paper was obtained from GE Healthcare Life Sciences (Pittsburgh, PA, USA).

2.2.3 Sample preparation and analysis

1. Urine was spiked at 100 ng/mL for atenolol, alprazolam, carbamazepine, diazepam, gabapentin, fentanyl, AB-CHMINACA, AM-2201, THC glucuronide and THC.
2. Paper strips were prepared by cutting Whatman 3MM chromatography paper into 5 mm × 40 mm strips, and spotting 2.5 µL of sesame oil at one end of each strip.
3. A second set of paper strips with no oil were also prepared and used for extraction as a comparison.
4. Samples were prepared and analysed in triplicate and the results averaged.
5. 50 µL of urine was applied using a glass capillary to the end of the strip containing the sesame oil and allowed to dry.

Fig. 5 Paper strip extraction. After drying, the extraction strip was cut into 5 mm increments and each segment analysed to determine change in concentration throughout the strip.

6. After drying, the strips were then cut into 5 mm increments (as shown in Fig. 5).
7. Each segment was spiked with 5 μL of a 500 ng/mL solution of the SILs: THC D3, THC-COOH D3, AB-CHMINACA D4, AM-2201 D5 and fentanyl D5 in methanol.
8. Samples were analysed by paper spray mass spectrometry, as described in the previous section, and the ratio between analyte and internal standard was determined for each analyte on each section.

2.2.4 Data analysis

After the ratio between analyte and internal standard was determined for all analytes in each 5 mm segment, trends were sought in the changes in ratio from the leading edge of the strip (where the urine was spiked and where the oil was if present for that sample). This was done by normalizing each successive 5 mm increment to the first 5 mm increment. Because the internal standard was spiked onto the 5 mm increments after the extraction, any decreasing trend should be an indication that the analyte had preconcentrated in the first 5 mm.

2.2.5 Summary

This protocol sought to determine which analytes would preconcentrate out of a urine matrix when passed through a spot of sesame oil on a strip of paper. By cutting the strip into 5 mm increments after passing the urine through, the amount of analyte in each segment should represent how likely the analyte is to preconcentrate in the oil or evenly distribute throughout the strip. Spiking the same amount of internal standard on each segment and measuring the ratio between analyte and internal standard allow trends in analyte distribution to be determined.

2.3 Measuring improvements to detection limits: Preconcentration and preservation

2.3.1 Rationale

Previous work had shown that cannabinoids could be preconcentrated and preserved using sesame oil. In order to determine whether these effects also lead to a corresponding improvement in detection limits, samples were prepared in a variety of ways (with and without sesame oil or with and without preconcentration) and stored for an extended period of time prior to analysis.

2.3.2 Materials and equipment

In addition to materials listed in previous sections, AB-CHMINACA metabolite 4,5-fluoro ADB, 5-fluoro ADB metabolite 2 were obtained from Cayman Chemical (Ann Arbor, MI, USA).

2.3.3 Sample preparation and analysis: Preservation

1. To evaluate effects of preservation on detection limits, a calibration curve was generated in urine spiked at 0.5, 5, 10, 100, 500 and 1000 ng/mL for THC, 11-OH-THC, THC-COOH, AB-CHMINACA, AM-2201, and fentanyl and 100 ng/mL SIL analogs for each analyte was spiked into each calibrator, except 11-OH-THC and THC-COOH.
2. In order to determine the extent of preservation, two sets of calibrators were prepared: one extracted using sesame oil as described earlier, and a second set extracted without sesame oil for comparison.
3. After extraction the first 5 mm segment of each paper strip (where the biofluid was deposited) was removed and stored in a plastic container inside a drawer at room temperature until analysis.
4. Two replicates were prepared and analysed at each calibration level and blanks (biofluid with internal standard) were analysed in triplicate.
5. Samples (the calibrators) were analysed by paper spray after 1, 7, and 27 days.

2.3.4 Sample preparation and analysis: Preconcentration

1. To evaluate the effects of preconcentration on detection limits, urine and oral fluid was spiked at 0.5, 5, 25, 100, 500 and 1000 ng/mL of THC, 11-OH-THC, THC-COOH, AB-CHMINACA, AM-2201, 5F-ADB and AB-CHMINACA metabolite 4 and 100 ng/mL SIL analogs for each analyte was spiked into each calibrator except the metabolites AB-CHMINACA metabolite 4, 11-OH-THC and THC-COOH.

2. In order to determine whether preconcentration improves detection limits, two sets of calibrators were prepared: one extracted using sesame oil as described earlier, and a second set in which a 5 μL aliquot of spiked biofluid was spiked onto a 5 mm square of chromatography paper containing sesame oil (in order to have a similar volume of oil to the extracted samples, a strip of paper was prepared with sesame oil as normal, then the first 5 mm segment was removed and spiked with biofluid).
3. After extraction, the first 5 mm segment of each paper strip (where the biofluid was deposited) was removed and stored in a plastic container inside a drawer at room temperature until analysis.
4. Two replicates were prepared and analysed at each calibration level and blanks (biofluid with internal standard) were analysed in triplicate.
5. Samples (the calibrators) were analysed by paper spray after 1 and 28 days to evaluate any links between preservation and preconcentration.

2.3.5 Data analysis
For both sets of experiments, results were evaluated based on changes to detection limits. For each blank and calibrator, the ratio between the area under the curve for the analyte and its internal standard was found. Metabolites used the stable isotopic label of the unmetabolized form for the internal standard. The slope and Y-intercept were determined by linear regression [33] with a weighting factor of $1/X^2$. Samples showing no signal or signal lower than three standard deviations above the average blank signal were omitted. Detection limits were calculated as three times the standard error of the Y-intercept divided by the slope.

2.3.6 Summary
Previous experiments showed that paper strip extraction using sesame oil can lead to preconcentration and preservation of cannabinoids. The two similar protocols described here are designed to evaluate how these effects impact the final limits of detection when analysing cannabinoids by paper spray. This is accomplished by creating calibration curves with and without sesame oil for preservation and with and without extraction to evaluate preconcentration.

2.4 Automated analysis
2.4.1 Rationale
Ideally the methods presented here could be incorporated into an easy to use disposable analytical device. To test this, a partially 3D printed device was

modelled and printed that was capable of holding the samples generated using these methods while being analysed by a paper spray autosampler.

2.4.2 Materials and equipment

In addition to the materials mentioned in previous sections, the bottom half of the autosampler cartridge and the Velox autosampler were produced by Prosolia (Indianapolis, IN, USA). The top half of the autosampler cartridge was modelled in Sketchup (Trimble, Sunnyvale, CA, USA) and 3D printed using a polypropylene filament using an Ultimaker 2+ extended (Geldermalsen, Gelderland, Netherlands).

2.4.3 Sample preparation and analysis

1. Calibration curves and blanks were prepared in oral fluid and extracted as described in the "Sample Preparation and Analysis: Preservation" section of the previous protocol.
2. Samples were analysed using an autosampler cartridge that consists of an injection moulded bottom half and a 3D printed polypropylene top half that snaps onto the bottom half and holds the 5 mm sample square as shown in Fig. 6.
3. Samples were analysed using a paper spray autosampler.

2.4.4 Data analysis

Detection limits were determined by linear regression as discussed in the previous protocol.

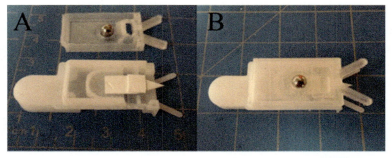

Fig. 6 Prototype paper strip extraction cartridge: (A) 3D printed polypropylene top half and injection moulded plastic bottom half, (B) assembled cartridge.

2.4.5 Summary
This body of work shows that using paper strip extraction with sesame oil can lower detection limits for cannabinoids. In order to test the ability of this method to be useful in a rapid screening technique closer to a real-life application, this protocol evaluates whether taking the samples generated by paper strip extraction and analysing them in a self-contained autosampler cartridge has a significant impact on detection limits relative to previously obtained results.

3. Results and discussion
3.1 Sesame oil preservation

$(-)$-*trans*-Δ^9-tetrahydrocannabinol (THC) and its metabolites can be unstable, making them difficult to analyse. THC, for example, will oxidize into cannabinol without preservation [30]. Additionally, due to their hydrophobicity, a significant portion of THC can be lost to plastic containers from aqueous matrices like urine [26]. Dronabinol, the synthetic version of THC sold for pharmaceutical purposes, uses sesame oil to preserve THC for extended periods of time at room temperature [30]. To determine whether this preservation could be extended to dried urine spots, sesame oil was added to the paper used to store the sample. Two additional oils were also tested as a comparison: oleic acid, a major component of sesame oil, and mineral oil, a petroleum-based oil without fatty acids. The amount of analyte remaining after storage was measured as a relative amount compared to the stable isotopic label (SIL) spotted immediately prior to analysis. Preservation is reported here as the percent analyte remaining in samples that had been dried for an hour compared to samples that had been dried for a day (Fig. 7).

Without oil, THC and its metabolites showed a significant drop over the course of a day while the two synthetic cannabinoids and fentanyl did not, indicating that the synthetic drugs were more stable than the natural cannabinoids in the dried urine spot. Of the three oils tested, only sesame oil effectively preserved THC. Sesame oil is a mixture of fatty acids, antioxidants, and other natural products [32,34]. A previous study on dronabinol indicated that the antioxidants within the oil likely helped THC preservation [30]. To test whether antioxidants were the reason THC was preserved in dried urine spots, a mixture of oleic, linoleic and palmitic acid in a similar ratio as sesame oil was tested. The synthetic sesame oil was also spiked either

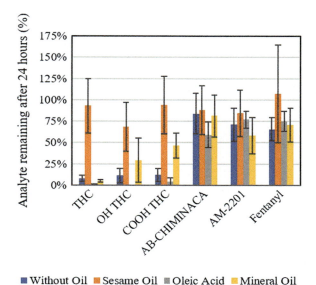

■ Without Oil ■ Sesame Oil ■ Oleic Acid ■ Mineral Oil

Fig. 7 Percent of analyte remaining (relative to freshly spotted internal standard) for dried urine samples after 24 h storage vs 1 h using different oils for preservation.

Fig. 8 Sesame oil (left) and synthetic sesame oil (right).

with the naturally occurring antioxidants sesamin and sesamolin or with the food preservative butylated hydroxytoluene (Fig. 8 shows a picture of sesame oil and synthetic sesame oil).

No combination of the naturally occurring antioxidants combined with synthetic sesame oil preserved THC or its metabolites to any significant amount. This could mean that there are other components of sesame oil that help preserve THC. As demonstrated in Fig. 8 there is a distinct colour to natural sesame oil that is not present in the synthetic, which suggests that the synthetic sesame oil is missing components to make it as similar as possible to natural sesame oil. An alternate option is that there could be an impurity in the oils that is preventing the antioxidants from functioning. The absolute signal intensity for all analytes using the synthetic sesame oil was notably lower than similar experiments using the nonsynthetic sesame oil, which likely indicates some impurity associated with one or more of the components that also inhibits paper spray ionization. What is out of the ordinary is that the metabolite OH-THC was preserved, at least partially (only about a 26% loss of OH-THC compared to the greater than 90% loss of THC and COOH-THC), in synthetic sesame oil with butylated hydroxytoluene (BHT). Given the complex nature of natural oils it would be a labour-intensive process to narrow down the exact source of sesame oil's preservation of THC, which is beyond the scope of the current work.

3.2 Evaluating preconcentration using paper strip extraction

Early experiments into analysing THC by paper spray were hindered by unexplained losses of analyte. The problem was that urine samples containing THC were being prepared in plastic vials. Because THC is extremely hydrophobic, it was depositing on the vial walls and being lost before being analysed. Later on, these problems prompted the idea that hydrophobic compounds like THC could be extracted from urine as it flows through a hydrophobic matrix, such as sesame oil on a strip of paper. In order to evaluate this idea, a urine sample containing both natural and synthetic cannabinoids was extracted. In addition, several pharmaceuticals with known logP values were also included to determine whether the technique was applicable to all drugs, only hydrophobic drugs, or only cannabinoids. After extraction, the strip was cut into 5 mm increments and spiked with SIL versions of each analyte.

The individual squares were analysed by paper spray and the ratio between the analyte and the SIL was plotted as a function of distance travelled through the paper strip. Three strips were analysed with and without sesame oil and the results were averaged. If the ratio for the analyte to internal standard was constant for each increment, then the analyte evenly distributed, whereas if the ratio decreased rapidly from the first 5 mm segment then the analyte preconcentrated. Data for four different drugs

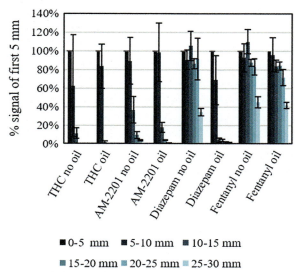

Fig. 9 Decrease in analyte concentration throughout a strip of paper with or without sesame oil at the end. Measurements expressed as a percentage of the analyte/SIL ratio for that segment relative to the ratio in the first 5 mm segment.

demonstrating the different behaviours are shown in Fig. 9, while the MS/MS data of the first four 5 mm segments for THC and its internal standard THC D3 are shown in Fig. 10.

The cannabinoids (THC, THC glucuronide, and both synthetic cannabinoids) decreased significantly by the third segment (10–15 mm) regardless of the presence of sesame oil on the paper as shown in Fig. 9. Looking at the mass spectra in Fig. 10 there is a noticeable decrease in the intensity for the MS/MS fragment of THC (193.1221) relative to the internal standard THC D3 (196.1412). This suggests that there is a specific interaction between these cannabinoids and cellulose that can lead to preconcentration from aqueous matrices regardless of the presence of sesame oil. All the other analytes, except for diazepam, had a more uniform distribution throughout the paper strip regardless of the presence of oil. Diazepam showed an even distribution without oil but was more concentrated in the first 10 mm with oil. The distribution of the drugs was not strictly determined by hydrophobicity. While all the molecules that concentrated at the head of the strip are hydrophobic, the fact that the more hydrophilic THC glucuronide also concentrated suggests additional interactions played a role. However, the fact that diazepam, the most hydrophobic of the noncannabinoids, did preconcentrate within the oil raises certain questions. Specifically, is there

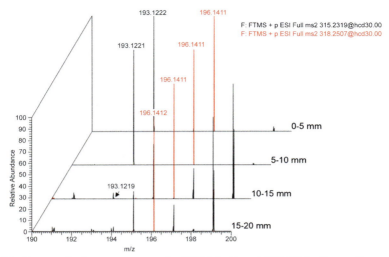

Fig. 10 Changes in the mass spectra for THC (black) and THC D3 (red) over the span of four 5 mm segments of extraction strip.

a threshold value that if an analyte's logP exceeds that value it will preconcentrate in the oil, or is there a specific interaction between diazepam and sesame oil like there is between cannabinoids and cellulose?

3.3 Measuring improvements to detection limits: Preservation and preconcentration

At this point in the work it had been shown that using sesame oil can preserve cannabinoids and that passing urine through a strip of paper (with or without sesame oil) caused the cannabinoids to preconcentrate at the head of the strip where the urine had been deposited. This meant that there were two potential mechanisms for improving detection limits using paper strip extraction. To evaluate whether these effects lead to decreased detection limits, a set of two experiments were performed to test each mechanism.

3.3.1 Preservation

To determine the effect of preservation on detection limits over time, calibration curves were prepared for THC in water by paper strip extraction with and without sesame oil. Extraction was carried out and the sample segments, the first 5 mm where the cannabinoids should have preconcentrated, were stored at room temperature in a drawer for 1, 7 and 27 days prior to analysis. The preservation effects of sesame oil are best illustrated by looking at the calibration curves obtained for THC without sesame oil before and

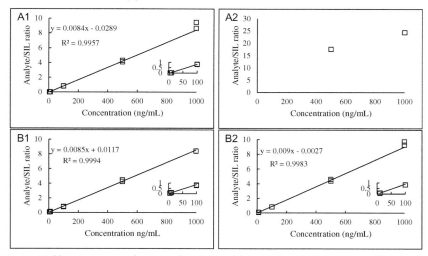

Fig. 11 Calibration curves for THC after 1 day (A1 and B1) and after 27 days (A2 and B2) using paper strip extraction without sesame oil (A1 and A2) and paper strip extraction with sesame oil (B1 and B2).

after 27 days of storage (Fig. 11A1 and A2, respectively) compared to with sesame oil over the same length of time (Fig. 11B1 and B2, respectively). By day 27, the samples without the preservation of the oil only had detectable THC in the two highest concentration standards (Fig. 11A2). The samples stored with sesame oil, on the other hand, still showed signal down to 10 ng/mL, demonstrating a marked improvement in stability for THC (Fig. 11B2). The high analyte/internal standard ratio in Fig. 11A2 was due to degradation of the internal standard (an isotopic label of THC) to undetectable levels. This caused the Y-intercept to be heavily skewed high relative to the other three calibration curves (the data points are still shown to give an estimate of the detection limit).

The lowest detectable concentrations at different time intervals for each analyte were determined based on the standard error of the Y-intercept and slope of the calibration curves. Calibration curve information is shown in Table 1; detection limits can be found in Table 2. Fentanyl was included as a noncannabinoid comparison. The synthetic cannabinoids showed minimal change over time, with or without the addition of oil. THC and 11-OH-THC show a measurable increase in the lowest detectable concentration over time without sesame oil. THC-COOH showed significant degradation over time even with sesame oil, albeit still improved over no preservative.

Table 1 Slope, Y-intercept, and R^2 values for the calibration curves after paper strip extraction with and without sesame oil after storage at room temperature.

	No oil			Oil		
	Slope	Y-intercept	R^2	Slope	Y-intercept	R^2
Day 1						
THC	0.0084 ± 0.0005	-0.029 ± 0.005	0.996	0.0085 ± 0.0004	0.012 ± 0.004	0.999
OH THC	0.0014 ± 0.0007	0.003 ± 0.007	0.916	0.00095 ± 0.00009	0.003 ± 0.001	0.942
COOH THC	0.0106 ± 0.0003	0.11 ± 0.04	0.998	0.0093 ± 0.0005	0.21 ± 0.01	0.997
AM-2201	0.047 ± 0.001	-0.07 ± 0.01	0.998	0.045 ± 0.002	-0.06 ± 0.02	0.999
AB-CHMINACA	0.00430 ± 0.00009	$-0.025 \perp 0.002$	0.997	0.0045 ± 0.0002	-0.011 ± 0.004	0.99
Fentanyl	0.0081 ± 0.0002	-0.046 ± 0.004	0.996	0.0078 ± 0.0004	-0.073 ± 0.009	0.996
Day 7						
THC	0.011 ± 0.009	-0.5 ± 0.1	0.992	0.0087 ± 0.0003	-0.003 ± 0.003	0.997
OH THC	0.0003 ± 0.0002	0.077 ± 0.003	0.937	0.0011 ± 0.0001	0.0005 ± 0.0012	0.863
COOH THC	0.20 ± 0.06	-15 ± 9	0.861	0.0114 ± 0.0007	0.1 ± 0.1	0.958
AM-2201	0.045 ± 0.002	-0.06 ± 0.02	0.999	0.031 ± 0.004	0.051 ± 0.006	0.988
AB-CHMINACA	0.0043 ± 0.0002	-0.003 ± 0.002	0.997	0.0043 ± 0.0002	-0.005 ± 0.002	0.999
Fentanyl	0.0079 ± 0.0001	-0.003 ± 0.002	0.999	0.0078 ± 0.0003	-0.038 ± 0.007	0.999

Continued

Table 1 Slope, Y-intercept, and R^2 values for the calibration curves after paper strip extraction with and without sesame oil after storage at room temperature.—cont'd

	No oil			Oil		
	Slope	Y-intercept	R^2	Slope	Y-intercept	R^2
Day 27						
THC	n/a	n/a	n/a	0.0090 ± 0.0004	-0.003 ± 0.008	0.998
OH THC	n/a	n/a	n/a	0.00093 ± 0.00005	0.005 ± 0.001	0.992
COOH THC	n/a	n/a	n/a	0.024 ± 0.005	-1.2 ± 0.9	0.839
AM-2201	0.033 ± 0.005	0.02 ± 0.05	0.914	0.044 ± 0.001	-0.06 ± 0.01	0.999
AB-CHMINACA	0.0044 ± 0.0002	0.008 ± 0.004	0.994	0.0042 ± 0.0001	0.012 ± 0.002	0.997
Fentanyl	0.0077 ± 0.0002	-0.023 ± 0.006	0.999	0.0068 ± 0.0008	-0.05 ± 0.02	0.994

n/a indicates there were insufficient data points to construct a calibration curve for that data set.

Table 2 Lowest detectable urine concentration (ng/mL) after paper strip extraction with and without sesame oil after storage at room temperature.

	Day 1		Day 7		Day 27	
Analyte	No oil	Oil	No oil	Oil	No oil	Oil
THC	2	1	30	1	$\geq 500^a$	3
11-OH-THC	20	4	30	4	$\geq 1000^a$	4
THC-COOH	10	5	130	30	$\geq 1000^a$	120
AM-2201	0.8	1	1	0.6	4	0.8
AB-CHMINACA	1	3	1	1	3	2
Fentanyl	2	3	0.8	3	2	7

[a]Detection limits significantly above the 100 ng/mL internal standard are rough estimates. Detection limits are three times the standard error of the Y-intercept divided by the slope.

3.3.2 Preconcentration

In the previous section, it was shown that using sesame oil to preserve natural cannabinoids could lower detection limits for THC. To independently determine the contributions of preconcentration vs preservation on the detection limits, urine samples were analysed with and without preconcentration. Oral fluid and urine were stored on paper with sesame oil, both with and without performing paper strip extraction. Urine samples were analysed after 1 day and 28 days of storage (Table 3). Oral fluid was added as a second, noninvasive, biofluid (Table 4). Both urine and oral fluid showed a noticeable improvement in detection limits of THC from preconcentration. THC-COOH likewise showed a significant improvement in the detection limits with preconcentration, more than an order of magnitude for both urine and oral fluid. As in the earlier experiments detailed in Table 2, THC-COOH did show some degradation after 28 days even with the oil. The metabolite AB-CHMINACA M4, which is more hydrophilic than AB-CHMINACA, showed elevated detection limits relative to AB-CHMINACA under all conditions. The calibration curves for 11-OH-THC and THC-COOH occasionally showed noticeably lower r^2 values than THC (as shown in Table 1) and higher limits of detection. This could be due to the lower hydrophobicity of the metabolites [35] or because the metabolites did not use an isotopically labelled analog as the internal standard. AM-2201 and AB-CHMINACA gave similar results regardless of the biofluid, preconcentration, or storage time. 5F-ADB in most cases showed about an order of magnitude improvement in its

Table 3 Lowest detectable urine concentration (ng/mL) with and without paper strip extraction after storage at room temperature with sesame oil.

Urine	Day 1		Day 28	
	Without preconcentration	With preconcentration	Without preconcentration	With preconcentration
THC	20	2	10	2
11-OH-THC	9	4	70	20
THC-COOH	200	10	$\geq 500^a$	40
AM-2201	0.2	0.2	0.2	0.2
AB-CHMINACA	1	0.2	1	2
AB-CHMINACA M4	130	60	70	80
5F-ADB	2	0.3	0.3	0.3

[a]Detection limits well above the 100 ng/mL internal standard are rough estimates.

Table 4 Lowest detectable oral fluid concentration (ng/mL) with and without paper strip extraction after storage at room temperature with sesame oil.

Oral fluid	Day 1		Day 28	
	Without preconcentration	With preconcentration	Without preconcentration	With preconcentration
THC	20	1	30	1
11-OH-THC	4	3	70	5
THC-COOH	10	3	$\geq 1000^a$	60
AM-2201	0.06	0.08	0.2	0.08
AB-CHMINACA	4	0.6	3	2
AB-CHMINACA M4	50	100	100	100
5F-ADB	2	0.2	2	0.2

[a]Detection limits well above the 100 ng/mL internal standard are rough estimates.

detection limit when performing paper strip extraction. These results suggest that paper strip extraction has the potential to improve the detection limits for synthetic cannabinoids relative to directly spotting the analyte on paper, but it is not consistent and is analyte dependent. Importantly, the detection limits for the synthetic cannabinoids did not increase with the addition of oil, meaning that both natural and synthetic cannabinoids can be detected with the same method.

3.4 Automated analysis

The work presented here is aimed at facilitating the use of paper spray in simultaneous screening for both natural and synthetic cannabinoids. Currently there are autosamplers and self-contained cartridges available to automate paper spray mass spectrometry. The techniques described here would require a modified version of these cartridges in order to carry out analysis. However, any modifications should maintain the speed and simplicity of the technique; otherwise, those modifications risk defeating the purpose of paper spray. A prototype paper spray cartridge was developed to demonstrate the potential for simple, fast, and automated implementation of paper strip extraction. The autosampler cartridge consisted of two main parts: an injection moulded bottom and a 3D printed top. In this work, the autosampler and the bottom part of the cartridge came from Prosolia, while the top part was 3D printed on an Ultimaker 2+ extended using polypropylene for the material. Polypropylene was selected for the top half due to its resistance to organic solvents. When the top half was snapped onto the bottom half, it made a seal around the sample square to flow solvent through the sample to the paper spray substrate (as shown in Fig. 6).

A calibration curve was generated using the prepared cartridges to determine the LODs (results in Table 5). THC and the three synthetic

Table 5 Detection limits in ng/mL in oral fluid using sesame oil and paper strip extraction in a half 3D printed autosampler cartridge.

Analyte	LOD
THC	4
5F–ADB	0.1
5F–ADB M2	0.5
AB–CHMINACA	6
AM–2201	0.1
AB–CHMINACA M4	20

cannabinoids and cannabinoid metabolites all gave roughly similar results to the manual experiments, considering the variability in calculating detection limits such as day-to-day variation in the mass spectrometer and lot-to-lot variability in the urine samples. AB-CHMINACA M4 gave a lower detection limit (20 ng/mL vs 100), but analysis of this compound was more variable on the whole due to the lack of an isotopically labelled analog. THC-COOH was also run in this experiment, but there were insufficient data points for a calibration curve. It is possible that some interferent eluting from the 3D printed plastic specifically affected THC-COOH ionization. Beyond this anomaly, the autosampler cartridge results showed that the paper strip extraction technique is compatible with automation.

4. Suitability for cannabinoid testing

The usefulness of paper strip extraction lies in having a technique capable of simultaneous and rapid detection of both synthetic and natural cannabinoids. To be a valid option for applications such as determining driving under the influence of drugs (DUID) or workplace drug screening, the technique must be sensitive enough to at least detect levels indicative of recent use in that specific biofluid. Current workplace drug testing methods typically focus on THC-COOH as a biomarker in urine with cut-offs at 50 ng/mL for the initial screen and 15 ng/mL for the subsequent confirmation [13]. Paper strip extraction detection of THC-COOH showed some problems. THC-COOH detection limits got noticeably higher over time relative to THC, indicating that THC-COOH was unstable in urine spots even with sesame oil (although the degradation was slower than without oil). Also, the detection limits for THC-COOH were significantly elevated in oral fluid and when using the automated cartridge. These results cast doubt on using paper strip extraction for routine detection of THC-COOH. However, THC-COOH is a metabolite that can be present for days in the body at variable levels over time depending on the frequency of use [36], and thus is not a good marker on its own for determination of recent use in cases of DUID [3].

For recent use, THC has been proposed as a better biomarker, as its presence in urine and oral fluid spikes within a few hours of use and drops off quickly [37–39]. For urine samples, the low ng/mL detection limits obtained by paper strip extraction are close to the 1.5 ng/mL cut-off proposed as an indication of recent use [38]. For oral fluid, the concentration of THC can vary quite a bit after recent use [37], but the cut-off concentration for THC from Substance Abuse and Mental Health Services

Administration (SAMHSA) is 2 ng/mL [40] and the cut-off for DUID from the European Union's Driving Under the Influences of Drugs, Alcohol, and Medicines (DRUID) is 1 ng/mL [41]. This also lies within the range detectable by paper strip extraction, although some further refinement of the technique and improvement in the detection limits are needed to decrease false negatives for samples near the cut-off value. In addition, as mentioned earlier and as shown in Fig. 11A2, the internal standard THC D3 is subject to degradation when unpreserved. Ideally, THC D3 would be sufficiently preserved by sesame oil, or if it does degrade to some extent during long-term storage, that it degrades at the same rate as THC to preserve the ratio. Considering that slope of the calibration line going from Fig. 11B1 and B2 does not change significantly over 27 days, this is likely the case. However, the potential for THC D3 to volatilize or degrade should be kept in mind during any further development of the method.

A significant limitation is also the inability of paper spray MS/MS to distinguish the isomers THC and cannabidiol (CBD) without chromatographic separations [42]. This would likely result in false positives for individuals using CBD for legal, medicinal purposes. Follow-up confirmatory testing would therefore be essential. Despite these limitations, paper spray coupled to paper strip extraction shows good promise for rapidly screening both synthetic and natural cannabinoids from urine or oral fluid samples.

5. Conclusion

A method of concentrating and preserving THC and its metabolites in urine or oral fluid spots was demonstrated and integrated into a cartridge compatible with a paper spray autosampler. By flowing urine or oral fluid through paper, synthetic and natural cannabinoids were retained at the head of the paper. THC, which is normally labile and difficult to analyse from a urine or oral fluid spot, was preserved for at least 27 days at room temperature with the addition of sesame oil to the paper. Combining these two techniques improved detection limits for THC to ng/mL levels in urine and oral fluid, close to current cut-off values for detection in cases of DUID were obtained. While differentiating isomers (such as THC and CBD) is still a problem for ambient ionization techniques, this method was also able to simultaneously detect natural and synthetic cannabinoids from a sample stored on paper.

Acknowledgements

Funding for this research was provided by a grant from the National Institute of Justice, Office of Justice Programs under award number 2016-DN-BX-007, by the National Institute on Drug Abuse of the National Institutes of Health under award number R21DA043037, and by Indiana University's Responding to the Addictions Crisis Grand Challenges initiative.

References

[1] B. Bills, N. Manicke, Using sesame seed oil to preserve and preconcentrate cannabinoids for paper spray mass spectrometry, J. Am. Soc. Mass Spectrom. 31 (3) (2020) 675–684.

[2] Marijuana Overview, http://www.ncsl.org/research/civil-and-criminal-justice/marijuana-overview.aspx. (accessed 11/1/2019).

[3] P. Bondallaz, B. Favrat, H. Chtioui, E. Fornari, P. Maeder, C. Giroud, Cannabis and its effects on driving skills, Forensic Sci. Int. 268 (2016) 92–102.

[4] K. Declues, S. Perez, A. Figueroa, A 2-year study of Δ 9-tetrahydrocannabinol concentrations in drivers: examining driving and field sobriety test performance, J. Forensic Sci. 61 (6) (2016) 1664–1670.

[5] Drugged Driving, Marijuana-Impaired Driving. http://www.ncsl.org/research/transportation/drugged-driving-overview.aspx (accessed 11/1/2019), 2019.

[6] S. Chavan, V. Roy, Designer drugs: a review, World J. Pharm. Pharm. Sci. 4 (8) (2015) 297–336.

[7] 2018 National Drug Threat Assessment, U.S. Department of Justice Drug Enforcement Administration, 2018.

[8] C. Rosenberg, Schedules of Controlled Substances: Temporary Placement of Six Synthetic Cannabinoids (5F-ADB, 5F-AMB, 5F-APINACA, ADB-FUBINACA, MDMB-CHMICA and MDMB-FUBINACA) into Schedule I. Justice, D. o., Ed, Drug Enforcement Administration, Department of Justice, 2017.

[9] M. Spaderna, P.H. Addy, D.C. D'Souza, Spicing things up: synthetic cannabinoids, Psychopharmacology 228 (4) (2013) 525–540.

[10] A.J. Adams, S.D. Banister, L. Irizarry, J. Trecki, M. Schwartz, R. Gerona, "Zombie"outbreak caused by the synthetic cannabinoid AMB-FUBINACA in New York, N. Engl. J. Med. 376 (3) (2017) 235–242.

[11] R. Kronstrand, M. Roman, M. Andersson, A. Eklund, Toxicological findings of synthetic cannabinoids in recreational users, J. Anal. Toxicol. 37 (8) (2013) 534–541.

[12] R.J. Tait, D. Caldicott, D. Mountain, S.L. Hill, S. Lenton, A systematic review of adverse events arising from the use of synthetic cannabinoids and their associated treatment, Clin. Toxicol. 54 (1) (2016) 1–13.

[13] K. Kulig, Interpretation of workplace tests for cannabinoids, J. Med. Toxicol. 13 (1) (2017) 106–110.

[14] F.P. Smith, Handbook of Forensic Drug Analysis, Elsevier Academic Press, Burlington, MA, 2005, p.46.

[15] A.N. Hoofnagle, M.H. Wener, The fundamental flaws of immunoassays and potential solutions using tandem mass spectrometry, J. Immunol. Methods 347 (1–2) (2009) 3–11.

[16] M.C. Hennion, Solid-phase extraction: method development, sorbents, and coupling with liquid chromatography, J. Chromatogr. A 856 (1+2) (1999) 3–54.

[17] A.H. Wu, R. Gerona, P. Armenian, D. French, M. Petrie, K.L. Lynch, Role of liquid chromatography-high-resolution mass spectrometry (LC-HR/MS) in clinical toxicology, Clin. Toxicol. 50 (8) (2012) 733–742.

[18] H.H. Maurer, Current role of liquid chromatography-mass spectrometry in clinical and forensic toxicology, Anal. Bioanal. Chem. 388 (7) (2007) 1315–1325.

[19] C.L. Feider, A. Krieger, R.J. DeHoog, L.S. Eberlin, Ambient ionization mass spectrometry: recent developments and applications, Anal. Chem. 91 (7) (2019) 4266–4290.

[20] R.B. Cody, J.A. Laramée, J.M. Nilles, H.D. Durst, Direct analysis in real time (DART) mass spectrometry, JEOL News 40 (1) (2005) 8–12.

[21] Introduction to DESI—Desorption Electrospray Ionisation Mass Spectrometry, http://www.npl.co.uk/science-technology/surface-and-nanoanalysis/surface-and-nanoanalysis-basics/introduction-to-desi-desorption-electrospray-ionisation-mass-spectrometry. (accessed 12/2/14).

[22] H. Wang, J. Liu, R.G. Cooks, Z. Ouyang, Paper spray for direct analysis of complex mixtures using mass spectrometry, Angew. Chem. Int. Ed. 49 (5) (2010) 877–880. (S877/1-S877/7).

[23] J. Liu, H. Wang, N.E. Manicke, J.-M. Lin, R.G. Cooks, Z. Ouyang, Development, characterization, and application of paper spray ionization, Anal. Chem. 82 (6) (2010) 2463–2471.

[24] C. Vega, C. Spence, C. Zhang, B.J. Bills, N.E. Manicke, Ionization suppression and recovery in direct biofluid analysis using paper spray mass spectrometry, J. Am. Soc. Mass Spectrom. 27 (4) (2016) 726–734.

[25] N.E. Manicke, B.J. Bills, C. Zhang, Analysis of biofluids by paper spray MS: advances and challenges, Bioanalysis 8 (6) (2016) 589–606.

[26] R.M. White, Instability and poor recovery of cannabinoids in urine, oral fluid, and hair, Forensic Sci. Rev. 30 (1) (2018) 33–49.

[27] C. Zhang, N.E. Manicke, Development of a paper spray mass spectrometry cartridge with integrated solid phase extraction for bioanalysis, Anal. Chem. 87 (12) (2015) 6212–6219.

[28] D.E. Damon, K.M. Davis, C.R. Moreira, P. Capone, R. Cruttenden, A.K. Badu-Tawiah, Direct biofluid analysis using hydrophobic paper spray mass spectrometry, Anal. Chem. 88 (3) (2016) 1878–1884.

[29] B.J. Bills, J. Kinkade, G. Ren, N.E. Manicke, The impacts of paper properties on matrix effects during paper spray mass spectrometry analysis of prescription drugs, fentanyl and synthetic canabinoids, Forensic Chem. 11 (2018) 15–22.

[30] M.F. Wempe, A. Oldland, N. Stolpman, T.H. Kiser, Stability of dronabinol capsules when stored frozen, refrigerated, or at room temperature, Am. J. Health Syst. Pharm. 73 (14) (2016) 1088–1092.

[31] Dronabinol, https://www.drugbank.ca/drugs/DB00470. (accessed 9/6/2019).

[32] T. Tashiro, Y. Fukuda, T. Osawa, M. Namiki, Oil and minor components of sesame (Sesamum indicum L.) strains, J. Am. Oil Chem. Soc. 67 (8) (1990) 508–511.

[33] A.M. Almeida, M.M. Castel-Branco, A.C. Falcao, Linear regression for calibration lines revisited: weighting schemes for bioanalytical methods, J. Chromatogr. B Analyt. Technol. Biomed. Life Sci. 774 (2) (2002) 215–222.

[34] G.S. Jamieson, W.F. Baughman, The chemical composition of sesame oil, J. Am. Chem. Soc. 46 (3) (1924) 775–778.

[35] Metabolite 11-Nor-9-carboxy-THC (THC-COOH), https://www.drugbank.ca/metabolites/DBMET02064. (accessed 9/6/2019).

[36] F. Musshoff, B. Madea, Review of biologic matrices (urine, blood, hair) as indicators of recent or ongoing cannabis use, Ther. Drug Monit. 28 (2) (2006) 155–163.

[37] A. Marsot, C. Audebert, L. Attolini, B. Lacarelle, J. Micallef, O. Blin, Comparison of cannabinoid concentrations in plasma, oral fluid and urine in occasional cannabis smokers after smoking cannabis cigarette, J. Pharm. Pharm. Sci. 19 (3) (2016) 411–422.

[38] J.E. Manno, B.R. Manno, P.M. Kemp, D.D. Alford, I.K. Abukhalaf, M.E. McWilliams, F.N. Hagaman, M.J. Fitzgerald, Temporal indication of marijuana use can be estimated from plasma and urine concentrations of $\Delta 9$-tetrahydrocannabinol, 11-hydroxy-$\Delta 9$-tetrahydrocannabinol, and 11-nor-$\Delta 9$-tetrahydrocannabinol-9-carboxylic acid, J. Anal. Toxicol. 25 (7) (2001) 538–549.

[39] D.A. Kidwell, J.C. Holland, S. Athanaselis, Testing for drugs of abuse in saliva and sweat, J. Chromatogr. B Biomed. Sci. Appl. 713 (1) (1998) 111–135.

[40] SAMHSA, Mandatory guidelines for federal workplace drug testing programs, Fed. Regist. 80 (2015) 28054–28101.

[41] M.J. Swortwood, M.N. Newmeyer, O.A. Abulseoud, M. Andersson, A.J. Barnes, K.B. Scheidweiler, M.A. Huestis, On-site oral fluid $\Delta 9$-tetrahydrocannabinol (THC) screening after controlled smoked, vaporized, and oral cannabis administration, Forensic Toxicol. 35 (1) (2017) 133–145.

[42] R.D. Espy, S.F. Teunissen, N.E. Manicke, Y. Ren, Z. Ouyang, A. van Asten, R.G. Cooks, Paper spray and extraction spray mass spectrometry for the direct and simultaneous quantification of eight drugs of abuse in whole blood, Anal. Chem. 86 (15) (2014) 7712–7718.

CHAPTER THIRTEEN

Quantitating cannabinoids in edible chocolates using heated ultrasonic-assisted extraction

James W. Favell[a,b], Ryan Hayward[a], Emily O'Brien[a], Seamus Riordan-Short[a], Nahanni Sagar[a], Rob O'Brien[a,c], Matthew Noestheden[a,b,]*

[a]Supra Research and Development, Kelowna, BC, Canada
[b]Department of Chemistry, I.K. Barber School of Arts and Sciences, University of British Columbia, Kelowna, BC, Canada
[c]Department of Biology, I.K. Barber School of Arts and Sciences, University of British Columbia, Kelowna, BC, Canada
*Corresponding author: e-mail address: matt@suprarnd.ca

Contents

1. Introduction		397
2. Materials and methods		400
	2.1 Apparatus	400
	2.2 Chemicals and solvents	401
	2.3 Preparation of solutions and calibration standards	402
	2.4 Preparation of chocolate extracts	403
	2.5 uHPLC analysis	403
	2.6 MS and source parameters	403
	2.7 Data processing	404
	2.8 Gravimetric analysis	405
	2.9 Single-laboratory validation with fortified samples	405
3. Results		406
	3.1 Chromatography	406
	3.2 Heated ultrasonic extraction	407
	3.3 Cold stabilisation	408
	3.4 Method performance	409
4. Conclusions		412
References		413

1. Introduction

With the recent legalisation of recreational *Cannabis sativa* L. (cannabis) use in Canada and many states in the USA, routes of administration have diversified beyond typical inhalation, with 16% of surveyed users in the USA reporting the consumption of edible cannabis products within the last 30 days in a 2014 study [1] . This trend extends to the commercial market, where edible sales reached $625 million USD in the two US states of California and Colorado in 2018 [2]. In Canada, 9.1% of respondents reported edible use as their primary method of cannabis consumption [3], and now that the production and sale of cannabis edibles has been legalised, this market is expected to surpass oils and tinctures. Underscoring this trend was a 2018 online survey of Canadians over 18, where 45.8% of respondents said they were willing to try edible cannabis products [4]. More specifically, respondents expressed interest in cannabis-based bakery and ready-made products (including candies); 46.1% and 26.6%, respectively said they would consider buying these products [4]. These results were consistent with a different survey of Canadians that indicated that the most commonly purchased cannabis edibles are baked goods (e.g. brownies and cookies) and candies (e.g. chocolate, hard candy, and gummies) [5].

The phytocannabinoids (i.e. the dominant pharmacologically active constituents) in cannabis are a class of plant secondary metabolites that act on CB_1 and/or CB_2 receptors in the endocannabinoid system [6]; CB_1 receptors have been linked to neurodegenerative disorders including multiple sclerosis and Huntington's disease [7], while CB_2 receptors are involved in inflammatory processes [8,9]. *Cannabis sativa* L. (cannabis) and its derivative products can contain an array of pharmacologically active secondary metabolites (Fig. 1A), beyond the familiar phytocannabinoids Δ^9-tetrahydrocannabinol (THC) and cannabidiol (CBD). Each phytocannabinoid is produced from acidic metabolic precursors (except CBN, which is a degradation product of THC and CBD) that have no documented effect on the endocannabinoid system [10]. The minor phytocannabinoids cannabigerol (CBG), cannabichromene (CBC), cannabinol (CBN), cannabidivarin (CBDV) and tetrahydrocannabivarin (THCV), each display a level of affinity towards CB_1 and CB_2 receptors, although, much less is known about their pharmacological effects [11].

Following the inhalation of cannabis smoke or vapour, psychoactive effects are experienced within minutes, with peak THC concentrations in plasma occurring within 3–10 min [12]. Contrastingly, following oral ingestion, effects are experienced 40–60 min after dosing, with peak THC concentrations occurring 1–5 h after ingestion [13,14]. Thus, cannabis intoxication differs in its mechanism of action between the two routes, resulting in a different user experience. Regulations on a standard dose of THC in edibles have generally been set at 10 mg per serving in the USA

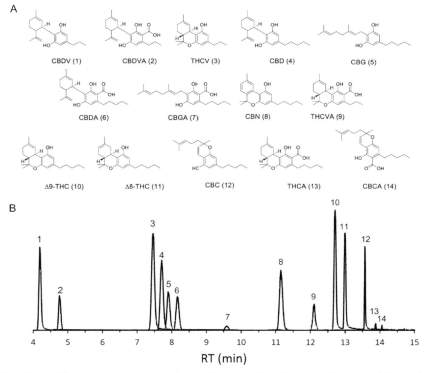

Fig. 1 (A) Chemical structures of the 14 cannabinoids examined in this study. (B) Chromatogram of a standard containing all 14 cannabinoids at 50 μg/mL.

and Canada (Oregon is 5 mg/serving). Dose control is essential for edible products, not only because of the delayed onset of intoxication, but also due to the potential for accidental consumption and cannabis poisoning. These problems have been exacerbated by quality control issues in consumer markets where edibles are legal. For instance, of 75 cannabis edibles tested in California, only 17% were properly labelled [15]. Determining the phytocannabinoid content of "THC free" products is also critical, since mislabelled products could result in accidental impairment; few products that claimed to contain a 1:1 ratio of THC to CBD actually contained this ratio [15]. Such inconsistencies in reported phytocannabinoid content represent a liability for producers and, more importantly, a major consumer safety and regulatory issue that needs to be addressed.

Dealing with such label claim issues requires the development of robust and accurate analytical methodologies to quantitate phytocannabinoids in edible matrices. However, edible matrices are complicated compared to cannabis flowers and extracts (where many excellent analytical methods have been published [16–18]), with many differing markedly in their macromolecular content, making a standard method for phytocannabinoid extraction and quantitation unlikely [19]. For example, a cannabis brownie may be comprised of 28.1% total fat, 49.3% carbohydrates, and 6.33% protein, while gummy bears are typically comprised of 77% carbohydrates, 6.9% protein, and 0% total fat [20]. Currently, few published methods exist for the extraction of major and minor phytocannabinoids from edible matrices. Previous studies have demonstrated the recovery of THC, CBD and CBN from cannabis edibles, but failed to quantitate minor phytocannabinoids or any acidic forms [21–25]. These omissions are noteworthy given that minor phytocannabinoids likely have synergistic effects to the major phytocannabinoids, and their concentrations may impact user intoxication, experience, and metabolism [11].

Given the high consumption of chocolate-based edibles [5] and their potential to be accidentally ingested by non-cannabis users, this matrix was identified as a critical need for the cannabis industry. Underscoring this was a call for methods issued by AOAC International in 2017, where they requested the quantitation of phytocannabinoids in milk, dark and white chocolates. Recently, the extraction of phytocannabinoids from chocolate by cryomilling samples, followed by a QuEChERS extraction was reported [25]. While this method has merit, the chosen detection method of thin-layer chromatography coupled to DESI-mass spectrometry prevented the detection of minor or acidic cannabinoids and DESI-mass spectrometry is not a widely implemented technique, so the uptake of the reported method is likely to be limited. Here we report the development of a robust analytical method to quantitate phytocannabinoids in chocolate using heated, ultrasonic-assisted extraction, followed by cold stabilisation and analysis using ultra-high-pressure liquid chromatography (UHPLC) coupled to either a photodiode array (PDA) or a Q Exactive Orbitrap mass spectrometer detector.

2. Materials and methods

2.1 Apparatus

(a) *Micropipettes*—Eppendorf Research Plus: 20, 200, 1000 and 5000 µL (Hamburg, Germany)
(b) *Analytical balance*—VWR-164AC 160 g Analytical Balance (VWR International, Radnor, PA)

Quantitating cannabinoids in edible chocolates 401

(c) *Top loading balance*—VWR-4502AC 4500 g top loading balance (VWR International, Radnor, PA)

(d) *Centrifuge*—Centrifuge 5804 R (Eppendorf, Hamburg, Germany)

(e) *uHPLC system*—Vanquish Horizon uHPLC system (ThermoFisher Scientific, Waltham, MA)

(f) *uHPLC column*—Accupore C_{18}, 150×4.6 mm, $2.6\,\mu$m (ThermoFisher Scientific, Waltham, MA)

(g) *MS system*—Q Exactive™ Hybrid Quadrupole-Orbitrap Mass Spectrometer with a HESI-II heated electrospray ionisation source in positive mode (ThermoFisher Scientific, Waltham, MA)

(h) *Software*—TraceFinder software Version 4.1 and FreeStyle Version 1.3 (ThermoFisher Scientific, Waltham, MA)

(i) *Polypropylene centrifuge tubes*—15 mL (Corning Inc., Corning, NY)

(j) *Microcentrifuge tubes*—1.5 mL (Eppendorf, Hamburg, Germany)

(k) *Heated sonicator*—Digital-Pro 6 L Professional Ultrasonic Cleaner

(l) *Vortex mixer*—Fisherbrand Digital Vortex Mixer (ThermoFisher Scientific, Waltham, MA)

(m) *HPLC autosampler vials*—Amber, 2 mL (Chromatographic specialities, Brockville, ON, Canada)

(n) *Pasteur pipets*—Fisherbrand disposable borosilicate glass pipets (ThermoFisher Scientific, Waltham, MA)

(o) *Vacuum concentrator*—Savant SPD121P-115 SpeedVac Concentrator (ThermoFisher Scientific, Asheville, NC)

Equivalent apparatus may be substituted.

2.2 Chemicals and solvents

(a) *Acetonitrile*—HPLC Grade, Fisher Scientific (Hampton, NH)

(b) *Methanol*—HPLC Grade, Fisher Scientific (Hampton, NH)

(c) *Water*—HPLC Grade, Fisher Scientific (Hampton, NH)

(d) *Formic acid*—Reagent Grade, $\geq 95\%$ (Honeywell Laboratory Solutions, Charlotte, NC)

(e) *Cannabichromene (CBC)*—1.0 mg/mL in methanol, Cerilliant Corp. (Round Rock, TX)

(f) *Cannabichromenic acid (CBCA)*—1.0 mg/mL in acetonitrile, Cerilliant Corp. (Round Rock, TX)

(g) *Cannabidiol (CBD)*—1.0 mg/mL in methanol, Cerilliant Corp. (Round Rock, TX)

(h) *Cannabidiolic acid (CBDA)*—1.0 mg/mL in acetonitrile, Cerilliant Corp. (Round Rock, TX)

(i) *Cannabidivarin (CBDV)*—1.0 mg/mL in methanol, Cerilliant Corp. (Round Rock, TX)

(j) *Cannabidivarinic acid (CBDA)*—1.0 mg/mL in acetonitrile, Cerilliant Corp. (Round Rock, TX)

(k) *Cannabigerol (CBG)*—1.0 mg/mL in methanol, Cerilliant Corp. (Round Rock, TX)

(l) *Cannabigerolic acid (CBGA)*—1.0 mg/mL in acetonitrile, Cerilliant Corp. (Round Rock, TX)

(m) *Cannabinol (CBN)*—1.0 mg/mL in methanol, Cerilliant Corp. (Round Rock, TX)

(n) *Δ8-tetrahydrocannabinol (Δ8-THC)*—1.0 mg/mL in methanol, Cerilliant Corp. (Round Rock, TX)

(o) *Δ9-tetrahydrocannabinol (Δ9-THC)*—1.0 mg/mL in methanol, Cerilliant Corp. (Round Rock, TX)

(p) *Δ9-tetrahydrocannabinolic acid (THCA)*—1.0 mg/mL in acetonitrile, Cerilliant Corp. (Round Rock, TX)

(q) *Tetrahydrocannabivarin (THCV)*—1.0 mg/mL in methanol, Cerilliant Corp. (Round Rock, TX)

(r) *Tetrahydrocannabivarinic acid (THCVA)*—1.0 mg/mL in acetonitrile, Cerilliant Corp. (Round Rock, TX)

Equivalent chemicals may be substituted.

2.3 Preparation of solutions and calibration standards

(a) *Mobile phase A*—(0.1% formic acid in 85:15 water: acetonitrile, v/v). Mobile phase A was prepared by diluting 1 mL formic acid in 850 mL HPLC-grade water and 150 mL HPLC-grade acetonitrile and mixed thoroughly. Solvents were added gravimetrically.

(b) *Mobile phase B*—(0.05% formic acid in 64:36 methanol: acetonitrile, v/v). Mobile phase B was prepared by diluting 0.5 mL formic acid in 640 mL HPLC-grade methanol and 360 mL HPLC-grade acetonitrile and mixed thoroughly. Solvents were added gravimetrically.

(c) *Sample diluent*—(60% mobile phase B, 40% mobile phase A). 9 mL of mobile phase B and 6 mL of mobile phase A were added to a 15 mL centrifuge tube and mixed thoroughly.

(d) *Cannabinoid standard mix (50 μg/mL)*—50 μL of each of the 14 cannabinoid standards (*vide supra*) was transferred to a 2 mL amber HPLC vial, in addition to 300 μL of the starting mobile phase and mixed thoroughly. This resulted in a stock solution where each of the cannabinoid standards was present at a concentration of 50 µg/mL.

(e) *Calibration standards*—A calibration curve was prepared at 0.05, 0.1, 0.25 µg/mL, 0.5, 1, 2.5, 5, 10 and 25 µg/mL by diluting the 50 µg/mL cannabinoid standard mix with the appropriate volume of the starting mobile phase.

2.4 Preparation of chocolate extracts

Dark chocolate, milk chocolate and white chocolate chips were purchased at local grocery stores and used as received. 500 mg of chocolate were accurately weighed using an analytical balance into a 15 mL centrifuge tube and 10.00 mL ACN was added. Samples were then manually mixed before sonication at 50 °C for 10 min, with additional manual mixing at 5 min and at the end of sonication. Extracts were placed in a freezer at -20 °C for 2 h. Afterwards, extracts were centrifuged at $3000 \times g$ for 5 min. A 1 mL aliquot of each extract was transferred to a 1.5 mL microcentrifuge tube and centrifuged at $13,000 \times g$ for a further 5 min. Extracts were diluted 100-fold, using the starting mobile phase as the diluent (see Section 2.3 step c), into 2 mL amber glass vials and analysed without further work-up.

2.5 uHPLC analysis

(a) *uHPLC System*—Vanquish Horizon uHPLC system (ThermoFisher Scientific, Waltham, MA)

(b) *uHPLC Column*—Accupore C_{18}, 150×4.6 mm, 2.6 µm (ThermoFisher Scientific, Waltham, MA)

(c) *Column/pre-heater temperature*—50 °C

(d) *Flowrate*—1.5 mL/min

(e) *Injection volume*—5 µL (MS) and 50 µL (DAD)

(f) *uHPLC gradient*—Settings in Table 1

2.6 MS and source parameters

(a) *Sheath/aux gas*—40/15

(b) *Sweep gas*—1

(c) *Spray voltage*—3.50 V

(d) *Capillary temperature*—320 °C

(e) *S-Lens RF*—50.0

(f) *Aux gas heater*—350 °C

(g) *Acquisition parameters*—settings in Table 2. Conducted in Full MS/dd-MS^2 (Top 5) mode.

Table 1 UHPLC gradient for the separation of phytocannabinoids.

Time (min)	%B
0.0	60
0.5	60
8.5	65
11.2	70
13.0	95
14.0	98
16.0	98
16.1	60
18.0	60

Table 2 Acquisition parameters for the full MS /dd-MS^2 (Top 5) scans.

MS parameter	Full MS	dd-MS2 (Top 5)
Resolution	70,000	17,500
AGC target	1e6	4e3
Maximum IT (ms)	250	50
Scan range (m/z)	200–400	2.0
Loop count	–	5
NCE	–	17.5, 35.0, 52.5
Dynamic exclusion (s)	–	2.0

2.7 Data processing

Data was acquired and processed with the TraceFinder (version 4.1) software package (Thermo Fisher Scientific). Data reduction and statistical calculations were performed using Microsoft Excel (Microsoft Corporation; Redmond, WA, USA). The ICIS detection algorithm was used to integrate all data, using 5-point Savitzky-Golay peak smoothing and ± 5 ppm mass accuracy tolerance for the precursor ion. All data was

Quantitating cannabinoids in edible chocolates

matched against a high-resolution accurate mass spectral library constructed using phytocannabinoid standards. Calibration curves were calculated as quadratic equations using inverse concentration weighting. Unless noted, data are reported as the mean ± the standard error of the mean (SEM).

2.8 Gravimetric analysis

Extracts ($n = 3$/condition) for each type of chocolate were prepared as above with the amount of time at $-20\,°C$ varied between 1, 2 and 4 h. Control extracts were kept at ambient conditions for the same time points. After cold stabilisation, samples were centrifuged at $13,000 \times g$ for 5 min. The resulting supernatant was evaporated to dryness using N_2. Dryness was evaluated by repeated weighing of samples until there was no more than a 0.25% difference between successive readings. The amount of residual material remaining after drying the supernatant was expressed relative to the original mass of the chocolate.

2.9 Single-laboratory validation with fortified samples

Blank extracts were fortified with the fourteen cannabinoid standards at 0.008% (low), 0.4% (mid), or 4% (high; w/w) after the final dilution to facilitate method validation using commercially available analytical standards. The absolute area response of all analytes in a blank sample analysed immediately after a calibrator were required to be $\leq 0.1\%$ of the absolute area of the same analytes in the high calibrator when carry-over was evaluated. Recovery was determined by fortifying blank samples at low, mid and high concentrations, with five replicates at each level in the three matrices of chocolate. Repeatability (RSD_r) was determined for each matrix at the mid concentration and intermediate precision (single analyst, RSD_R) was calculated using three separate days with five replicates at the mid concentration for each matrix. The method detection limit (MDL), defined as the minimum concentration of analyte that can be reported with 99% confidence that a measured concentration is distinguishable from a blank sample [26], was found by fortifying a total of seven replicates of each matrix across 3 days of validation at the concentration, with two analysts contributing to the preparation of extracts. The MDLs were calculated with the following equation:

$$MDL = t \times s_s$$

Where t is 3.134, the Student's t-value for a single-tailed 99th percentile t statistic and standard deviation estimate with 6 degrees of freedom, and S_s is the standard deviation of the seven fortified replicate samples for each matrix. The method reporting limit (MRL) [26], defined as the lowest concentration level reported for a given test method (which must be equal to or greater than the MDL), was determined from seven replicates in each matrix prepared on a single day by a single analyst. The half range for the prediction interval of results (HR$_{PIR}$) is calculated from this data:

$$HR_{PIR} = 3.963 \times s$$

Where 3.963 is a constant for the seven extraction replicates and s is the standard deviation. This value is used to calculate the upper and lower prediction interval of results (PIR):

$$PIR = \frac{x \pm HR_{PIR}}{Fortified\ Concentration} \times 100\%$$

If the upper PIR is less than 150% and the lower PIR is greater than 50%, the MRL is verified.

3. Results

3.1 Chromatography

The final chromatographic method was able to resolve all 14 cannabinoids contained within the standard mix. (Fig. 1B). CBDV was the first cannabinoid to elute, at 4.20 min. The final cannabinoid to elute was CBCA at 14.06 min. Resolution of these analytes was critical, as compatibility with optical detection and MS (*vide infra*) would not be possible given the limited specificity of optical detectors for phytocannabinoids. The neutral cannabinoids exhibited higher response factors than their acidic counterparts using MS, which was expected as the ESI system was operated in the positive ionisation mode; CBGA (9.58 min), THCA (13.88 min), and CBCA (14.06 min) were particularly impacted in this regard. This did not present a problem in quantitating these compounds given the sensitivity of the developed method relative to the analyte concentrations typically found in cannabis edibles (e.g. 10 mg THC/serving). Performing this analysis using a triple quadrupole mass spectrometer with polarity switching would mitigate the response factor disparity reported here.

It is also worth noting that the developed chromatographic method can accommodate up to 50 μL injections without appreciable changes to chromatographic performance (Fig. 2), which facilitates using optical detectors that will inherently have lower sensitivity than MS-based approaches.

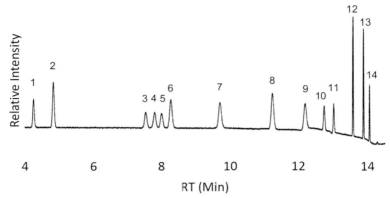

Fig. 2 PDA data (230 nm) at 50 μL (vs 5 μL for MS) demonstrating the ability of the chromatographic method to accommodate for the sensitivity differences between the reported MS-based detection and optical detectors.

Similar to the use of cost-effective heated ultrasonic extraction in lieu of cryomilling (*vide infra*), the use of optical detectors will encourage adoption of the developed method using instrumentation that are economically and technically more accessible.

3.2 Heated ultrasonic extraction

Initial method development investigations focused on ball milling and cryomilling to homogenise the chocolate samples prior to extraction (data not shown). Due to the fat content of the chocolate matrix, standard ball milling did not achieve suitable sample homogeneity. Cryomilling was more successful but was deemed unsuitable for routine quantitation due to the limited throughput potential and the expense required to purchase and operate typical cryomills.

Given that chocolate will, depending on the type, melt above 30 °C, the potential to melt the chocolate into a homogenous liquid was explored. To aid in the extraction process a heated ultrasonic bath was utilised. Setting the sonic bath to 50 °C ensured that the chocolate (milk, dark and white) melted during the extraction process. While a higher temperature could have been used ($ACN_{bp} = 82\,°C$), the temperature was kept low to ensure that decarboxylation of the acidic cannabinoids was not a concern. This conservative approach was necessary since analytical standards for acidic cannabinoids were not available in suitable quantities to make chocolates in support of method development and validation.

3.3 Cold stabilisation

The extraction of cannabinoids from chocolate using ACN will result in the co-extraction of waxes and macromolecules present in the chocolate (solvent extraction has been used in the quantitation of fatty acids in chocolate [25]). These co-extracted matrix components can have detrimental effects on the performance of analytical instruments—most notably a reduction in the operational lifetime of analytical columns. Cold stabilisation (i.e. winterisation) was proposed as an approach to sample preparation that could address this problem. At colder temperatures, waxes and macromolecules will be less soluble in the organic solvent and are therefore likely to precipitate out of solution. This would allow for their removal via centrifugation. In the present study, $-20\,°C$ was selected as the cold stabilisation temperature due to the ubiquity of freezers with this temperature setting.

As the first step towards quantitating the impact that cold stabilisation may have, a series of experiments were designed to determine the amount of co-extracted matrix components removed during this process. As depicted in Fig. 3, there was less residual material present in the supernatant of samples that had been cold stabilised at $-20\,°C$ for 1, 2 or 4 h, when compared to a sample that had been held under ambient conditions for the same

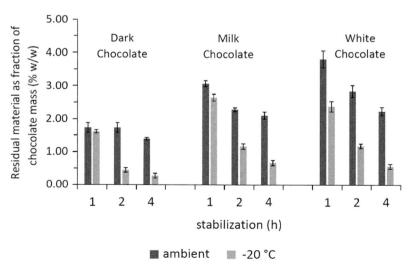

Fig. 3 Mass of residual material in supernatant expressed as a percentage of the original mass of chocolate, with and without cold stabilisation at $-20\,°C$.

time. This trend was the same across dark, milk and white chocolate matrices. The discrepancy in absolute losses following cold stabilisation in the different matrices was attributed to the relative fat content of each, with dark chocolate < milk chocolate < white chocolate. Notably, there was a decrease in the amount of residual material under ambient conditions too. This result left little doubt regarding the importance of cold stabilisation, as it suggested that chocolate extracts left sitting in a thermostated autosampler (waiting for analysis) have the potential to foul analytical columns due to the precipitation of co-extracted matrix components. Based on these data, 2h was selected as the time allotted to cold stabilisation, since it provided a suitable balance between sample clean-up and throughput within an eight-hour workday.

3.4 Method performance

For method validation, extracts were fortified at 0.008%, 0.4%, and 4% (w/w; cannabinoid/chocolate) after extraction (*vide supra*). While this approach is not an ideal way to perform method validation, the cost of infusing chocolates with the phytocannabinoids of interest at levels representative of the concentration ranges in cannabis edibles was prohibitively high. Notwithstanding this, all analytes were quantitated over a simulated range of 0.008–4% (w/w) with quadratic fit and inverse concentration weighting. The same calibration functions and dynamic ranges were evaluated for each day of validation ($n = 3$), with all correlation coefficients (R^2) being >0.99. Carry-over was determined to be $\leq 0.1\%$ by area (data not shown) for all analytes evaluated after a high calibration sample at 4% (w/w).

3.4.1 Accuracy

For the purposes of this study, recoveries were determined accurate if they varied by less than or equal to $\pm 20\%$ at the low and mid concentrations and by less than or equal to $\pm 10\%$ at the high concentration (Tables 3–5). In the dark chocolate matrix (Table 3) all compounds fall within the prescribed region except for CBG and CBN. In milk chocolate (Table 4) it is apparent that the phytocannabinoids were biassed high at the low fortification level. Similarly, recoveries are biassed low at the low fortification level in white chocolate (Table 5) with five cannabinoids (CBDV, CBDVA, CBDA, CBGA, and THCA) outside of the target recoveries.

Table 3 Method validation summary for cannabinoids in a dark chocolate matrix.

ID	Analyte	Recovery			Conducted at midlevel		Method limits (% w/w)	
		0.008% (w/w)	0.4% (w/w)	4% (w/w)	% RSD_r	% RSD_R	MDL	MRL
1	CBDV	93.5	96.1	94.0	6.52	15.9	0.005	0.008
2	CBDVA	90.0	92.4	97.8	6.05	15.4	0.003	0.008
3	THCV	102	97.6	97.3	5.69	13.2	0.004	0.008
4	CBD	101	97.4	103	6.29	11.4	0.004	0.008
5	CBG	101	92.8	87.3	6.40	14.9	0.005	0.008
6	CBDA	99.3	91.3	95.8	6.79	12.2	0.003	0.008
7	CBGA	102	92.5	93.9	3.64	9.80	0.003	0.008
8	CBN	113	93.7	89.3	7.63	12.1	0.004	0.008
9	THCVA	104	93.3	91.6	7.66	10.4	0.004	0.008
10	Δ9-THC	101	94.8	91.3	5.96	9.14	0.005	0.008
11	Δ8-THC	96.8	97.6	92.6	6.51	17.6	0.004	0.008
12	CBC	103	94.8	90.4	5.94	10.3	0.004	0.008
13	THCA	92.3	88.7	92.1	6.15	8.64	0.003	0.008
14	CBCA	107	85.4	98.1	6.57	7.63	0.003	0.008

3.4.2 Repeatability and intermediate precision

Repeatability, expressed here as relative standard deviation (RSD_r), was determined across five replicate samples prepared at a fortified concentration of 0.4% (w/w). In all three matrices, repeatability was acceptable, with an average RSD_r across all compounds of 6.3% in dark chocolate, 6.6% in milk chocolate, and 3.9% in white chocolate (Tables 3–5). Similarly, intermediate precision was assessed as the relative standard deviation (RSD_R) of 15 replicates over the course of 3 days (five replicates per matrix per day). As might be expected, RSD_R was greater than RSD_r. Across all 14 analytes in a given matrix, the average RSD_R was found to be 12.1% in dark chocolate, 8.0% in milk chocolate and 9.5% in milk chocolate. For all compounds in all three

Quantitating cannabinoids in edible chocolates 411

Table 4 Method validation summary for cannabinoids in a milk chocolate matrix.

		Recovery			Conducted at mid level		Method limits (% w/w)	
ID	Analyte	0.008% (w/w)	0.4% (w/w)	4% (w/w)	% RSD$_r$	% RSD$_R$	MDL	MRL
1	CBDV	112	95.5	98.8	6.60	7.26	0.003	0.008
2	CBDVA	117	92.5	101	6.03	7.05	0.003	0.008
3	THCV	116	95.4	99.0	6.03	7.59	0.003	0.008
4	CBD	116	97.9	99.6	7.05	7.75	0.003	0.008
5	CBG	114	97.8	99.2	6.00	7.44	0.003	0.008
6	CBDA	115	93.8	99.6	5.51	7.27	0.003	0.008
7	CBGA	119	97.9	101	6.79	6.89	0.003	0.008
8	CBN	114	97.6	96.3	6.39	7.76	0.003	0.008
9	THCVA	123	97.3	98.4	5.90	7.44	0.004	0.008
10	Δ9-THC	114	102	97.6	5.97	7.07	0.003	0.008
11	Δ8-THC	115	101	95.0	5.96	7.82	0.003	0.008
12	CBC	109	97.6	97.0	5.63	8.64	0.003	0.008
13	THCA	118	98.0	97.7	9.50	11.5	0.003	0.008
14	CBCA	109	100	103	9.26	10.3	0.002	0.008

matrices, both RSD$_r$ and RSD$_R$ yielded Horwitz ratios (HorRat) [27] between 0.3 and 1.3, indicating suitable method repeatability and intermediate precision.

3.4.3 Method detection limit and method reporting limit

The MDL and MRL were determined in a similar fashion, the primary difference being that the verification of the MDL was conducted across three separate days of analysis while the MRL verification was conducted on a single day. For this method, the successfully verified MRL was 0.008% (w/w) for all analytes in all matrices, whereas MDLs ranged from 0.002% to 0.006% (w/w; Tables 3–5).

Table 5 Method validation summary for cannabinoids in a white chocolate matrix.

		Recovery			Conducted at midlevel		Method limits (% w/w)	
ID	Analyte	0.008% (w/w)	0.4% (w/w)	4% (w/w)	% RSD$_r$	% RSD$_R$	MDL	MRL
1	CBDV	77.6	94.7	91.2	4.13	12.0	0.006	0.008
2	CBDVA	76.1	92.5	95.8	4.12	10.2	0.004	0.008
3	THCV	86.1	96.6	97.1	3.42	9.60	0.005	0.008
4	CBD	86.1	96.0	104	3.74	8.56	0.005	0.008
5	CBG	84.8	91.9	82.8	3.79	12.9	0.006	0.008
6	CBDA	77.4	91.5	95.3	4.17	10.9	0.005	0.008
7	CBGA	76.0	90.8	88.9	3.18	8.66	0.006	0.008
8	CBN	98.0	92.0	84.7	3.16	10.2	0.005	0.008
9	THCVA	85.0	89.8	83.4	3.49	14.4	0.006	0.008
10	Δ9-THC	86.9	94.9	90.5	4.44	7.08	0.006	0.008
11	Δ8-THC	84.0	97.4	92.1	3.80	16.1	0.006	0.008
12	CBC	86.5	93.8	92.2	5.09	7.59	0.005	0.008
13	THCA	78.0	87.5	89.3	3.55	4.73	0.005	0.008
14	CBCA	93.5	85.7	91.6	4.06	5.24	0.004	0.008

4. Conclusions

This approach to extracting phytocannabinoids from chocolates for potency testing is simple, amenable to high-throughput and is economically accessible for producers to implement. Moreover, the developed method allows for the reliable quantitation of cannabinoids in chocolate matrices, filling a pressing need of regulatory bodies and cannabis producers. The developed method provides several advantages in addition to its quantitative capabilities, such as cold stabilisation. This technique proved to be effective at removing co-extracted waxes from the supernatant extracts, thereby reducing both matrix interferences and wear on the analytical instrument. The method's compatibility with multiple detectors lends itself to broad uptake within the cannabis industry. Adaption and expansion to encompass other challenging matrices, such as gummies, will serve to increase the applicability and usefulness of this method, providing greater accuracy in edible product labelling and support the safe consumption of cannabis.

References

[1] G.L. Schauer, B.A. King, R.E. Bunnell, G. Promoff, T.A. Mcafee, Toking, aping, and eating for health or fun: marijuana use patterns in adults, U.S., Am. J. Prev. Med. 50 (2016) (2014) 1–8.

[2] A. Blake, I. Nahtigal, The evolving landscape of cannabis edibles, Curr. Opin. Food Sci. 28 (2019) 25–31.

[3] Statistics Canada, Method of consumption most often used by type of cannabis user, household population aged 15 or older, Canada, fourth quarter 2018 [data set], in: National Cannabis Survey, Fourth Quarter 2018, Canadian Federal Government, 2019.

[4] S. Charlebois, J. Music, B. Sterling, S. Somoyogi, Edibles and Canadian Consumers' Willingness to Consider Recreational Cannabis in Food or Beverage Products: A Second Assessment, Halifax, Dalhousie University, 2019.

[5] Deloitte, Nurturing New Growth: Canada Gets Ready for Cannabis 2.0, https://doi.org/10.1109/RCIS.2013.6577673, 2019.

[6] A.C. Howlett, M.E. Abood, CB1 and CB2 receptor pharmacology, Adv. Pharmacol. 80 (2017) 169–206.

[7] D.A. Kendall, G.A. Yudowski, Cannabinoid receptors in the central nervous system: their signaling and roles in disease, Front. Cell. Neurosci. 10 (2017) 294.

[8] C.J. Lucas, P. Galettis, J. Schneider, The pharmacokinetics and the pharmacodynamics of cannabinoids, Br. J. Clin. Pharmacol. 84 (2018) 2477–2482.

[9] R. Pertwee, The diverse CB1 and CB2 receptor pharmacology of three plant cannabinoids: D9-tetrahydrocannabinol, cannabidiol and D9-tetrahydrocannabivarin, Br. J. Pharmacol. 153 (2008) 199–215.

[10] C.M. Andre, J.F. Hausman, G. Guerriero, Cannabis sativa: the plant of the thousand and one molecules, Front. Plant Sci. 7 (2016) 1–17.

[11] E.B. Russo, Taming THC: potential entourage effects, Br. J. Pharmacol. 163 (2011) 1344–1364.

[12] P. Sharma, P. Murthy, M.M.S. Bharath, Chemistry, metabolism, and toxicology of cannabis: clinical implications, Iran. J. Psychiatry 7 (2012) 149–156.

[13] R.M. Kaufmann, B. Kraft, R. Frey, D. Winkler, S. Weiszenbichler, C. Bäcker, S. Kasper, H.G. Kress, Acute psychotropic effects of oral cannabis extract with a defined content of δ9-tetrahydrocannabinol (THC) in healthy volunteers, Pharmacopsychiatry 43 (2010) 24–32.

[14] M.A. Huestis, Human cannabinoid pharmacokinetics, Chem. Biodivers. 4 (2007) 1770–1804.

[15] R. Vandrey, J.C. Raber, M.E. Raber, B. Douglass, C. Miller, M.O. Bonn-Miller, Cannabinoid dose and label accuracy in edible medical cannabis products, JAMA 313 (2015) 2491–2493.

[16] E.M. Mudge, S.J. Murch, P.N. Brown, Chemometric analysis of cannabinoids: chemotaxonomy and domestication syndrome, Sci. Rep. 8 (2018) 13090.

[17] C. Agarwal, K. Máthé, T. Hofmann, L. Csóka, Ultrasound-assisted extraction of cannabinoids from Cannabis sativa L. optimized by response surface methodology, J. Food Sci. 83 (2018) 700–710.

[18] L.J. Rovetto, N.V. Aieta, Supercritical carbon dioxide extraction of cannabinoids from Cannabis sativa L, J. Supercrit. Fluids 129 (2017) 16–27.

[19] J.T. Tanner, W.R. Wolf, W. Horwitz, Nutritional metrology: the role of reference materials in improving quality of analytical measurement and data on food components, in: H. Greenfield (Ed.), Quality and Accessibility of Food-Related Data, AOAC International, Arlington, 1995, pp. 99–104.

[20] A. Leghissa, Z.L. Hildenbrand, K.A. Schug, The imperatives and challenges of analyzing Cannabis edibles, Curr. Opin. Food Sci. 28 (2019) 18–24.

[21] D.W. Lachenmeier, L. Kroener, F. Musshoff, B. Madea, Determination of cannabinoids in hemp food products by use of headspace solid-phase microextraction and gas chromatography-mass spectrometry, Anal. Bioanal. Chem. 378 (2004) 183–189.

[22] M. Pellegrini, E. Marchei, R. Pacifici, S. Pichini, A rapid and simple procedure for the determination of cannabinoids in hemp food products by gas chromatography-mass spectrometry, J. Pharm. Biomed. Anal. 36 (2005) 939–946.

[23] J.R. Stenzel, G. Jiang, Identification of Cannabinoids in Baked Goods by UHPLC/MS, https://www.thermofisher.com/document-connect/document-connect.html?url=https%3A%2F%2Fassets.thermofisher.com%2FTFS-Assets%2FCMD%2FApplication-Notes%2FAN433_62876_MSQ_ForTox(1).pdf&title=QXBwbGljYXRpb24gTm90ZTogSWRlbnRpZmljYXRpb24gb2YgQ2FubmFiaW5vaWRzIGluIEJha2VkIEdvb2RzIGJ5IFVIUExDL01T, 2008. Accessed February 20, 2020.

[24] X. Wang, D. Mackowsky, J. Searfoss, M.J. Telepchak, LC-GC, in: Determination of cannabinoid content and pesticide residues in cannabis edibles and beverages, LC-C 34 (10) (2017) 20–27.

[25] M. Yousefi-Taemeh, D.R. Ifa, Analysis of tetrahydrocannabinol derivative from cannabis-infused chocolate by QuEChERS-thin layer chromatography-desorption electrospray ionization mass spectrometry, J. Mass Spectrom. 54 (2019) 834–842.

[26] Office of Water, United States Environmental Protection Agency, Definition and Procedure for the Determination of the Method Detection Limit, Revision 2, US Federal Government via the US EPA, 2016.

[27] W. Horwitz, R. Albert, The Horwitz ratio (HorRat): a useful index of method performance with respect to precision, J. AOAC Int. 89 (2006) 1095–1109.

CHAPTER FOURTEEN

Analyses of cannabinoids in hemp oils by LC/Q-TOF-MS

Imma Ferrer*

Center for Environmental Mass Spectrometry, Department of Environmental Engineering, University of Colorado, Boulder, CO, United States
*Corresponding author: e-mail address: imma.ferrer@colorado.edu

Contents

1. Introduction		415
2. Methods		417
2.1	Definition	417
2.2	Rationale	417
2.3	Materials, equipment and reagents	417
2.4	Protocols	418
2.5	Precursor techniques	419
2.6	Safety considerations and standards	419
2.7	Analysis and statistics	419
2.8	Related techniques	419
2.9	Pros and cons	419
2.10	Alternative methods/procedures	420
2.11	Troubleshooting and optimization	420
2.12	Summary	420
3. Results and discussion		420
3.1	Chromatographic separation of cannabinoids	420
3.2	Accurate mass detection of cannabinoids	420
3.3	Fragmentation patterns of cannabinoids	423
3.4	Analyses of hemp oils	445
4. Conclusions		450
Acknowledgements		450
References		451

1. Introduction

Cannabis products, especially those derived from hemp plants, have been gaining popularity in the last few years. It is important to take into account that their safety and health properties are directly related to their

Comprehensive Analytical Chemistry, Volume 90
ISSN 0166-526X
https://doi.org/10.1016/bs.coac.2020.04.014

© 2020 Elsevier B.V.
All rights reserved.

415

chemical composition, which can vary widely depending on the type of crop or manufacturing method. For this reason, it is important that sound and robust methodologies are developed for the analysis of the main active constituents of these products, which are cannabinoids [1,2].

The method of choice for the determination of cannabinoids in cannabis plants or cannabis derived products has been liquid chromatography coupled with ultraviolet detection or diode array detection (LC/UV or LC/DAD) due to the high content of these compounds in this type of products [3,4]. While this might be true for the main components, delta-tetrahydrocannabinol (THC) and cannabidiol (CBD), other components, such as cannabigerol (CBG) or cannabinol (CBN) might be present in trace amounts and a more sensitive methodology might be needed. In this sense tandem mass spectrometry techniques, using LC/MS/MS with defined transitions for each compound are more appropriate [4–6]. In fact, the use of mass spectrometry in the cannabis industry is becoming more popular, since it can tackle more widespread analysis [7]. However, the use of accurate mass techniques, such as time-of-flight, for the determination of cannabinoids has not been described much in detail; a few papers exist describing this methodology for the analysis of these compounds [8–10]. With high resolution mass spectrometry, it is possible to figure out the cannabinoid profile in commercial hemp derived products. Because all cannabinoid compounds have very similar chemical structures and are inter-correlated it is important to find small differences in ionization and/or fragmentation patterns that aid to the unequivocal identification of each one of these compounds.

Up to recent times, the attention on the cannabinoid class focused mainly on the major cannabinoids, such as CBD, THC and CBN [11]. Minor cannabinoids have been studied to a smaller extent [12,13], and comprehensive cannabinoid profiles in commercial products are scarce [8]. But some of these minor cannabinoids are gaining popularity in light of the new pharmacological properties attributed to them [14,15]. For this reason, it is very important to evaluate and quantitate their presence in commercial products, especially for pet and human consumption.

In this work we developed a method based on LC/Q-TOF-MS for the characterization and chemical profiling of 17 different cannabinoids. The detailed MS-MS fragmentation for each one of them is shown and discussed. Finally, we applied the methodology to the analysis of commercial hemp derived oil samples.

2. Methods

2.1 Definition

Detection of cannabinoids derived from hemp plants by liquid chromatography coupled with quadrupole time-of flight mass spectrometry (LC/Q-TOF-MS).

2.2 Rationale

Most methods reported in the literature use LC-UV or GC-MS for the determination of cannabinoids [16]. However, GC-MS techniques are not recommended since some inter-conversions between THC and CBD can occur due to elevated inlet temperatures. Our protocol uses LC-MS with accurate mass detection for the unequivocal identification of all the cannabinoids studied.

2.3 Materials, equipment and reagents

Analytical standards for 17 cannabinoids were purchased from Cerilliant (Round Rock, TX, USA). The compounds were cannabichromene (CBC), cannabidiol (CBD), cannabigerol (CBG), cannabicyclol (CBL), cannabinol (CBN), cannabichromenic acid (CBCA), cannabidiolic acid (CBDA), cannabigerolic acid (CBGA), cannabinolic acid (CBNA), cannabidivarin (CBDV), cannabidivarinic acid (CBDVA), delta-8-tetrahydrocannabinol (Δ8-THC), delta-9-tetrahydrocannabinol (Δ9-THC), tetrahydrocannabinolic acid (THCA), tetrahydrocannabivarin (THCV), tetrahydrocannabivarinic acid (THCVA), cannabicitran (CBCT), and 11-hydroxy-tetrahydrocannabinol (HTHC). Three deuterated labelled standards for CBD, CBN and THC were also purchased for QA/QC purposes: cannabidiol d-3 (CBD-d3), cannabinol d-3 (CBN-d3) and delta-9-tetrahydrocannabinol d-3 (THC-d3).

HPLC grade acetonitrile, methanol and water were obtained from Burdick and Jackson (Muskegon, MI, USA). Formic acid was obtained from Sigma-Aldrich (St. Louis, MO, USA).

HPLC system (consisting of vacuum degasser, temperature-controlled autosampler and column compartments, and a binary pump) (Agilent Series 1290, Agilent Technologies, Santa Clara, CA, USA).

Reversed phase C8 analytical column of 150 mm × 4.6 mm and 3.5 µm particle size (Zorbax Eclipse XDB-C8).

The injected sample volume: 5 µL.

Ultra-high definition quadrupole time-of-flight mass spectrometer Model 6545 Agilent (Agilent Technologies, Santa Clara, CA, USA) equipped with electrospray Jet Stream Technology, operating in positive ion mode.

A calibrant solution was delivered by an external quaternary pump. This solution contains the internal reference masses (purine ($C_5H_4N_4$ at m/z 121.0509 and HP-921 [hexakis-(1H,1H,3H-tetrafluoro-pentoxy) phosphazene] ($C_{18}H_{18}O_6N_3P_3F_{24}$) at m/z 922.0098).

2.4 Protocols

2.4.1 Chromatographic separation

- Mobile Phase A: water (0.1% formic acid).
- Mobile phase B: acetonitrile.
- Flow rate: 0.6 mL/min.
- Gradient: 10%B held for 5 min, then to 100%B in 30 min, held at 100% for 3 min.
- Postrun time: 10 min.

2.4.2 TOF mass detection

- Mass range: 100–1000 m/z at 2 GHz high dynamic range mode.
- Operation parameters: capillary voltage: 3500 V; nebulizer pressure: 45 psi; drying gas: nitrogen at 10 L/min; drying gas temperature: 250°C; sheath gas flow: 11 L/min; sheath gas temperature: 350°C; nozzle voltage: 0 V, fragmentor voltage: 175 V; skimmer voltage: 65 V; octopole RF: 750 V.

2.4.3 MS-MS conditions

- Mass range: 40–700 m/z.
- Isolation window: 1.3 Da using "Abundance Dependent Acquisition" mode.
- Collision energies: 15 V and 30 V.

2.4.4 Sample preparation for hemp oils

- Weigh 20 mg of hemp oil in a glass test tube.
- Dissolve with 1 mL of methanol.
- Dilute solution to an approximate concentration of 20 µg/mL (based on initial concentration of CBD in the commercial hemp oil).
- Inject 5 µL onto LC/Q-TOF-MS instrument.

2.5 Precursor techniques

Concentrations of the standards purchased were 1 mg/mL in either methanol or acetonitrile (depending on solubility). Standards were kept stored at $-18\,°C$. From these solutions, working standard solutions were prepared by dilution with acetonitrile and water.

Calibration curves were prepared from a matrix of olive oil spiked with the mix of the 17 cannabinoids ranging from 0.1 ppm to 50 ppm. Internal standards were used for QA/QC purposes.

2.6 Safety considerations and standards

All safety considerations and analytical chemistry standard practices were followed for all the procedures described above.

2.7 Analysis and statistics

Analysis was carried out with MassHunter version 7.00.

Chemdraw was used for fragmentation pathways and for calculation of accurate masses of each ion.

2.8 Related techniques

Liquid chromatography with tandem mass detection (LC-MS-MS).

2.9 Pros and cons

Pros	Cons
Unequivocal technique with accurate mass using time-of-flight detection	Lengthy method development
Profiling of cannabinoids can be obtained for several hemp oils and commercial products	High scientific skill for the determination of fragmentation patterns
Quantitation can be carried out with this technique	
No degradation or inter-conversion is observed for the compounds studied	

2.10 Alternative methods/procedures

Alternative methods include the use of triple quadrupole coupled with mass spectrometry (LC-MS-MS) and it has been described in some previous works [4,5].

2.11 Troubleshooting and optimization

Problem	Solution
Matrix interferences	Use of internal standards is recommended
Complex MS-MS fragmentation spectra	Mass fragmentation interpretation with accurate mass

2.12 Summary

Chromatographic separation for 17 cannabinoids was achieved.

Fragmentation patterns for all cannabinoids were obtained under MS-MS conditions.

Analyses of commercial hemp oils revealed the presence of CBD and other major cannabinoids.

3. Results and discussion

3.1 Chromatographic separation of cannabinoids

Chromatographic separation for all the cannabinoids studied was accomplished with a reversed phase column (C8). In general, hydrophilic compounds such as cannabidivarin (CBDV) and tetrahydrocannabivarin (THCV), and their corresponding acids CBDVA and THCVA, eluted at shorter retention times as expected when compared to the other cannabinoids. The most hydrophobic compounds, cannabichromene (CBC), cannabicyclol (CBL) and cannabicitran (CBCT) eluted at a much later retention time, making necessary an extra 3 min of 100% of acetonitrile elution time. Fig. 1 shows the separation of all the compounds studied here.

3.2 Accurate mass detection of cannabinoids

Table 1 lists the elemental composition, accurate mass and main ion (base peak ion) observed for the cannabinoids studied here. The main ions

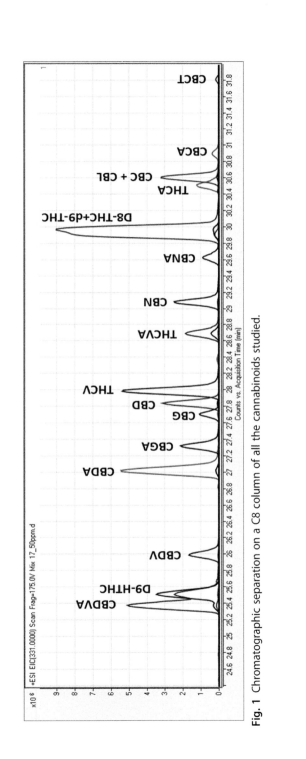

Fig. 1 Chromatographic separation on a C8 column of all the cannabinoids studied.

Table 1 Cannabinoids studied in this work with their elemental composition, accurate mass of the base peak ion and retention time.

Cannabinoids	Elemental composition	Accurate mass	Main ion	Ret. time (min)
Cannabidiol (CBD)	$C_{21}H_{30}O_2$	315.2319	M + H	27.9
Cannnabidivarin (CBDV)	$C_{19}H_{26}O_2$	287.2006	M + H	26
Cannabidiolic acid (CBDA)	$C_{22}H_{30}O_4$	341.2111/403.1856	M-H_2O/M + 2Na	27
Cannabidivarinic acid (CBDVA)	$C_{20}H_{26}O_4$	313.1798	M-H_2O	25.4
Δ8-Tetrahydrocannabinol (Δ8-THC)	$C_{21}H_{30}O_2$	315.2319	M + H	30
Δ9-Tetrahydrocannabinol (Δ9-THC)	$C_{21}H_{30}O_2$	315.2319	M + H	29.9
Δ9-Tetrahydrocannabinolic acid (THCA)	$C_{22}H_{30}O_4$	341.2111/403.1856	M-H_2O/M + 2Na	30.5
Tetrahydrocannabivarin (THCV)	$C_{19}H_{26}O_2$	287.2006	M + H	28
Tetrahydrocannabivarinic acid (THCVA)	$C_{20}H_{26}O_4$	313.1798/375.1543	M-H_2O/M + 2Na	28.7
11-Hydroxy-Δ9-THC (Δ9-HTHC)	$C_{21}H_{30}O_3$	331.2268	M + H	25.5
Cannabinol (CBN)	$C_{21}H_{26}O_2$	311.2006	M + H	29.1
Cannabinolic acid (CBNA)	$C_{22}H_{26}O_4$	355.1904/399.1543	M + H/M + 2Na	29.7
Cannabigerol (CBG)	$C_{21}H_{32}O_2$	317.2475	M + H	27.7
Cannabigerolic acid (CBGA)	$C_{22}H_{32}O_4$	343.2268/405.2012	M-H_2O/M + 2Na	27.3
Cannabichromene (CBC)	$C_{21}H_{30}O_2$	315.2319	M + H	30.6
Cannabichromenic acid (CBCA)	$C_{22}H_{30}O_4$	359.2217/403.1856	M + H/M + 2Na	30.9
Cannabicyclol (CBL)	$C_{21}H_{30}O_2$	315.2319	M + H	30.6
Cannabicitran (CBCT)	$C_{21}H_{30}O_2$	315.2319	M + H	31.7

obtained by positive ion electrospray are usually the protonated molecules $[M+H]^+$. However, the acid forms, such as CBDVA, THCVA, CBDA, THCA and CBGA, all experienced a loss of water in the source. This phenomenon has been reported for these types of molecules. The cannabinoids CBCA and CBNA were the exception to this phenomenon, probably due to the higher hydrophobicity of these molecules. Furthermore, double sodium adducts were observed for all the acids forms. Because the acids might be deprotonated in the pH of the mobile phase, a sodium adduct can bond to the oxygen in the carboxylic moiety, and then a second sodium forms an adduct to give rise to a positive ion. All the acid forms would work very well under negative ion electrospray conditions, however for this methodology developed here we decided to keep everything in positive ion mode to have a general detection method for all the cannabinoids. Triple quadrupoles can usually alternate between positive and negative ion detection, but QTOF instruments do not perform well under alternating voltage conditions.

3.3 Fragmentation patterns of cannabinoids

Each one of the cannabinoids studied was analysed under MS-MS conditions (see Methods section) and the detailed fragmentations obtained were proposed based on accurate mass assignments for each fragment ion. In general, moderate fragmentation occurred at 15 V, but better signal for the fragment ions were obtained at 30 V and that is the reason this collision energy was chosen for all the results presented in the next subsections.

Cannabinoids can be classified into 8 main different types: cannabigerol, cannabidiol, tetrahydrocannabinol, cannabinol, cannabichromene, cannabielsoin, cannabicitran, and cannabicyclol types. In this work, we classified them into these groups when discussing the fragmentation patterns, since many similarities were observed within the groups studied. The cannabielsoin group was not studied since the standards were not available at the time of the study.

3.3.1 Cannabidiol type
Cannabidiol (CBD)

The MS-MS spectrum at 30 V is shown in Fig. 2 for cannabidiol (CBD). The chemical structure is shown as well and the numbering of carbons will follow the rest of the cannabinoids discussed here. The protonated molecule $[M+H]^+$ is 315.2319 and it fragments extensively at this collision energy. The most relevant fragments are at m/z 259.1674 (F1), corresponding to

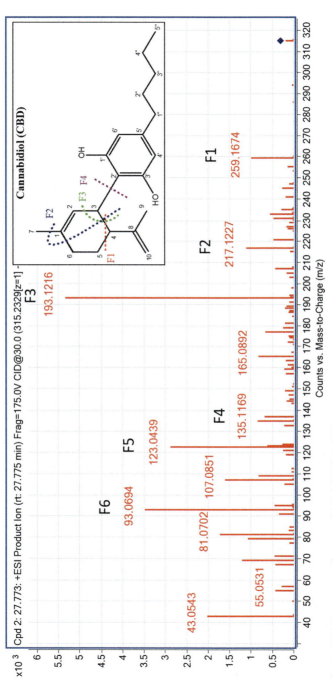

Fig. 2 MS-MS spectrum for cannabidiol (CBD) at 30V.

the loss of four carbon units from the terpene moiety (methylpropene unit); m/z 217.1227 (F2), which is the sequential loss of a propylene unit; m/z 193.1216 (F3), which corresponds to the olivetol moiety with the carbon unit attached to C2' of the benzene ring; m/z 135.1169 (F4) corresponding to the terpene moiety (limonene); m/z 123.0439 (F5) corresponding to resorcinol with the carbon unit attached to C2'; and m/z 93.0694 (F6) which corresponds to the 1-methylcyclohexa-1, 4-diene structure.

Cannabidivarin (CBDV)

Cannabidivarin is the analog to CBD with a shorter alkyl chain in the aromatic ring, so instead of five carbons it has 3 carbons, which comes from the divarinolic acid structure. The protonated molecule $[M+H]^+$ is 287.2006, and it also fragments extensively at 30 V. Therefore, as expected, the analogy to the main ion at m/z 193 for CBD, it would be the ion at m/z 165.0903 (F3), which is indeed the main fragment ion obtained for CBDV, as shown in Fig. 3. Other characteristic fragment ions are those at m/z 219.1370 (F1) and m/z 189.0906 (F2), which are unique to this structure with the 3-carbon alkyl chain. Other similar ions to CBD were obtained here as well since they involve the resorcinol and terpene structure, also present in this compound: m/z 135.1165 (F4), m/z 123.0444 (F5) and m/z 93.0692 (F6).

Cannabidiolic acid (CBDA)

Cannabidiolic acid is the carboxylated form of CBD. The protonated molecule $[M+H]^+$ is 359.2217. The corresponding MS-MS spectrum is shown in Fig. 4. This type of acids experiment a rapid water loss of 18 units, leading to the ion at m/z 341.2105 (F1). Another small, but important fragment is that at m/z 285.1481 (F2), which corresponds to the breakage between the bonds at C4–C5 and C3–C4 atoms. The other relevant ions are at m/z 261.1481 (F3), which comes from the breakage of the terpene moiety at C1–C6 bond and C3–C4 bonds, and the fragment at m/z 219.1010 (F4), which forms from the terpene loss (with C3 left in the molecule) and represents the main fragment ion. All of these fragmentations occur after the water loss.

Cannabidivarinic acid (CBDVA)

Cannabidivarinic acid is the carboxylated form of CBDV. Similarly to CBDA, this compound experiments a rapid water loss from the protonated molecule at m/z 331.1904, leading to the fragment ion at m/z 313.1798 (F1),

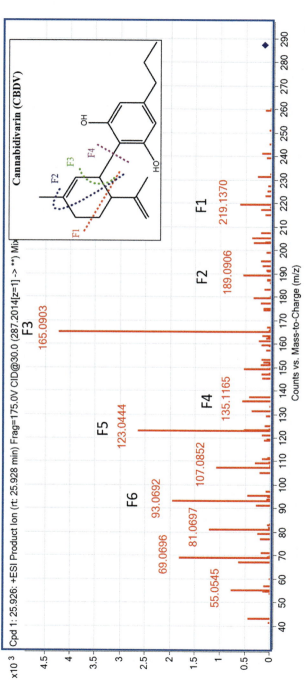

Fig. 3 MS-MS spectrum for cannabidivarin (CBDV) at 30 V.

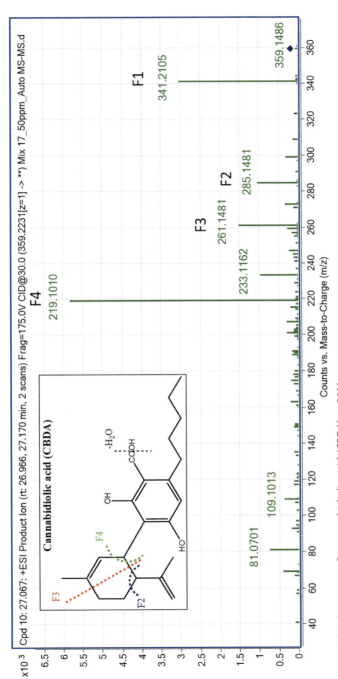

Fig. 4 MS-MS spectrum for cannabidiolic acid (CBDA) at 30V.

as shown in Fig. 5. The loss of a methylpropene occurs (similarly to CBD and CBDA) resulting in m/z 257.1163 (F2). Moreover, the same loss between C1–C6 and C3–C4 bonds are seen for this molecule, giving rise to the ion at m/z 233.1171 (F3). And again, the terpene loss (with C3 left in the molecule) occurs as well, leading to the main ion at m/z 191.0698 (F4).

3.3.2 Tetrahydrocannabinol type

Δ9-Tetrahydrocannabinol (Δ9-THC)

The tetrahydrocannabinol type is characteristic of the presence of a dihydropyran ring (instead of a free hydroxyl group in C3'), which confers this type of molecules higher lipophilicity, hence the later elution times. Tetrahydrocannabinol is the main component in *Cannabis indica* and a minor component (usually established at <0.3%) in hemp. The protonated molecule $[M+H]^+$ for THC is 315.2319 (same exact mass as CBD). Interestingly, the MS–MS spectra of THC is also practically identical to that of CBD, as seen in Fig. 6. The same exact fragments seen for CBD are obtained for THC, except a slight difference in ion intensities, which are generally lower for THC. In addition, one of the fragments seems more prevalent in THC when compared to CBD, that is the ion at m/z 231.1376 (F2), which involves the breaking of the ether bond and then a similar breakage between bonds C1–C6 and C3–C4, re-forming a benzopyran structure. All the other fragments (F1, F3, F4, F5 and F6) are identical to those of CBD. Δ8-Tetrahydrocannabinol (Δ8-THC) was also studied under MS–MS conditions, and it was identical to that of Δ9-THC (spectrum not shown here).

Tetrahydrocannabivarin (THCV)

Tetrahydrocannabivarin is the molecule related to THC but with a 3-carbon alkyl chain (see Fig. 7). The protonated molecule $[M+H]^+$ is 287.2006 (identical to CBDV). The fragment at m/z 231.1375 (F1) is similar to the fragment at m/z 259 (F1) from THC. The fragment ion at m/z 205.1222 (F2) is again similar to the F2 fragmentation in THC. The main fragment ion at m/z 165.0905 (F3) is equivalent to the m/z 193 fragment in THC and corresponds to the "olivetol" moiety with a 3-carbon alkyl chain and the C3 present (same ion observed for CBDV as well). All the other fragments (F4, F5, and F6) are identical to the ones discussed for CBD and THC.

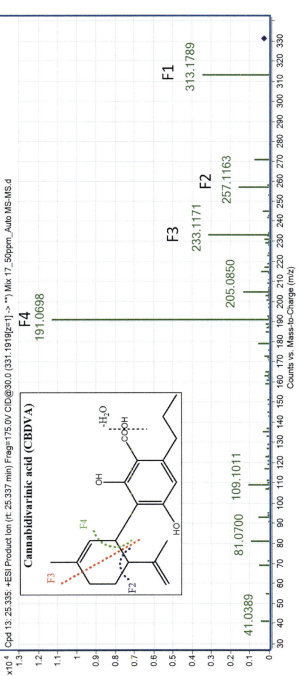

Fig. 5 MS-MS spectrum for cannabidivarinic acid (CBDVA) at 30V.

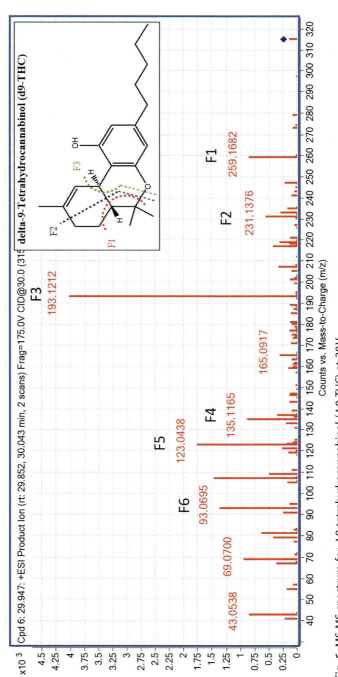

Fig. 6 MS-MS spectrum for Δ9-tetrahydrocannabinol (Δ9-THC) at 30V.

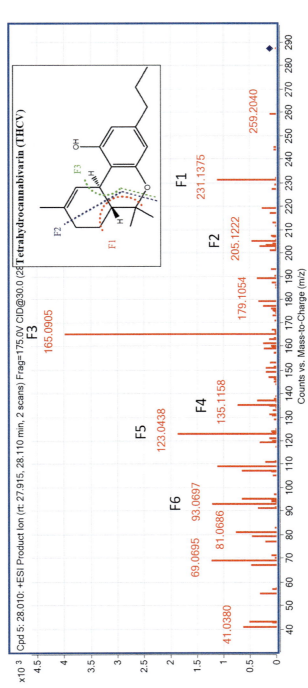

Fig. 7 MS-MS spectrum for tetrahydrocannabivarin (THCV) at 30 V.

Tetrahydrocannabinolic acid (THCA)

Tetrahydrocannabinolic acid protonated molecule $[M+H]^+$ is 359.2217. Its MS-MS spectrum (Fig. 8) is again identical to that of CBDA. The same exact fragments at m/z 341.2109 (F1) (loss of water), m/z 285.1481 (F2), m/z 261.1475 (F3), and m/z 219.1011 (F4) are observed for this molecule. In this case, no differences in ion intensity were observed when compared to CBDA.

Tetrahydrocannabivarinic acid (THCVA)

Tetrahydrocannabivarinic acid protonated molecule $[M+H]^+$ is 331.1904. Its corresponding MS-MS fragmentation spectrum is shown in Fig. 9. The same exact losses as CBDVA were observed, with slightly differences in ion intensity. The loss of water at m/z 313.1786 appears to be more intense in THCVA when compared to CBDVA. Moreover, one additional fragment ion at m/z 271.1327 (F2') was observed at a higher intensity, which corresponds to the loss of propene, probably occurring at the terpene moiety. All the other fragments are the same ones as discussed for CBDVA.

11-Hydroxy-tetrahydrocannabinol (HTHC)

HTHC is one of the major human metabolites of THC [17]. The protonated molecule $[M+H]^+$ is 331.2268. The corresponding MS-MS spectra is shown in Fig. 10. A loss of water is observed by the fragment at m/z 313.2152 (F1). Interestingly, this molecule losses a second water forming the ion at m/z 295.2045 (F2). Other fragments of interest are at m/z 257.1165 (F3) and m/z 233.1169 (F4), which are depicted in Fig. 10 and involve structure rearrangements. The main fragment ion at m/z 191.0698 (F5) corresponds to the same fragmentation as F3 but with an additional loss of the aliphatic 5-carbon chain on the aromatic ring.

3.3.3 Cannabinol type

Cannabinol (CBN)

Cannabinol is the transformation product of THCA found in aged cannabis (via oxidation process), it is mainly the aromatized form which can also be formed by UV light. If a cannabis plant for example is exposed to sunlight for a prolonged period of time, THCA will convert to CBNA, which by decarboxylation process will lead to CBN [17]. The protonated molecule $[M+H]^+$ is 311.2006. The fragmentation MS-MS spectrum for CBN is shown in Fig. 11. Because of the stability of the aromatic ring, this molecule fragments much differently than the other cannabinoids. The C-C bond

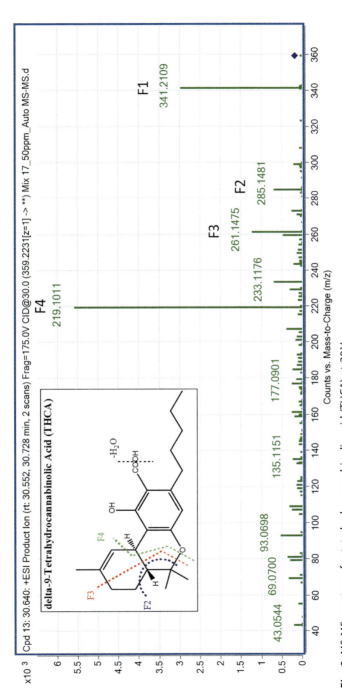

Fig. 8 MS-MS spectrum for tetrahydrocannabinolic acid (THCA) at 30V.

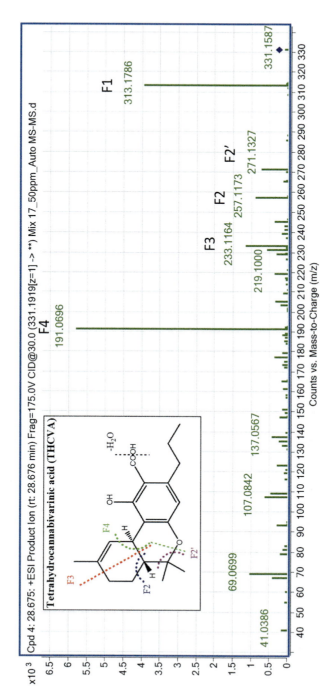

Fig. 9 MS-MS spectrum for tetrahydrocannabivarinic acid (THCVA) at 30V.

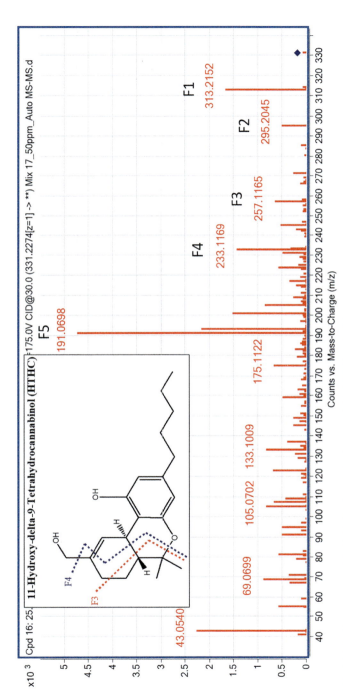

Fig. 10 MS-MS spectrum for 11-hydroxy-tetrahydrocannabinol (HTHC) at 30V.

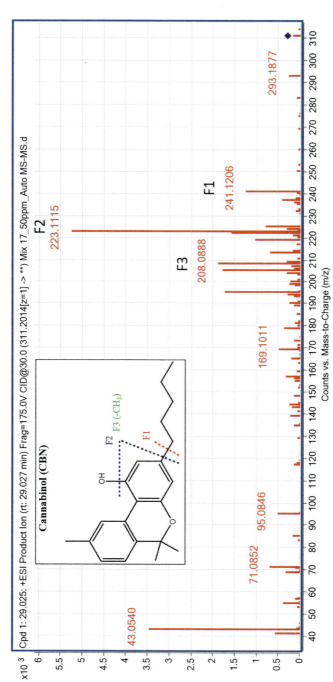

Fig. 11 MS-MS spectrum for cannabinol (CBN) at 30 V.

between two benzene rings is stronger (more difficult to break) than the C-C bond between a benzene ring and a terpene moiety. A small fragment at m/z 293.1877 is observed corresponding to a water loss. A higher intensity fragment at m/z 241.1206 (F1) is formed by cleavage of the aliphatic 5-carbon chain. The main fragment ion at m/z 223.1115 (F2) forms via the same fragmentation and an additional water loss. And finally, the fragment at m/z 208.0888, which is an odd electron ion, is formed by an additional methyl loss.

Cannabinolic acid (CBNA)

Cannabinolic acid is the precursor of CBN and, as mentioned before, it is a degradation product from THCA either in the cannabis plant or in produced cannabis, after long times of exposure to solar light. The protonated molecule $[M+H]^+$ is 355.1904. The fragmentation spectrum for CBNA is shown in Fig. 12. As expected, the major loss for this compound is a loss of water, common to all the acidic forms of the cannabinoids, to form the ion at m/z 337.1793 (F1). A second loss of water, similar to that observed for CBN, also occurs for this compound giving rise to the ion at m/z 319.1697 (F2). Finally, the ion at m/z 235.1115 (F3) forms via a re-arrangement of the molecule after losing the three molecules of water and the 5-carbon chain.

3.3.4 Cannabigerol type

Cannabigerol (CBG)

Cannabigerol comes from cannabigerolic acid (CBGA) after decarboxylation via heat or light. The protonated molecule $[M+H]^+$ is 317.2475. The fragmentation MS–MS spectra of cannabigerol is probably the simplest and most straightforward of all the cannabinoids studied here. As seen in Fig. 13, only two major fragments are observed: the ion at m/z 193.1215 (F1), which is common to all the other major cannabinoids and corresponds to the olivetol moiety with the ortho methyl group, and the ion at m/z 123.0438 (F2), which again corresponds to the resorcinol moiety with the ortho methyl group. A smaller ion is seen at m/z 207.1369, which probably corresponds to the breakage of the isoprenoid moiety between atoms C1-C2.

Cannabigerol acid (CBGA)

Cannabigerol acid is consider the "mother" of all the cannabinoids in the plant. It is the precursor for CBDA, THCA and CBCA, among others, via different enzymatic reactions. The protonated molecule $[M+H]^+$ is

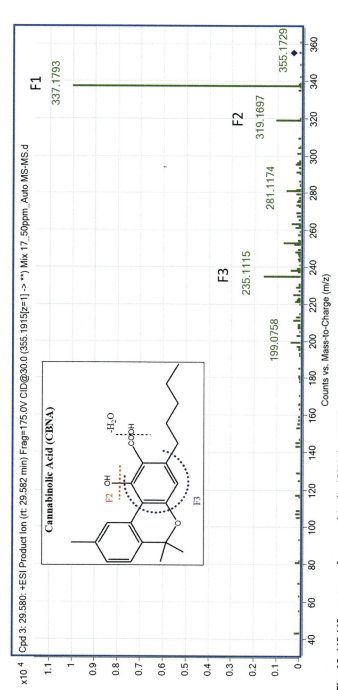

Fig. 12 MS-MS spectrum for cannabinolic (CBNA) at 30 V.

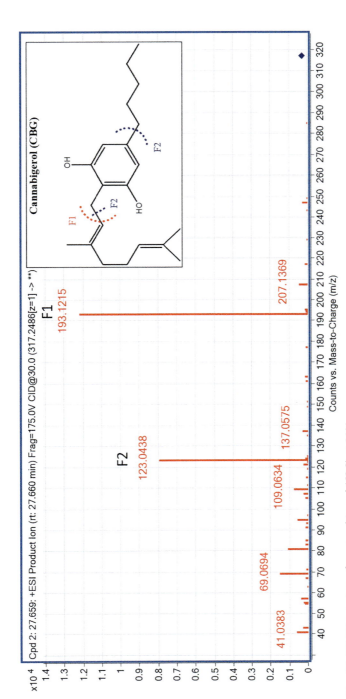

Fig. 13 MS-MS spectrum for cannabigerol (CBG) at 30V.

261.2373. Again, the MS–MS fragmentation spectrum is the simplest of all (Fig. 14), showing only one fragment ion at m/z 219.1011 (F1), corresponding to the loss of the isoprenoid moiety (breaking at C2-C3), after the loss of water. It is worth mentioning that this molecule (like all the other acid forms studied) losses water rapidly in the electrospray source (before MS–MS conditions), forming the ion at m/z 343.2237, which is slightly visible in the MS–MS fragmentation spectrum as well.

3.3.5 Cannabichromene type
Cannabichromene (CBC)

Cannabichromene protonated molecule $[M+H]^+$ is identical to that of CBD and THC, m/z 315.2319. This molecule presents a very similar MS–MS fragmentation spectrum to THC, in spite of the apparent different chemical structure (Fig. 15). The fragments at m/z 259.1685 (F1), m/z 231.1381 (F2), m/z 193.1216 (F3) and m/z 123.0434 (F5) are the same ones observed for THC. Two other fragments at m/z 137.0595 (F4) and m/z 69.0700 (F6) were also prominent and are discussed here. The fragment F4 is a loss of the 5-carbon aliphatic chain and an opening of the benzopyran ring. And the small fragment F6 corresponds to 2-methylbut-2-ene.

Cannabichromenic acid (CBCA)

Cannabichromenic acid is the precursor of CBC and the protonated molecule $[M+H]^+$ is 359.2217. Its fragmentation spectrum is shown in Fig. 16. Again, it is an almost identical spectrum to that of THCA. The fragment at m/z 341.2106 (F1) is the classic loss of water. And all the other ions at m/z 285.1478 (F2), m/z 261.1476 (F3) and m/z 219.1010 (F4) are the same exact ones than those for THCA. A unique fragment that appears for this molecule is at m/z 273.1484 (F2'), which corresponds to the cleavage of yet another C–C bond in the aliphatic chain (the isoprene unit), similar to the F6 fragmentation for CBC.

3.3.6 Cannabicyclol type
Cannabicyclol (CBL)

Cannabicyclol is one of the less common cannabinoids, it forms by degradation of CBC through natural irradiation or acid conditions. The protonated molecule $[M+H]^+$ is 315.2319 (identical to CBD, THC and CBC). The fragmentation spectrum is shown in Fig. 17. The fragments observed for this molecule are at m/z 235.1689 (F1), m/z 193.1218 (F2), m/z 165.0906 (F3), which is the same fragmentation from F1 plus the loss of

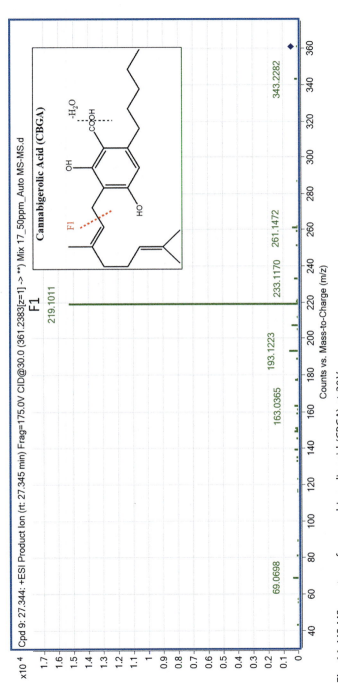

Fig. 14 MS-MS spectrum for cannabigerolic acid (CBGA) at 30V.

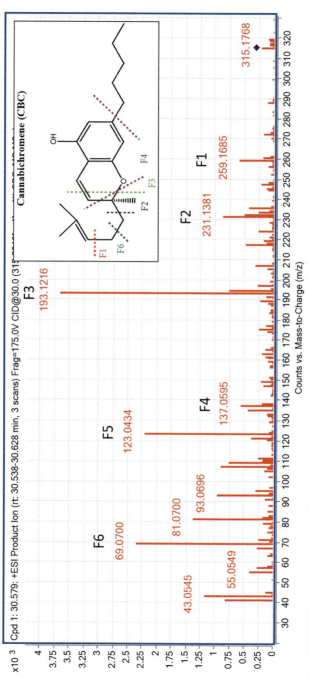

Fig. 15 MS-MS spectrum for cannabichromene (CBC) at 30V.

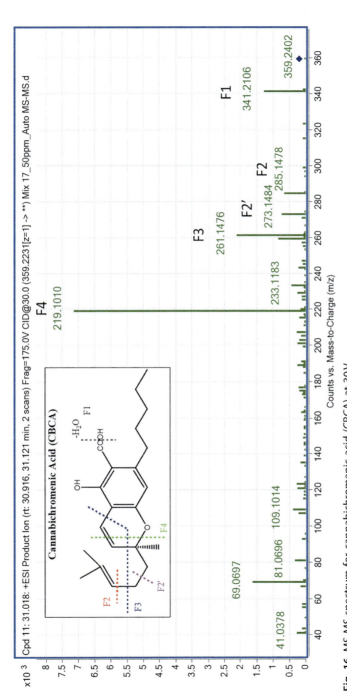

Fig. 16 MS-MS spectrum for cannabichromenic acid (CBCA) at 30V.

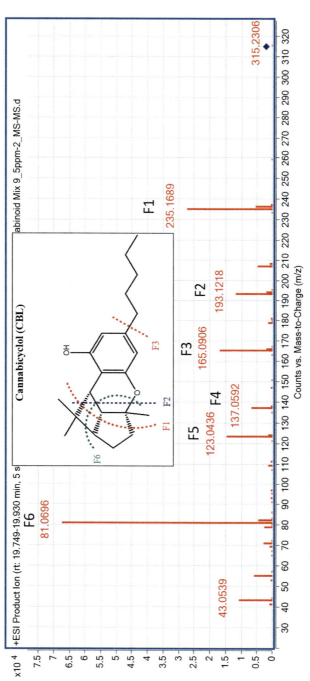

Fig. 17 MS-MS spectrum for cannabicyclol acid (CBL) at 30V.

Analyses of cannabinoids in hemp oils 445

the 5-carbon aliphatic chain, m/z 137.0592 (F4) which is an additional ethylene loss, m/z 123.0436 (F5) corresponding to the resorcinol moiety with one carbon, and the main fragment ion at m/z 81.0696 (F6), which corresponds to the methylcyclopentene moiety, which is very characteristic of this molecule.

3.3.7 Cannabicitran type
Cannabicitran (CBCT)
Cannabicitran is found in highest concentrations in the distillates from the Type III flowers of the cannabis plant, which have high levels of CBD and low levels of THC [18]. However, compared to most other cannabinoids, it is found in relatively low concentrations. Chemically it has an extra ring compared to all the other cannabinoids, but it has exactly the same formula composition and same exact mass to CBD, THC, CBC and CBL. This compound is the most hydrophobic of all the cannabinoids studied, thus having a very late retention time. The protonated molecule $[M+H]^+$ is 315.2319. The MS-MS fragmentation spectrum is shown in Fig. 18, and, surprisingly it is almost identical to that of THC, with a slightly difference in ion fragment intensities. All fragments obtained F1, F2, F3, F4, F5 and F6 are identical to those discussed earlier for THC. If one assumes the ether bond between the C and O is labile, then the molecule is exactly identical to THC, so it makes sense that all the fragment ions obtained are also identical.

Finally, we have represented some of the common ions found for most of the cannabinoids in Fig. 19, with their corresponding accurate masses.

3.4 Analyses of hemp oils

Twelve different hemp oils containing various compositions of cannabinoids (from pure CBD to full-spectrum) were analysed by this methodology. All of the commercial oils obtained came from local stores in Colorado, some of them were purchased at a regular grocery store. The oils were analysed by LC/Q-TOF-MS and a composition cannabinoid profile was obtained for each one of them. As expected, all the samples contained cannabidiol (CBD) as advertised in the label (frequency of detection was 100%). However, a high percentage of samples (75%) also contained THC amounts, but those concentrations were in all cases below 0.3%. Some of the cannabinoids studied were not detected, those were: CBG, THCVA, CBNA, CBL, CBCT and HTHC. Table 2 shows the average concentrations for each cannabinoid in the 12 hemp oil samples.

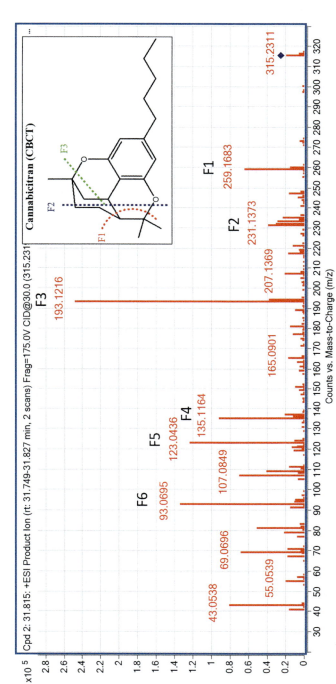

Fig. 18 MS-MS spectrum for cannabicitran (CBCT) at 30V.

Analyses of cannabinoids in hemp oils

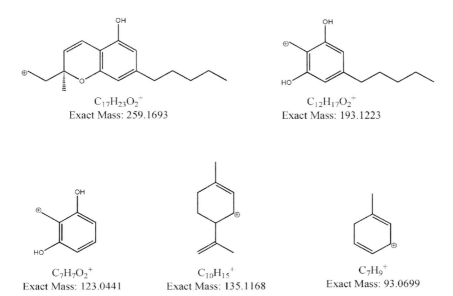

Fig. 19 Common ions obtained from the MS-MS spectra of the majority of the compounds studied here.

Fig. 20 shows as an example the cannabinoid profile for three different hemp oils. Hemp oil #1 was a pet product labelled as CDB (with supposedly no THC present), however as seen in this figure the CBD content was rather low, and it was mainly its acid form the one detected at highest concentration (CBDA). Surprisingly, THCA was also detected at a higher percentage in this product (10%). Hemp oil #4 was a full spectrum oil for human consumption; CBD and CBDA were the main components for this product, but THC and CBCA were also detected at lower percentages (2% and 4%, respectively). Finally, hemp oil #7 was labelled as a product containing CBDV and THCV (relatively new minor cannabinoid with some pharmacological properties, such as appetite suppression). In this case the relative percentage was mainly from CBD and CBDV, with smaller percentages of THC and THCV present as well.

Fig. 21 shows the relative frequency of detection (normalized to CBD since all products contained this main ingredient) obtained for each cannabinoid for the 12 hemp oils analysed. As we can see in this figure, the second most detected cannabinoid was THC, indicating that either some

Table 2 Average concentrations obtained for all the cannabinoids studied in 12 hemp oil samples.

Compound	Average concentration (mg/mL)
CBD	7.89
CBDV	0.60
CBDA	0.64
CBDVA	0.03
THC	0.15
THCV	0.46
THCA	0.19
THCVA	n.d.
HTHC	n.d.
CBN	0.13
CBNA	n.d.
CBG	n.d.
CBGA	0.08
CBC	0.24
CBCA	0.15
CBL	n.d.
CBCT	n.d.

n.d., not detected.

of the commercial manufacturing for these oils are not separating efficiently CBD from THC, or in the case of full-spectrum oils, that was one of the major components extracted from the hemp plants. A third component that was present at a higher percentage of samples was CBC. This compound is again formed in the plant through an enzymatic reaction from CBC synthase, so its presence in the samples is expected [19]. CBDA was also detected quite a few times, indicating some of the processes involved in the extraction of oils use the plant as such, without heating or drying. CBDV was also found in a similar percentage of samples. Finally, all the other compounds were found only in a relative small number of samples.

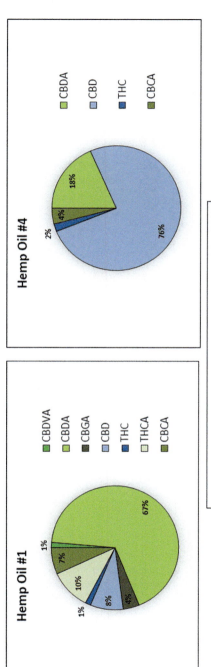
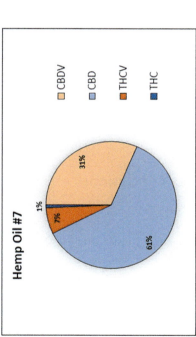

Fig. 20 Cannabinoid profile (as a percentage of detected cannabinoids) for 3 hemp oils.

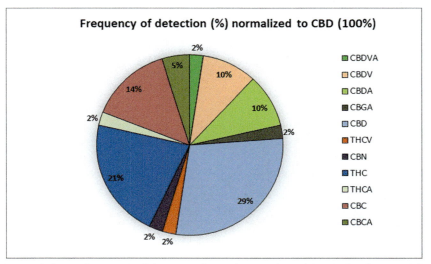

Fig. 21 Relative frequency of detection for each cannabinoid (normalized to CBD) for the 12 hemp oils analysed.

4. Conclusions

A methodology for the analysis of 17 major cannabinoids was developed by liquid chromatography coupled to quadrupole time-of-flight mass spectrometry (LC/Q-TOF-MS). The individual mass fragmentation spectra for each one of the compounds analysed was presented and major ions were identified and assigned to chemical structures. Fragmentation patterns for each one were described and accurately reported. Finally, 12 commercial samples of hemp oils were analysed and the presence of all the cannabinoids was evaluated, quantitated and reported.

Acknowledgements

Agilent Technologies is acknowledged for instrument and technical support. No other financial support was received for this work. All products analysed were purchased at local stores in CO.

This chapter is dedicated to our beloved Kitty (2003–2017) who inconspicuously got us started on the analyses of cannabinoids. We will never know if hemp oil #1 helped, but hopefully in the next few years a large amount of science and testing will be done in order to understand the pharmacological benefits of these type of compounds for human and living species. And we hope this paper can contribute even if a tiny amount to the basics of this type of research.

References

[1] P. Morales, D.P. Hurst, P.H. Reggio, Molecular targets of the phytocannabinoids: a complex picture, Prog. Chem. Org. Nat. Prod. 103 (2017) 103–131. https://doi.org/10.1007/978-3-319-45541-9_4.

[2] C.E. Turner, M.A. Elsohly, E.G. Boeren, Constituents of *cannabis sativa* l. xvii. A review of the natural constituents, J. Nat. Prod. 43 (1980) 169–234. https://doi.org/10.1021/np50008a001.

[3] C. Citti, B. Pacchetti, M.A. Vandelli, F. Forni, G. Cannazza, Analysis of cannabinoids in commercial hemp seed oil and decarboxylation kinetics studies of cannabidiolic acid (CBDA), J. Pharm. Biomed. Anal. 149 (2018) 532–540. https://doi.org/10.1016/j.jpba.2017.11.044.

[4] O. Aizpurua-Olaizola, U. Soydaner, E. Öztürk, D. Schibano, Y. Simsir, P. Navarro, N. Etxebarria, A. Usobiaga, Evolution of the cannabinoid and terpene content during the growth of Cannabis sativa plants from different chemotypes, J. Nat. Prod. 79 (2016) 324–331. https://doi.org/10.1021/acs.jnatprod.5b00949.

[5] P. Berman, K. Futoran, G.M. Lewitus, D. Mukha, M. Benami, T. Shlomi, D. Meiri, A new ESI-LC/MS approach for comprehensive metabolic profiling of phytocannabinoids in Cannabis, Sci. Rep. 8 (2018) 14280. https://doi.org/10.1038/s41598-018-32651-4.

[6] I. Di Marco Pisciottano, G. Guadagnuolo, V. Soprano, M. De Crescenzo, P. Gallo, A rapid method to determine nine natural cannabinoids in beverages and food derived from Cannabis sativa by liquid chromatography coupled to tandem mass spectrometry on a QTRAP 4000, Rapid Commun. Mass Spectrom. 32 (2018) 1728–1736. https://doi.org/10.1002/rcm.8242.

[7] B. Nie, J. Henion, I. Ryona, The role of mass spectrometry in the cannabis industry, J. Am. Soc. Mass Spectrom. 30 (2019) 719–730. https://doi.org/10.1007/s13361-019-02164-z.

[8] C. Citti, P. Linciano, S. Panseri, F. Vezzalini, F. Forni, M.A. Vandelli, G. Cannazza, Cannabinoid profiling of hemp seed oil by liquid chromatography coupled to high-resolution mass spectrometry, Front. Plant Sci. 10 (2019) 120. https://doi.org/10.3389/fpls.2019.00120.

[9] M. Hädener, M.Z. Kamrath, W. Weinmann, M. Groessl, High-resolution ion mobility spectrometry for rapid cannabis potency testing, Anal. Chem. 90 (2018) 8764–8768. https://doi.org/10.1021/acs.analchem.8b02180.

[10] A. Macherone, J. Stevens, M. Adams, A. Roth, K. Kaikaris, S.D. Antonio, R. Steed, Liquid chromatography-time-of-flight mass spectrometry for cannabinoid profiling and quantitation in hemp oil extracts, www.chromatographyonline.com LC-GC 35 (2017) 23–26. http://www.chromatographyonline.com/liquid-chromatography-time-flight-mass-spectrometry-cannabinoid-profiling-and-quantitation-hemp-oil.

[11] C.N. Pegoraro, D. Nutter, M. Thevenon, C.L. Ramirez, Chemical profiles of cannabis sativa medicinal oil using different extraction and concentration methods, Nat. Prod. Res. (2019) 1–4. https://doi.org/10.1080/14786419.2019.1663515.

[12] S.A. Ahmed, S.A. Ross, D. Slade, M.M. Radwan, I.A. Khan, M.A. Elsohly, Minor oxygenated cannabinoids from high potency Cannabis sativa L, Phytochemistry 117 (2015) 194–199. https://doi.org/10.1016/j.phytochem.2015.04.007.

[13] M.M. Radwan, M.A. ElSohly, A.T. El-Alfy, S.A. Ahmed, D. Slade, A.S. Husni, S.P. Manly, L. Wilson, S. Seale, S.J. Cutler, S.A. Ross, Isolation and pharmacological evaluation of minor cannabinoids from high-potency Cannabis sativa, J. Nat. Prod. 78 (2015) 1271–1276. https://doi.org/10.1021/acs.jnatprod.5b00065.

[14] L.O. Hanuš, S.M. Meyer, E. Muñoz, O. Taglialatela-Scafati, G. Appendino, Phytocannabinoids: a unified critical inventory, Nat. Prod. Rep. 33 (2016) 1357–1392. https://doi.org/10.1039/c6np00074f.

[15] E.B. Russo, Taming THC: potential cannabis synergy and phytocannabinoid-terpenoid entourage effects, Br. J. Pharmacol. 163 (2011) 1344–1364. https://doi.org/10.1111/j.1476-5381.2011.01238.x.

[16] C.L. Ramirez, M.A. Fanovich, M.S. Churio, Cannabinoids: Extraction Methods, Analysis, and Physicochemical Characterization, first ed., Elsevier B.V., 2018. https://doi.org/10.1016/B978-0-444-64183-0.00004-X.

[17] M.A. Huestis, Pharmacokinetics and metabolism of the plant cannabinoids, $\Delta 9$-tetra-hydrocannibinol, cannabidiol and cannabinol, Handb. Exp. Pharmacol. 168 (2005) 657–690. https://doi.org/10.1007/3-540-26573-2-23.

[18] Extract Labs, (n.d.) https://www.extractlabs.com/cbd-guides/what-is-cbt.

[19] C.E. Turner, M.A. Elsohly, Biological activity of Cannabichromene, its homologs and isomers, J. Clin. Pharmacol. 21 (1981) 283S–291S. https://doi.org/10.1002/j.1552-4604.1981.tb02606.x.

CHAPTER FIFTEEN

The estimation of cannabis consumption through wastewater analysis

Lubertus Bijlsma[a],*, Daniel A. Burgard[b], Frederic Been[c], Christoph Ort[d], João Matias[e], Viviane Yargeau[f]

[a]Research Institute for Pesticides and Water, University Jaume I, Castellón, Spain
[b]Department of Chemistry, University of Puget Sound, Tacoma, WA, United States
[c]KWR Water Research Institute, Nieuwegein, The Netherlands
[d]Eawag, Urban Water Management, Swiss Federal Institute of Aquatic Science and Technology, Dübendorf, Switzerland
[e]European Monitoring Centre for Drugs and Drug Addiction, Lisbon, Portugal
[f]Department of Chemical Engineering, McGill University, Montreal, QC, Canada
*Corresponding author: e-mail address: bijlsma@uji.es

Contents

1. Wastewater-based epidemiology	453
2. Cannabis biomarkers	457
3. Analytical methodology	461
4. Wide-scale use of WBE for cannabis estimates in Canada and US.	472
5. Future research and current asset	476
Acknowledgements	477
References	477

1. Wastewater-based epidemiology

Wastewater-based epidemiology (WBE) profits from raw influent wastewater samples that are collected routinely at the inlet of most sewage treatment plants (STPs). In these "pooled urine samples", the concentrations of illicit drugs and their metabolites can be quantified to estimate the total amount of drugs consumed by a community. This provides a non-invasive, near-real-time analysis of drug use within the area of a sewer network connected to a STP.

In the 1990s, liquid-chromatography coupled to mass spectrometry (LC-MS) was initially used as a technique to monitor micropollutants, such as pharmaceuticals, personal care products and pesticides, and to study the impact of treated and untreated liquid household waste in the environment. It was, at that time, suggested to use the quantified data of pharmaceuticals to estimate its consumption by the general public or in hospitals from a catchment of a STP, and WBE had its first "official" mention in 2001 [1]. The approach has rapidly developed over the last decade to monitor a myriad of illicit drug residues in near-real-time with numerous methodological and applied publications (>250, Keywords "sewage epidemiology" OR "wastewater-based epidemiology" SCOPUS 14.2.2020). WBE has established itself as a useful routine application in estimating temporal and geographical trends in illicit drug use and strongly complements the various sources of information on the drug situation, where each source has different uses and strengths. In comparison with survey methods, wastewater analysis is not subject to response and non-response bias. By formally testing wastewater, this method is expected to be closer to the true spectrum of drugs being consumed rather than relying on individual recollection or belief. It also has the potential to provide timely information, within short time frames, on geographical and temporal trends. However, it cannot provide information on prevalence and frequency of use, numbers of users, types of user and purity of the drugs. Triangulation of data from wastewater analysis with data obtained through other indicators is an important area of continuing work that will help establish the merits and validity of both.

WBE consists of five steps (see Fig. 1). (1) Representative daily composite samples of raw wastewater are collected. (2) Concentrations of selected substances are quantified. (3) The concentrations are multiplied with the wastewater volumes measured over the period of sampling to obtain loads of drug residues in sewers. (4) Daily loads are divided by the number of people present in the catchment area of the STP to facilitate comparison among cities based on these population–normalized estimates. Finally, (5) the total daily consumption of a drug is estimated by applying a specific correction factor to the daily sewer loads. The correction factor considers the average excretion rate of a given drug residue and the molecular mass ratio of the parent drug to its metabolite, but can also take the stability or purity of a drug into account. This final step is, however, optional and not needed if trends are the desired outcome, or if for example excretion rates are not available or reliable. Typically, results are reported up to step 4 or 5 and sometimes a sixth step is also performed, i.e., number of doses.

To minimize uncertainties, it is recommended to follow best practice in all steps:

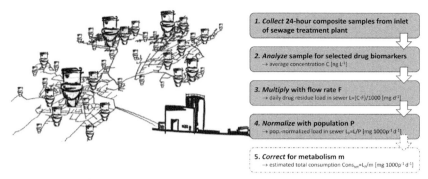

Fig. 1 Key steps for estimating drug consumption at the community level based on raw wastewater collected at the inlet of a sewage treatment plant ($Cons_{tot} = \frac{C \cdot F}{P} \cdot \frac{1}{m}$ [$mg\ 1000p^{-1}\ d^{-1}$], where C is the concentration [ng/L], F the flow rate [$m^3\ d^{-1}$], P the population [–] and m the metabolism [excretion rate in %]. *Modified from* K. V. Thomas, L. Bijlsma, S. Castiglioni, A. Covaci, E. Emke, R. Grabic, F. Hernández, S. Karolak, B. Kasprzyk-Hordern, R.H. Lindberg, M. Lopez de Alda, A. Meierjohann, C. Ort, Y. Pico, J.B. Quintana, M. Reid, J. Rieckermann, S. Terzic, A.L.N. van Nuijs, P. de Voogt, *Comparing illicit drug use in 19 European cities through sewage analysis*, Sci. Total Environ. 432 (2012) 432–439. https://doi.org/10.1016/j.scitotenv.2012.06.069.

Sampling of wastewater. In most modern STPs, a sampling scheme to collect samples on a routine basis is in place. It is typically used to quantify removal rates for traditional compounds such as nutrients or commonly used micropollutants. Ideally, samples are collected at least every 10 min in a flow-proportional mode over a 24-h period to obtain a composite sample adequately representing the daily average concentration [2,3]. Alternatively, a volume-proportional mode can be applied with a similar number of individual samples collected over a day. Time-proportional modes may imply more systematic or random uncertainties depending on the—typically unknown—intra-day variations of concentration profiles [4]. Relevant information on the catchment area and STP under investigation should be collected and documented [5]. For small catchments or even outlets of individual premises, e.g., schools or prisons, requirements are more stringent because an individual toilet flush containing the substance of interest may pass the monitoring station in less than 1 min and are likely being missed with traditional sampling equipment. Also measuring flow to calculate loads and technical implementation of sampling is much more demanding [4,6]. Besides, especially when sampling small communities, ethical principles should be considered to evade stigmatization of a particular group [7].

Selection of biomarkers. The selection of specific drug biomarkers is not an easy task, since an ideal target drug residue needs to fulfil specific requirements to ensure the reliable application of WBE. An appropriate biomarker is (i) excreted as a high fraction of the consumed dose, (ii) a human metabolite specific for consumption to differentiate from unconsumed drugs (e.g. disposal), (iii) stable during in-sewer transport and (iv) detectable in raw wastewater.

Estimation of population size. An important uncertainty when estimating per capita consumption of illicit drugs by means of wastewater analysis relates to the size and variability of the de facto population in a catchment [5,8]. Several methods based on census data, mobile device data and measuring hydrochemical parameters or specific substances in wastewater are currently employed to estimate the population [5,8,9]. The most reliable procedure would be to use all possible information and weight it according to certainty of individual estimates. With the refinement of the other contributors, population size has become one of the largest uncertainties in wastewater analysis [5,10].

Analytical methods. The chemical analysis of urinary biomarkers of illicit drugs in untreated wastewater is an analytical challenge. Drug residues are often present at very low concentration levels (ng/L) and wastewater is a complex sample matrix that contains particulate matter and compounds that may interfere with the analysis of the target analytes. Hence, a sample treatment step, which consists in a filtration step and solid phase extraction (SPE) is typically applied prior to analysis, in order to remove matrix interferences and to pre-concentrate target biomarkers.

LC coupled to tandem MS (LC-MS/MS) with triple quadrupole (QqQ) mass analyzers is currently the most popular analytical technique for the quantitative determination of illicit drug residues in wastewater samples. These instruments have a wide dynamic range, reach high sensitivity and selectivity, are relatively easy to operate, and allow multi-residue analysis in a single run, which permits reducing analysis time and costs. The QqQ instruments operate in MS/MS mode, where at least two specific precursor-to-product ion transitions for each target analyte should be monitored [5]. A quantification transition, most often the most sensitive one to favour the quantification at lower concentrations, and a confirmation transition. It is generally accepted that for a reliable positive finding in the sample, both transitions need to show a chromatographic peak at the same retention time as the reference standards as well as the compliance of the ion ratio between the two transitions [11]. Moreover, the use of isotope-labelled internal standards (ILIS) for each target analyte is pivotal for wastewater analysis and added to the sample prior to sample treatment (i.e. as surrogate), to correct for matrix effects and compensate for potential errors associated with sample preparation. The performances of the analytical methodologies need to be fully validated for all target analytes in terms of linearity, accuracy, precision and limits of quantification (LOQ), and it is imperative to analyse internal quality controls (QCs) for daily method variations and perform regular checks of external QCs to guarantee the reporting of reliable WBE data. The latter can be done by participating in inter-laboratory exercises, such as those that are yearly organized by SCORE (www.score-network.eu) [12,13].

Nowadays, applying WBE for estimating cocaine, amphetamine, methamphetamine and MDMA use is well established and the related uncertainties are well known [5]. However, when applying it for estimating the use of cannabis, several specific challenges need to be carefully considered. In the following sections, an overview of the applications will be given, the utility and potential of WBE for cannabis, but also the limitations and bottlenecks with particular emphasis on the analytical methodology applied.

2. Cannabis biomarkers

Δ^9-Tetrahydrocannabinol (THC), the psychoactive ingredient in cannabis can be absorbed by diverse routes of administration such as smoking, oral, oromucosal, rectal, transcutaneous, and intravenous, while elimination from the body is equally diverse such as faeces, urine, sweat, oral fluid, and hair [14]. THC pharmacokinetic processes are dynamic and may be affected by a person's frequency and magnitude of use. THC is metabolized by microsomal hydroxylation to 11-hydroxy-THC (THC-OH) which is both a potent psychoactive metabolite and an intermediate for further metabolism to 11-Nor-9-carboxy-THC (THC-COOH) by liver alcohol-dehydrogenase enzymes (Fig. 2) [15]. THC is a highly lipophilic compound with storage in the body in adipose tissue, the extent to which appears to be determined by the frequency of use, leading to different detection windows in biological matrices among types of user [16]. Although there are more positive urine tests for cannabinoids than for any other drug class in workplace drug testing, a scarcity of urinary excretion data from controlled clinical studies of cannabis exist [17]. As noted above, there are a variety of routes of administration, and major differences exist in the ratio of the concentration of metabolites depending on these routes [15].

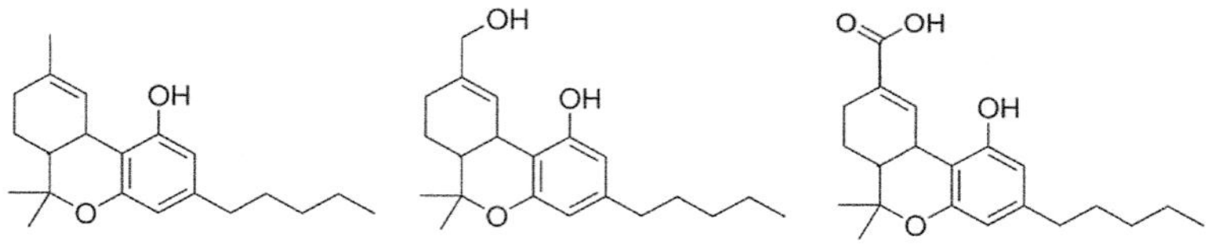

THC THC-OH THC-COOH

Fig. 2 Chemical structures of THC, THC-OH and THC-COOH.

Khan and Nicell identified 19 suitable studies that reported on various relevant aspects of THC excretion [18]. These studies involved a large range of doses, frequency of use, routes of administration, and numbers of participants, yielding a large range in the excretion profiles of the original THC dose in either/both urine and faeces. THC-COOH is the primary urinary metabolite and final point in the metabolization process and although a significant fraction of urinary THC-COOH is excreted as the glucuronide conjugate, conversion back to free THC-COOH in wastewater is expected due to sewer conditions and microbial activity in the conveyance system [18–20]. Hence, from a WBE perspective, THC-COOH is most frequently selected and determined in wastewater (see also Section 3 and Table 3). THC and THC-OH have been detected at a high rate in sewage sludge [21] owing to their more lipophilic nature, but THC-COOH is not expected to sorb onto wastewater influent suspended solids, and almost the entire faecal load of THC-COOH is expected to partition into the aqueous phase [18]. As will be discussed in later sections, THC-COOH can still be problematic to quantify compared to other drugs of abuse, because even in the carboxylated form, its lipophilicity in terms of Log D is still greater (Table 1) than drugs such as amphetamine $LogD = -0.79$ at pH 7.4 [22].

Several criteria have to be met for a biomarker to be suitable for WBE studies [23]. Besides that, a biomarker must be unique to human activity and excreted in substantial amounts, it must also be stable in wastewater during in-sewer transport and during storage until analysis.

The hydraulic residence times in the sewers can be from tens of minutes to nearly 24 h. The conveyance system consists of gravity sewers with an air column above the water surface and/or pressurized sewers. These contain biofilms with different microbial communities. Once the wastewater reaches the entrance to the treatment plant, a sample is collected and may spend up to 24 h in a container as a representative composite is obtained. Auto sampler containers are typically maintained at 4 °C. An aliquot is obtained from the composite and then either analysed immediately or more

Table 1 Predicted LogD values for THC, its metabolites and other illicit drugs as reference.

Compound	pH = 2	Log D (pH = 7.4)	Log D (pH = 8.0)
THC	5.94	5.94	5.92
THC-OH	4.66	4.66	4.64
THC-COOH	5.13	1.98	1.72
Methamphetamine	−1.1	−0.39	0.06
Amphetamine	−1.2	−0.79	−0.17
Cocaine	−1.3	1.8	2.1
Benzoylecgonine	−1.3	−0.7	−0.81

https://chemaxon.com/

often frozen at −20 °C and processed days, weeks or even months later. McCall et al. provide a literature review and found only a handful of studies that have investigated the stability of THC-COOH either in-sewer or in-sample [24]. THC-COOH has been shown to have <20% transformation in-sewer with variable in-sample results over five different studies [24]. The variety of stability results are due to differences in storage temperatures (20 °C, 4 °C, −20 °C) and length of periods studied. However, the larger issue with THC-COOH appears to come from the pH of the samples. Many WBE multi-residue analysis procedures call for acidification to pH 2 as a preservative technique, which has little effect on most illicit drugs. However, for THC-COOH, pH 2 results in a protonation of the carboxylic acid and a neutral molecule, which in theory leads to a much higher tendency towards sorption to particles or container walls.

One strength of the WBE approach is the estimation of parent drug consumption through back-calculation using wastewater loads (grams/day), parent/metabolite molar mass ratio, and pharmacokinetic excretion factors. This was first applied for THC-COOH by Zuccato et al. [25], using a correction factor of 152, which assumed a 0.6% of the THC dose excreted as THC-COOH and taking into account the molar mass ratio of parent drug to metabolite. Eq. 1 shows the calculation of a correction factor.

$$cor.factor = \frac{mass_{consumed.parentdrug}}{mass_{excreted.metabolite}} \times \frac{molarmass_{parent.drug}}{molarmass_{metabolite}} \tag{1}$$

This excretion factor seems to come primarily from a study involving six healthy male participants with a history of cannabis use, smoking two different amounts [17]. The results from this study showed the percent of the administered dose excreted in urine was 0.54 ± 0.12, but no results were published concerning the THC-COOH fraction from the participant's faeces. The 0.6% value has been widely cited in the field as an excretion fraction, which can be used to back-calculate THC consumption from THC-COOH loads [26-28,25,29]. However, Postigo et al. [30] proposed instead 2.5% (correction factor of 36.4), a larger excretion factor for the aqueous THC-COOH available in wastewater. This larger factor was established due to this particular study finding no THC-OH in the wastewater yet 2% had been reported as the fraction of THC-OH excreted in urine [30,31]. Since no THC-OH was detected it was assumed that there was rapid oxidation to THC-COOH and thus the sum of THC-COOH (0.5%) and oxidized THC-OH (2%) was used. Gracia-Lor et al. [32] published a refined correction factor for the back-calculation of THC. This study again references 18 studies reporting THC metabolite excretion in urine from various routes and 14 studies concerning THC metabolite excretion in faeces. After reviewing the available literature, the "refined" mean excretion rate of THC-COOH was reported lower at 0.5% and a correction factor of 182. This new, smaller excretion factor appears to be dominated by urine THC-COOH levels from smoked cannabis. Furthermore, Been et al. [33] used the previously mentioned clinical studies to derive an average excretion rate for THC-COOH. In this study, authors combined all the available study results in a Bayesian hierarchical model, which included prevalence data from surveys and THC-COOH loads measured in wastewater. Combining all the data in a single model, the authors calculated posterior distributions of user prevalence, total cannabis consumption and excretion rates. While they assumed smoking as the most predominant administration route for Switzerland (according to available survey data), the authors estimated an excretion rate for THC-COOH of 7.0% (95%-CI range 4.2–10.6%) and 0.04% (95%-CI range 0.01–0.08%) through faeces and urine, respectively. Finally, Burgard et al. [10] monitored THC-COOH loads in wastewater over a 3 year period corresponding to the new legalized recreational market in Washington State, USA. This study specifically did not perform back-calculations to report THC consumed owing to the discrepancy among reported excretion rates in urine and faeces, the types of users, and the 440% difference that comes with different routes of administration. Smoked cannabis reportedly yields 0.5% urinary excretion while oral doses yield $2.2 \pm 1.2\%$ [18]. In addition to wastewater loads, Burgard et al. [10] tracked recreationally purchased THC mass and compared these two indicators during the same time period. While only an estimate of the contribution from the residual black market could be made, they concluded that the excretion estimate of 2.5% [30] seems much more reasonable for use with WBE calculations. The obtained excretion rate is also in the same order of magnitude as the combined faeces and urine excretion rate estimated by Been et al. [33]. Furthermore, it is important to realize the possible shifting cannabis markets and the changes in desirable routes of administration. During the Washington State study, initially 90% of the sold cannabis was smokable products, but in under 3 years that had dropped to 83% of the market. Changing user demographics may lead to less smoking of cannabis i.e., A total of 48% of eighth grade survey respondents reported routes other than smoking preferred (eating, drinking, vaping, etc.) [10]. Less smoking and more oral use of cannabis will lead to larger excretion rates of the consumed THC dose. Moreover, an important distinction needs to be made with regard to the back-calculated amounts, namely whether the goal is to estimate total THC use or total cannabis use. In the latter case, an additional factor taking into account the average purity of sold cannabis needs to be included in the calculations.

Finally, cannabidiol (CBD) has gained in popularity due to its ability to alleviate certain medical conditions and is used as a medicine, an ingredient in foods, and as a dietary supplement. Numerous CBD products marketed may also contain small percentages of THC and could thus contribute to the total load of THC in wastewater. However, this contribution is uncertain, but seems small compared to THC used for recreational purposes. Therefore, at this point, it might be assumed that the small fractions of THC in CBD products is not a significant component of the overall THC load. Nevertheless, when monitoring THC, the additional fraction coming from CBD products adds to the overall uncertainty.

3. Analytical methodology

The analytical methodologies applied in WBE studies commonly include multiple substances. The quantitative determination of several illicit drug biomarkers in a single analysis is more practical, faster and thus cost-effective. Multi-residue methods do not only regard to the analysis, but also concern the sample collection, storage and sample treatment. However, when measuring multiple compounds, a compromise of the experimental conditions is often required.

Cannabis biomarkers have lower polarity, i.e., higher lipophilicity, compared to other illicit drugs and metabolites (see also Table 1). Despite the different physico-chemical properties, biomarkers of cannabis use have been included in multi-residue methodologies for the analysis of influent wastewater. Specifically, its main human urinary metabolite THC-COOH, has been the most reported biomarker of cannabis consumption in WBE publications. Yet, results of the inter-laboratory exercises performed by SCORE revealed difficulties related to the chemical analysis of THC-COOH in wastewater suggesting that concentrations found might be underestimated [12]. In fact, the authors illustrated that whilst laboratories performed well when analysing THC-COOH in methanol (i.e. relative standard deviation (RSD) <25% from the group's mean), underreporting of up to 90% (from the group's mean) were reported for tap water samples. In the latter case, laboratories were asked to process the samples according to their established procedures, which generally involved SPE and, in some cases, also sample acidification (as will be discussed further on, this can dramatically reduce the recovery of THC-COOH from water samples). Moreover, other non-instrumental factors, such as possible sorption to sample containers, partition on biofilms and particulate matter and excretion rates should be better understood to provide more accurate back-calculation estimates of the total amount of cannabis/THC used by a population [11,34]. Nevertheless, monitoring consumption trends using excreted THC-COOH in wastewater can give unique and timely information [10], but the data and methods used to generate these data should be carefully evaluated and interpreted.

In this section, focus has been put on the analytical methodology. In total 29 peer-reviewed articles were published since 2006 describing validated analytical methodologies for the determination of cannabis biomarkers in wastewater. Furthermore, important progress has been made in relation to the initial aspects of the analytical procedure, where it has been highlighted that the order of the initial sample preparation steps after sample collection is crucial [34]. Tables 2 and 3 show the sample preparation steps and instrumental parameters of the reported analytical methods, respectively.

The collection of representative 24 h composite samples is pivotal in the WBE approach (see Section 1). Different sample container materials have been used to collect and store samples such as amber glass, high-density polyethylene (HDPE), polypropylene (PP) and polyethylene terephthalate (PET) (Table 2). After sample collection and prior to SPE, a filtration and/or centrifuged step is often performed to prevent the SPE cartridge sorbent from clogging. A wide range of different filter types with pore sizes from

Monitoring cannabis through wastewater analysis

Table 2 Sample preparation steps.

Year, Ref	Sample container	Filtration/ Centrifugation	Acidification (Y/N)	ILIS addition	SPE	CF
2006, [35]	Glass	F: GF/A 1.6 μm	Y, prior to SPE	After F, prior to SPE	Oasis MCX (3 mL, 60 mg)	250
2007, [36]	n/a	F: GF/A	N	After F, prior to SPE	Oasis HLB (6 mL, 200 mg)	400
2008, [37]	Glass	F: GF 1 μm, Nylon 0.45 μm	N	After F, prior to SPE	Online Oasis HLB Prospekt-2 (10.4 mg)	–
2009, [38]	HDPE	C: 5 min, 4500 rpm	Y, prior to SPE	After C, prior to SPE	Oasis MCX (6 mL, 150 mg)	10
2010, [39]	Glass	F: GF BF 85/70	Y, prior to F	After F	Direct injection	–
2010, [40]	Glass	F: GF, NC 0.45 μm	Y, prior to SPE	After F, prior to SPE	Oasis HLB (6 mL, 200 mg)	500
2011, [41]	Glass	F: n/a	Y, prior to F	After F, prior to SPE	Oasis MCX (6 mL, 150 mg)	n/a
2011, [30]	PET	F: GF 1 μm, Nylon 0.45 μm	N	After F, prior to SPE	Online Oasis HLB Prospekt-2 (10.4 mg)	–
2012, [42]	n/a	F: GF, NC 0.45 μm	Y, prior to SPE	After F, prior to SPE	Oasis MCX (6 mL, 150 mg)	200
2012, [43]	Glass	F: NC 0.45 μm	Y, prior to SPME	After F, prior to SPME	SPME: DVB-CAR-PDMS fibre	–
2013, [44]	Glass	F: GF/A 1 μm, PES 0.45 μm	N	After F, prior to SPE	Oasis HLB (6 mL, 150 mg)	200
2013, [45]	n/a	F: RC 0.45 μm	N	After F, prior to SPE	Inline C_{18} Hypersil gold 20×2.1 mm 12 μm	–

Continued

Table 2 Sample preparation steps.—cont'd

Year, Ref	Sample container	Filtration/ Centrifugation	Acidification (Y/N)	ILIS addition	SPE	CF
2013, [26]	PP	F: GF/B 1 mm	N	After F, prior to SPE	Oasis HLB (6 mL, 500 mg)	500
2013, [46]	n/a	F: GF/D 2.7 µm	Y, prior to SPE	After F, prior to SPE	Oasis MCX (6 mL, 150 mg) + Strata NH_2 (3 mL, 200 mg)	250
2014, [47]	PET	F: RC 0.45 µm	N	Y, 15 min before F	Strata X (6 mL, 500 mg)	250
2014, [48]	HDPE	F: CEM 0.45 µm	N	Before F, prior to SPE	Oasis HLB (3 mL, 60 mg)	25
2014, [27]	PP	F: GF/D 2.7 µm	N	After F, prior to SPE	Oasis HLB (6 mL, 500 mg)	200
2014, [49]	HDPE	F: GF 0.7 µm	N	Before F, prior to SPE	Oasis HLB (3 mL, 60 mg)	250
2015, [28]	PET	F: GFC 1 µm, GF 0.5 µm	N	After F	Online Hypersep RetainPEP 20 × 2.1 mm 12 µm	–
2016, [50]	n/a	F: n/a	Y, prior to LLE	After F, prior to LLE	LLE	175
2017, [51]	n/a	F: GF/F	N	No ILIS	Chromabond HR-X (6 mL, 500 mg)	n/a
2018, [52]	Glass	C: 5 min, 4500 rpm, F: GF	N	After F, prior to SPE	Oasis HLB (6 mL, 200 mg)	100
2018, [53]	n/a	F: GF/A 0.7 µm	N	Before F, prior to SPE	Oasis MCX (6 mL, 150 mg)	1000
2018, [54]	PP	–	Y, with LLE	n/a	LLE	100

Table 2 Sample preparation steps.—cont'd

Year, Ref	Sample container	Filtration/ Centrifugation	Acidification (Y/N)	ILIS addition	SPE	CF
2018, [55]	Glass	C: 5 min, 4500 rpm, F: GF	N	After F, prior to SPE	Oasis HLB (n/a)	100
2019, [10]	HDPE	F: RC syringe 0.2 μm	N	Before F, prior to SPE	Strata-CX (3 mL, 60 mg)	n/a
2019, [56]	HDPE	F: GF 1.6 mm	Y, prior to SPE	Before F, prior to SPE	Strata-XC 33 μm (6 mL, 200 mg)	250
2019, [57]	Glass	F: GF 0.7 μm	N	After F, prior to SPE	Oasis HLB (6 mL, 500 mg)	100
2019, [58]	HDPE	F: GF 0.7 μm, Nylon 0.45 μm	N	Before F, prior to SPE	Oasis HLB (3 mL, 60 mg)	50

CF, Concentration Factor; n/a, information not available.

0.2 μm to 1.6 mm have been reported: glass fibre (GF), cellulose ester membrane (CEM), regenerated cellulose (RC), nitro cellulose (NC) and nylon membranes (Table 2). For multi-residue drug analysis, it was recommended to adjust the pH to acidic conditions after sample collection to decrease possible degradation and increase the in-sample stability for the majority of the illicit drug biomarkers [5]. However, at acidic pH, THC-COOH is present in its non-charged hydrophobic form, which means that it has potential to sorb to sample container or filter materials. Causanilles et al. [34] demonstrated that sorption of THC-COOH to container walls occurred more rapidly and to a higher extend at pH 2.5 for glass and PP containers. Furthermore, THC-COOH recovery after filtration was highly dependent on the pH [46,34]. When filtering large volumes of wastewater at neutral pH, recovery losses around 30% were reported independent of the filter material, but at acidic pH losses during filtration were above 75% [34]. Therefore, it is not advisable to acidify the samples, if it is not required by the selected protocol. Hence, a best-practice protocol for the initial sample preparation steps has been proposed, i.e., first addition of ILIS, second filtration and third acidification (if required) [34]. Table 2 shows that five out of the eight (62.5%) articles published in the years 2018 and 2019, thus

Table 3 Analytical parameters.

Year, Ref	Biomarkers	Instrument	Ionization	Quantification	Confirmation
2006, [35]	THC-COOH	RP-LC-MS/MS (QqQ)	ESI -	343 > 299	343 > 245
2007, [36]	THC-COOH, THC	RP-LC-MS/MS (QqQ)	ESI +	345 > 327; 315 > 193	345 > 193; 315 > 123
2008, [37]	THC-COOH, THC, THC-OH	RP-LC-MS/MS (QqLIT)	ESI -	343 > 299; 313 > 245; 329 > 311	343 > 191; 313 > 191; 329 > 268
2009, [38]	THC-COOH	RP-LC-MS/MS (QqQ)	ESI +	345 > 193	345 > 299, 345 > 327
2010, [39]	THC-COOH	RP-LC-MS/MS (QqQ)	ESI +	345 > 327	345 > 299
2010, [40]	THC-COOH, THC	GC–MS (Ion Trap)	EI	473 > 355; 386 > 371	-; 386 > 330, 386 > 315
2011, [41]	THC-COOH, THC	RP-LC-MS/MS (QqQ)	ESI +/−	343 > 299 (−); 315 > 193 (+)	343 > 245 (−); 315 > 123 (+)
2011, [30]	THC-COOH, THC, THC-OH	RP-LC-MS/MS (QqLIT)	ESI -	343 > 299; 313 > 245; 329 > 311	343 > 191; 313 > 191; 329 > 268
2012, [42]	THC-COOH, THC	RP-LC-HRMS (QTOF)	ESI -	343.1915; 313.2173	299.2017; 245.1547
2012, [43]	THC-COOH, THC	GC–MS (Q)	EI	371; 386	473,488; 303, 371
2013, [44]	THC-COOH, THC, THC-OH	RP-LC-HRMS (LTQ-Orbitrap)	ESI +	345.2060; 315.2319; 331.2268	327, 299; 259, 193; 313

2013, [45]	THC-COOH	RP-LC-HRMS (Q-Orbitrap)	ESI +	345.2060 > 299.2006	345.2060 > 327.1953
2013, [26]	THC-COOH	RP-LC-MS/MS (QqQ)	ESI -	343 > 299	343 > 245
2013, [46]	THC-COOH, THC-OH	RP-LC-MS/MS (QqQ)	ESI -	343 > 245; 329 > 311	343 > 191; 329 > 173
2014, [47]	THC-COOH	RP-LC-MS/MS (QqQ)	ESI +	345 > 41	345 > 327
2014, [48]	THC-COOH	RP-LC-MS/MS (QqQ)	ESI +	345 > 193	345 > 299, 345 > 327
2014, [27]	THC-COOH	RP-LC-MS/MS (QqQ)	ESI -	343 > 299	343 > 245
2014, [49]	THC-COOH	RP-LC-MS/MS (QqQ)	ESI -	343 > 299	343 > 245
2015, [28]	THC-COOH, THC	RP-LC-HRMS (Q-Orbitrap)	ESI +	345.2060; 315.2319	RT ILIS ±0.03 min, isotopic fit
2016, [50]	THC-COOH	RP-LC-MS/MS (QqQ)	ESI -	343 > 299	343 > 245
2017, [51]	THC-COOH	RP-LC-MS/MS (QqQ)	ESI +	345 > 299	–
2018, [52]	THC-COOH, THC, THC-OH	RP-LC-MS/MS (QqQ)	ESI +	345 > 327; 315 > 193; 331 > 313	345 > 299; 315 > 123; 331 > 193

Continued

Table 3 Analytical parameters.—cont'd

Year, Ref	Biomarkers	Instrument	Ionization	Quantification	Confirmation
2018, [53]	THC-COOH, THC-OH	RP-LC-MS/MS (QqQ)	ESI +	345 > 327; 331 > 313	345 > 299; 331 > 193
2018, [54]	THC-COOH, THC, THC-OH, THC(OH)$_2$	SFC-MS/MS (QqQ)	ESI +	345 > 299; 315 > 193; 331 > 313; 315 > 193	345 > 193; 315 > 123; 331 > 201; 315 > 123
2018, [55]	THC-COOH, THC, THC-OH	RP-LC-MS/MS (QqQ)	ESI +	345 > 327; 315 > 193; 331 > 313	345 > 299; 315 > 123; 331 > 193
2019, [10]	THC-COOH	RP-LC-MS/MS (QqQ)	ESI −	343 > 299	343 > 245
2019, [56]	THC-COOH, THC	RP-LC-MS/MS (QqQ)	ESI +	345 > 299; 315 > 193	345 > 193; 315 > 123
2019, [57]	THC-COOH, THC, THC-OH	RP-LC-MS/MS (QqQ)	ESI +	345 > 299; 315 > 123; 331 > 313	345 > 193; 315 > 193; 331 > 193
2019, [58]	THC-COOH	RP-LC-MS/MS (QqQ)	ESI +	345 > 299	345 > 193

after the recommendations were made, applied this best-practice protocol, whereas only three of 19 (15.8%) applied the correct order of steps in earlier publications. Two articles applied either centrifugation [38], or the order of initial steps was unknown [51].

Concentration levels of most illicit drug biomarkers in wastewater are in the ng/L–mg/L range, and despite that the sensitivity of modern instruments is excellent, a pre-concentration step is generally needed in order to reach the required quantification limits. In addition, matrix components that might co-elute with the analyte and interfere with the analytical measurement leading to a suppression or enhancement of the analyte response (i.e. matrix effects) may also be removed. Off-line SPE is most often applied for sample pre-concentration and clean-up, but also fully automated large volume injections [39] and on-line SPE applications [37,30,45,28] have been reported. Polymeric-based SPE sorbents with reversed phase (RP) properties built of generic hydrophilic and lipophilic balanced (Oasis HLB) monomers or strong cation-exchange mixed mode sorbents built upon RP copolymers (Oasis MCX or Strata-XC) were most popular in multi-residue methods (Table 2). Although cation-exchange mode cartridges allow improved selectivity towards basic analytes, THC-COOH in its neutral form at low pH may also be retained by the mixed RP characteristics resulting in satisfactory recoveries. Furthermore, some alternative sample preparation protocols can be found in the literature: liquid-liquid extraction (LLE) and solid-phase micro extraction (SPME) were both proposed for extracting anionic THC-COOH from wastewater [43,50,54]. The main drawback, however, is the limited applicability to measure multiple compounds. Yet, for the determination of cannabis only in wastewater, these specific sample preparation protocols could be good alternatives.

The use of internal standards, preferably the labelled form of the analyte of interest, is essential and compulsory in WBE studies. ILIS should be added to the sample as surrogate, i.e., just after sample collection, to correct for matrix effects and for potential errors associated with sample manipulation and storage. All reported methodologies used deuterated analogues of the cannabis biomarkers as ILIS for more accurate quantification, except one study that did not use any ILIS [51].

THC-COOH was always included as biomarker of cannabis consumption when analysing wastewater (Table 3) and the majority use LC-MS/MS with QqQ for its determination (22 of 29, 76%). Reversed phase (RP) analytical columns based on C18 were mostly used for chromatographic separation of drug biomarkers including THC-COOH. Acid was often added to mobile phases to favour the formation of protonated molecules, but acid can deteriorate the sensitivity and chromatographic performance of THC-COOH [48]. QqQ analyzers are known for their robustness and excellent sensitivity and selectivity, but hybrid systems with high-resolution mass spectrometry (HRMS) such as time-of-flight or Orbitrap analyzers have also shown good performances for both qualitative and quantitative analysis [42,44,45,28]. However, HRMS instrument are more expensive and require experienced operators. Furthermore, gas chromatography coupled to mass spectrometry (GC–MS) has been applied for the analysis of THC-COOH and THC in wastewater providing high levels of selectivity and sensitivity [40,43], yet a derivatization step is required to make them compatible with GC. Consequently, sample treatment is more laborious and time-consuming. An alternative could be supercritical-fluid chromatography (SFC) coupled to tandem mass spectrometry [54], but these instruments are often not standard available in laboratories.

For a reliable positive finding of THC-COOH in wastewater using tandem MS instruments, it is recommended that a minimum of at least two specific MS/MS transitions is monitored, whereas for HRMS instruments, at least two ions need to be monitored [5,11]. THC-COOH has been measured in both negative-ion and positive-ion mode. In principle, more abundant ionization would be expected in negative mode, due to higher trend towards the ionization of the acidic group. However, sensitivity seems manufacturer dependent, for example, Waters instruments seem to perform better in positive electrospray ionization (ESI) mode. Thus, a better ionization towards the basic group [38]. In positive ESI mode, m/z 345 > 327 has been selected as quantification transition in five occasions (Table 3). Although this transition probably was the most sensitive one, and would thus favour the quantification of THC-COOH at lower concentrations, it also corresponds to the non-specific loss of water. Therefore, this transition might be more problematic and prone to be interfered when analysing wastewater [11]. Similar is the selection of m/z 345 > 299 (or m/z 343 > 299 in negative ESI mode) corresponding to a non-specific loss of the carboxylic acid group. Hence, the selection of a less abundant, but more specific transition m/z 345 > 193 can be beneficial, presenting better signal-to-noise (s/n) ratios and thus sensitivity (Fig. 3). Chromatographic separation might not be an important issue when using MS/MS for detection, although it can be essential to avoid or minimize matrix effects, especially in complex influent wastewater samples. LC separation becomes, however, a crucial issue when

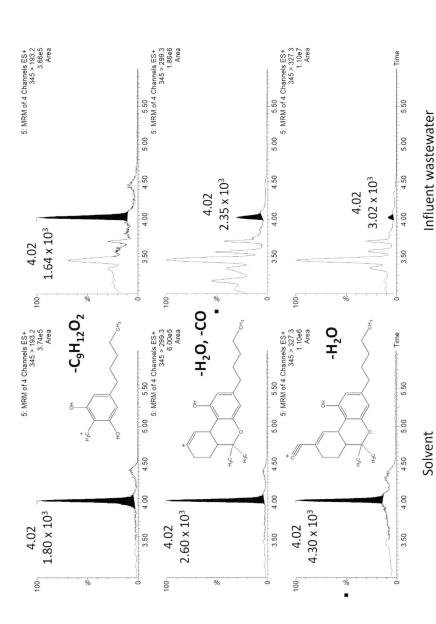

Fig. 3 Selectivity of THC-COOH transitions. From F. Hernandez, S. Castiglioni, A. Covaci, P. de Voogt, E. Emke, B. Kasprzyk-Hordern, C. Ort, M. Reid, J. V Sancho, K. V Thomas, A.L.N. van Nuijs, E. Zuccato, L. Bijlsma, Mass spectrometric strategies for the investigation of biomarkers of illicit drug use in wastewater, Mass Spectrom. Rev. 37 (2018) 258–280.

compounds have the same MS/MS transitions [38]. THC and CBD present common transitions, but more importantly for monitoring cannabis consumption through WBE, their metabolites THC-COOH and CBD-7-COOH also share their MS/MS transitions. Therefore, co-elution needs to be avoided, especially since CBD has become more popular and can thus be present in wastewater.

Limits of detection (LODs) and limits of quantification (LOQs) of THC-COOH in wastewater are generally in the ng/L—µg/L range and usually estimated based on s/n ratios of 1:3 and 1:10, respectively [11]. The estimation of realistic and comparable LODs and LOQs is, however, complicated in wastewater, as notable variations in chemical composition between samples occur. LOQs would need to be estimated from samples containing biomarkers at low level, where the s/n in the chromatograms and matrix effects can be taken into account. Furthermore, LOQs should be estimated at s/n 10 for the quantification transition, but also at s/n 3 for the confirmation transition to ensure not only the quantification of the compound but also its reliable identification [48]. The ultimate and most homogeneous approach is to estimate LOD and LOQ from inter-laboratory exercises where all participants analyse the same samples with their own analytical methodology [11].

4. Wide-scale use of WBE for cannabis estimates in Canada and US.

In the context of an increasing proportion of the population considering that consuming cannabis should not be a criminal offence, going from 51% in 2001 [59] to 70% in 2016 [60] in Canada, of a search for strategies to defeat the black market and to serve public health goals, several countries have adopted a policy of decriminalization to make simple possession a non-criminal offence and other countries or jurisdictions, such as Canada, Uruguay, the USA states Colorado, California and Illinois, and the Australian Capital Territory in Australia have legalized possession and use of recreational cannabis. In 2013, for the first time, the majority of United States population polled favoured legalizing cannabis, which followed the first legalization of a recreational cannabis market in the states of Washington and Colorado in 2012 [61]. Now 11 states and Washington D.C. have legal recreational markets. The changes in legislation increased the needs for the monitoring of cannabis consumption in the populations affected by these changes. This led to innovative strategies to collect data to track the prevalence of cannabis prior to and following the legalization.

One objective of implementing new methods to collect statistics was to identify sources of data with a lower risk of relying on underestimated consumption levels prior to legalization, which might be significant prior to legalization due to stigma associated with use of illegal drugs and the reluctance to disclose purchases from non-regulated suppliers. One strategy that was implemented in Canada and in Washington State, USA, was the use of WBE.

The implementation of the WBE approach at the scale of a country, which in the case of Canada involved 15 STPs up to about 6000 km apart, collecting wastewater monthly, over a period of a year, in five large urban centres across the country, and representing nearly 8.4 million people, was associated with few challenges, such as shortage and type of container material, sample storage and logistics. Tests using sampling bottles made of PET, the material recommended by SCORE [5,13] and HDPE (the most widely available) demonstrated that no significant difference was observed between the two bottles materials over a period of 7 days of storage. The amount of THC-COOH in the wastewater collected in this study in Canada was quantified applying an extraction [62] and LC-MS method [63], which are based on a procedure previously reported by Rodayan et al. [64].

The deployment of WBE at large scale provided data over extended periods of time for both Canada and Washington State, USA. For Canada, the results of the monitoring for cannabis and other drugs are available online [65]. Fig. 4A shows the average weekly loads of THC-COOH per capita over a year for each geographical area included in the pilot project and Fig. 4B shows the annual trend of the average weekly loads of THC-COOH per capita over the whole country (combined sites) of the pilot project. Unfortunately, significant variability of the monthly results was observed (Fig. 4A) and at the scale of the country, it was not possible to clearly identify a trend that might be associated with the legalization of cannabis in 2018 (Fig. 4B).

Focusing on a smaller geographical area can provide a different perspective. To evaluate the significance of the trend in consumption observed over time for a given population, the results obtained using WBE were compared to the self-reported data collected over the same period of time by Statistics Canada [66] for the corresponding population. Based on self-reported consumption (Table 4), a slight increase in prevalence was observed for the quarter during which cannabis was legalized (Q4 2018), but it decreased

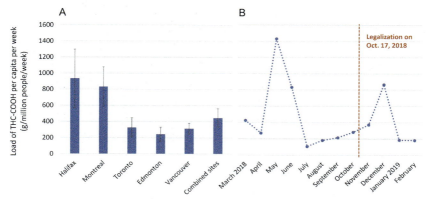

Fig. 4 (A) Average weekly loads of THC-COOH per capita (g/million people/week) by city over a one-year period of the pilot study (March 2018 to February 2019) ($n = 12$, error bars = lower and upper limits over the 12 months) and (B) Average weekly loads of THC-COOH per capita (g/million people/week) by month over the whole country (combined sites), adapted from Statistics Canada [65], 26/08/2019. *This does not constitute an endorsement by Statistics Canada of this product.*

Table 4 Prevalence of cannabis use in the selected Canadian province using self-reported information and the corresponding estimates of consumption for the corresponding metropolitan area included in the WBE pilot project.

Time period	Percentage of people reporting consumption (95% confidence interval)	WBE estimates of consumption based on THC-COOH loads
Q3 2018	10.1% (7.7–13.3)	July 2018: 411 g/week August 2018: 445 g/week September 2018: 1561 g/week
Q4 2018[a]	13.6% (10.9–16.8)	October 2018: 920 g/week November 2018: 768 g/week December 2018: 4396 g/week
Q1 2019	11.0% (8.8–13.8)	January 2019: 665 g/week February 2019: 588 g/week March 2019: *Not available*

[a]Legalization was on October 17, 2018, at the beginning of Q4 2018.
Data extracted from Statistics-Canada, Prevalence of Cannabis Use in the Past Three Months, Self-Reported, (2019). Table 13-10-0383-01. https://doi.org/https://doi.org/10.25318/1310038301-eng.

to very similar percentages of prevalence afterwards. The trend of consumption based on the load of THC-COOH estimated using WBE (Table 4), also suggests that there was an increase in consumption, just before legalization and during the fourth quarter of 2018. The peak of consumption might be explained by an increase interest due to legalization as well as the holiday season. However, the data for the beginning of 2019 do not suggest a sustained increase in consumption.

In Washington State, two STPs in a city of 200,000 people were monitored using WBE for THC-COOH for 7 months prior to the first legal recreational cannabis stores opening then followed by 29 months after the new marketplace began. Fig. 5 shows that while the legal THC dispensed in the catchment area of the two STPs increased at nearly 70% per quarter, the THC-COOH metabolite only increased at 9% per quarter. This indicated a much lower increase in cannabis use than new cannabis being introduced into the market and thus a decrease in cannabis from the black market is

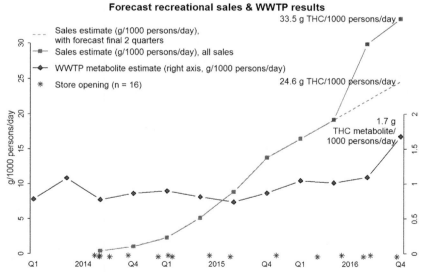

Fig. 5 Quarterly per capita estimates for THC consumption by sales (grey line) and by wastewater THC-COOH metabolite analysis (black line). The dashed line is the forecast continued recreational sales whereas the Q3 and Q4 in 2016 included all sales, including the State tracked medical market that came online starting Q3 2016. *From D.A. Burgard, J. Williams, D. Westerman, R. Rushing, R. Carpenter, A. LaRock, J. Sadetsky, J. Clarke, H. Fryhle, M. Pellman, C.J. Banta-Green, Using wastewater-based analysis to monitor the effects of legalized retail sales on cannabis consumption in Washington State, USA, Addiction 114 (2019) 1582–1590. https://doi.org/10.1111/add.14641.*

concluded. Total THC consumption did indeed increase but the stated goal of legalization to reduce the black market seems to have been achieved. This finding is an example of where a wastewater approach has shown its true value as a complementary approach to traditional used metrics. These other metrics have not yet been able to make any estimates as to an effect on the black market with a direct data source.

5. Future research and current asset

Monitoring cannabis use is highly relevant and interesting, due to its widespread use and ongoing policy changes. The estimation of community drug use through the chemical analysis of specific human biomarkers in wastewater has demonstrated its potential as a useful complementary approach to established drug monitoring tools such as general population surveys on drug use, treatment registry and law enforcement data. The application of WBE to monitor cannabis, however, has been challenging. Therefore, research has recently been centered on refining cannabis estimates, which resulted in a protocol for the storage, handling, and analysis of wastewater samples [34]. Yet, the main focus in this study was on the dissolved phase and did not include suspended solids, which seem to be of particular relevance for lipophilic cannabis biomarkers. Although THC may partition to particulates ultimately accumulating in the biosolids, THC-COOH is expected to partition into the aqueous phase and only predicted to adsorb to suspended solids between 1.1% and 8.5% [33]. However, it is not clear if the fraction excreted in faeces, which is supposedly higher than in urine, remains in the faeces even if they disaggregate in wastewater and are in contact with a high volume of wastewater. Furthermore, differences in wastewater characteristics—i.e., content and type of suspended solids—operation and design of sewer systems and the material of the sampling container may result in variable losses due to sorption of biomarkers and thus lead to variable amounts of chemical loadings measured in the liquid phase [67–69]. Moreover, transformation of cannabis biomarkers (i.e., THC-OH to THC-COOH) in sewer biofilms and during in-sewer transport may also significantly affect the data reported [18,30]. The information currently available is still limited and not conclusive. A better understanding of the possible role of suspended solids in raw wastewater needs attention and ongoing research moves in this direction.

While uncertainties exist around the measurement and quantification of the cannabis active ingredient THC and its metabolites in wastewater, the method does provide useful insights. Wastewater provides one of the only direct measures to estimate community level drug consumption. As shown in this chapter, wide-scale and community level use is a valuable input when assessing major changes to a drug's legal status. Sewer catchments and wastewater properties can vary widely among locations, depending on type of sewer system, special industrial discharges and weather conditions. However, if longitudinal monitoring occurs (i) within the same catchment, (ii) during similar conditions (i.e. dry weather), then relative trends in use can be evaluated even without knowing (iii) the exact sorption to particles, and degradation factors and (iv) average excretion rates since these are expected to remain relatively constant over time. In such situations, trends in use before and after legislative changes are helpful in assessing the effect of the legislation. The Canadian government sees wastewater as a complementary tool to traditional metrics and has invested in its use beyond the inaugural year presented here. Research is currently conducted to identify the potential sources of variability and facilitate the deployment of the method at large scale over extended periods of time. The Washington State case study provides lawmakers with evidence that one of the goals of cannabis legalization, a decrease in the cannabis black-market, appears to be working. This ability to understand the entire cannabis market through community consumption in relation to sales data is a result that would have been difficult to demonstrate applying traditional drug use indicators.

Acknowledgements

This publication was funded by the European Union's Justice Programme—Drugs Policy Initiatives, EuSeME (project number 861602). The Washington State pilot program was supported by the National Institute on Drug Abuse of the National Institutes of Health (R03DA038806). The pilot project in Canada was funded by Statistics Canada and conducted with the support of managers and staff at wastewater treatment plants in the five participating metropolitan areas.

References

[1] C.G. Daughton, Illicit drugs in municipal sewage: proposed new non-intrusive tool to heighten public awareness of societal use of illicit/abused drugs and their potential for ecological consequences, in: C.G. Daughton, T.L. Jones-Lepp (Eds.), Pharmaceuticals and Personal Care Products in the Environment: Scientific and Regulatory Issues, American Chemical Society, 2001, pp. 348–364.

[2] C. Ort, M.G. Lawrence, J. Reungoat, J.F. Mueller, Sampling for PPCPs in wastewater systems: comparison of different sampling modes and optimization strategies, Environ. Sci. Technol. 44 (2010) 6289–6296. https://doi.org/10.1021/es100778d.

[3] C. Ort, M.G. Lawrence, J. Rieckermann, A. Joss, Sampling for pharmaceuticals and personal care products (PPCPs) and illicit drugs in wastewater systems: are your conclusions valid? A critical review, Environ. Sci. Technol. 44 (2010) 6024–6035. https://doi.org/10.1021/es100779n.

[4] C. Ort, Quality assurance/quality control in wastewater sampling, in: C. Zhang, J.F. Mueller, M. Mortimer (Eds.), Quality Assurance & Quality Control of Environmental Field Sampling, Future Science, London, UK, 2014, pp. 146–168.

[5] S. Castiglioni, L. Bijlsma, A. Covaci, E. Emke, F. Hernandez, M. Reid, C. Ort, K.V. Thomas, A.L.N. van Nuijs, P. de Voogt, E. Zuccato, Evaluation of uncertainties associated with the determination of community drug use through the measurement of sewage drug biomarkers, Environ. Sci. Technol. 47 (2013) 1452–1460.

[6] C. Ort, J.M. Eppler, A. Scheidegger, J. Rieckermann, M. Kinzig, F. Sörgel, Challenges of surveying wastewater drug loads of small populations and generalizable aspects on optimizing monitoring design, Addiction 109 (2014) 472–481. https://doi.org/10.1111/add.12405.

[7] J. Prichard, W. Hall, P. de Voogt, E. Zuccato, Sewage epidemiology and illicit drug research: the development of ethical research guidelines, Sci. Total Environ. 472 (2014) 550–555. https://doi.org/10.1016/j.scitotenv.2013.11.039.

[8] J.W. O'Brien, P.K. Thai, G. Eaglesham, C. Ort, A. Scheidegger, S. Carter, F.Y. Lai, J.F. Mueller, A model to estimate the population contributing to the wastewater using samples collected on census day, Environ. Sci. Technol. 48 (2014) 517–525. https://doi.org/10.1021/es403251g.

[9] K.V. Thomas, A. Amador, J.A. Baz-Lomba, M. Reid, Use of mobile device data to better estimate dynamic population size for wastewater-based epidemiology, Environ. Sci. Technol. 51 (2017) 11363–11370. https://doi.org/10.1021/acs.est.7b02538.

[10] D.A. Burgard, J. Williams, D. Westerman, R. Rushing, R. Carpenter, A. LaRock, J. Sadetsky, J. Clarke, H. Fryhle, M. Pellman, C.J. Banta-Green, Using wastewater-based analysis to monitor the effects of legalized retail sales on cannabis consumption in Washington State, USA, Addiction 114 (2019) 1582–1590. https://doi.org/10.1111/add.14641.

[11] F. Hernandez, S. Castiglioni, A. Covaci, P. de Voogt, E. Emke, B. Kasprzyk-Hordern, C. Ort, M. Reid, J.V. Sancho, K.V. Thomas, A.L.N. van Nuijs, E. Zuccato, L. Bijlsma, Mass spectrometric strategies for the investigation of biomarkers of illicit drug use in wastewater, Mass Spectrom. Rev. 37 (2018) 258–280.

[12] A.L.N. van Nuijs, F.Y. Lai, F. Been, M.J. Andres-Costa, L. Barron, J.A. Baz-Lomba, J.D. Berset, L. Benaglia, L. Bijlsma, D. Burgard, S. Castiglioni, C. Christophoridis, A. Covaci, P. de Voogt, E. Emke, D. Fatta-Kassinos, J. Fick, F. Hernandez, C. Gerber, I. González-Mariño, R. Grabic, T. Gunnar, K. Kannan, S. Karolak, B. Kasprzyk-Hordern, Z. Kokot, I. Krizman-Matasic, A. Li, X. Li, A.S.C. Löve, M.L. de Alda, A.K. McCall, M.R. Meyer, H. Oberacher, J. O'Brien, J.B. Quintana, M. Reid, S. Schneider, S.S. Simoes, N.S. Thomaidis, K. Thomas, V. Yargeau, C. Ort, Multi-year inter-laboratory exercises for the analysis of illicit drugs and metabolites in wastewater: development of a quality control system, TrAC, Trends Anal. Chem. 103 (2018) 34–43. https://doi.org/10.1016/j.trac.2018.03.009.

[13] C. Ort, K. V. Thomas, S. Castiglioni, L. Bijlsma, M.J. Reid, E. Zuccato, A. Covaci, A. van Nuijs, B. Kasprzyk-Hordern, F. Hernandez, P. de Voogt, E. Emke, Sewage Analysis CORe group Europe (SCORE), accessed February 26, 2020. http://www.score-network.eu/.

[14] M.A. Huestis, Human cannabinoid pharmacokinetics, Chem. Biodivers. 4 (2007) 1770–1804. https://doi.org/10.1002/chin.200747256.

[15] M.E. Wall, B.M. Sadler, D. Brine, H. Taylor, M. Perez-Reyes, Metabolism, disposition, and kinetics of delta-9-tetrahydrocannabinol in men and women, Clin. Pharmacol. Ther. 34 (1983) 352–363. https://doi.org/10.1038/clpt.1983.179.

[16] E.L. Karschner, E.W. Schwilke, R.H. Lowe, W.D. Darwin, R.I. Herning, J.L. Cadet, M.A. Huestis, Implications of plasma Δ9-tetrahydrocannabinol, 11-hydroxy-THC, and 11-nor-9-carboxy-THC concentrations in chronic cannabis smokers, J. Anal. Toxicol. 33 (2009) 469–477. https://doi.org/10.1093/jat/33.8.469.

[17] M.A. Huestis, J.M. Mitchell, E.J. Cone, Urinary excretion profiles of 11-nor-9-carboxy-Δ9- tetrahydrocannabinol in humans after single smoked doses of marijuana, J. Anal. Toxicol. 20 (1996) 441–452. https://doi.org/10.1093/jat/20.6.441.

[18] U. Khan, J.A. Nicell, Sewer epidemiology mass balances for assessing the illicit use of methamphetamine, amphetamine and tetrahydrocannabinol, Sci. Total Environ. 421–422 (2012) 144–162. https://doi.org/10.1016/j.scitotenv.2012.01.020.

[19] G. D'Ascenzo, A. Di Corcia, A. Gentili, R. Mancini, R. Mastropasqua, M. Nazzari, R. Samperi, Fate of natural estrogen conjugates in municipal sewage transport and treatment facilities, Sci. Total Environ. 302 (2003) 199–209.

[20] A.L.N. van Nuijs, S. Castiglioni, I. Tarcomnicu, C. Postigo, M.L. de Alda, H. Neels, E. Zuccato, D. Barcelo, A. Covaci, Illicit drug consumption estimations derived from wastewater analysis: a critical review, Sci. Total Environ. 409 (2011) 3564–3577. https://doi.org/10.1016/j.scitotenv.2010.05.030.

[21] N. Mastroianni, C. Postigo, M.L. De Alda, D. Barcelo, Illicit and abused drugs in sewage sludge: method optimization and occurrence, J. Chromatogr. A 1322 (2013) 29–37. https://doi.org/10.1016/j.chroma.2013.10.078.

[22] F. Mack, H. Bönisch, Dissociation constants and lipophilicity of catecholamines and related compounds, Naunyn Schmiedebergs Arch. Pharmacol. 310 (1979) 1–9. https://doi.org/10.1007/BF00499868.

[23] E. Gracia-Lor, S. Castiglioni, R. Bade, F. Been, E. Castrignanò, A. Covaci, I. González-Mariño, E. Hapeshi, B. Kasprzyk-Hordern, J. Kinyua, F.Y. Lai, T. Letzel, L. Lopardo, M.R. Meyer, J. O'Brien, P. Ramin, N.I. Rousis, A. Rydevik, Y. Ryu, M.M. Santos, I. Senta, N.S. Thomaidis, S. Veloutsou, Z. Yang, E. Zuccato, L. Bijlsma, Measuring biomarkers in wastewater as a new source of epidemiological information: current state and future perspectives, Environ. Int. 99 (2017) 131–150. https://doi.org/10.1016/j.envint.2016.12.016.

[24] A.-K. McCall, R. Bade, J. Kinyua, F.Y. Lai, P.K. Thai, A. Covaci, L. Bijlsma, A.L.N. van Nuijs, C. Ort, Critical review on the stability of illicit drugs in sewers and wastewater samples, Water Res. 88 (2016) 933–947.

[25] E. Zuccato, C. Chiabrando, S. Castiglioni, R. Bagnati, R. Fanelli, Estimating community drug abuse by wastewater analysis, Environ. Health Perspect. 116 (2008) 1027–1032. https://doi.org/10.1289/ehp.11022.

[26] T. Nefau, S. Karolak, L. Castillo, V. Boireau, Y. Levi, Presence of illicit drugs and metabolites in influents and effluents of 25 sewage water treatment plants and map of drug consumption in France, Sci. Total Environ. 461–462 (2013) 712–722. https://doi.org/10.1016/j.scitotenv.2013.05.038.

[27] D.A. Devault, T. Néfau, H. Pascaline, S. Karolak, Y. Levi, First evaluation of illicit and licit drug consumption based on wastewater analysis in Fort de France urban area (Martinique, Caribbean), a transit area for drug smuggling, Sci. Total Environ. 490 (2014) 970–978. https://doi.org/10.1016/j.scitotenv.2014.05.090.

[28] N.V. Heuett, C.E. Ramirez, A. Fernandez, P.R. Gardinali, Analysis of drugs of abuse by online SPE-LC high resolution mass spectrometry: communal assessment of consumption, Sci. Total Environ. 511 (2015) 319–330. https://doi.org/10.1016/j.scitotenv.2014.12.043.

[29] S. Terzic, I. Senta, M. Ahel, Illicit drugs in wastewater of the city of Zagreb (Croatia)—estimation of drug abuse in a transition country, Environ. Pollut. 158 (2010) 2686–2693. https://doi.org/10.1016/j.envpol.2010.04.020.

[30] C. Postigo, M.L. de Alda, D. Barceló, Evaluation of drugs of abuse use and trends in a prison through wastewater analysis, Environ. Int. 37 (2011) 49–55. https://doi.org/10.1016/j.envint.2010.06.012.

[31] C. Postigo, M.J. López de Alda, D. Barceló, Drugs of abuse and their metabolites in the Ebro River basin: occurrence in sewage and surface water, sewage treatment plants removal efficiency, and collective drug usage estimation, Environ. Int. 36 (2010) 75–84. https://doi.org/10.1016/j.envint.2009.10.004.

[32] E. Gracia-Lor, E. Zuccato, S. Castiglioni, Refining correction factors for back-calculation of illicit drug use, Sci. Total Environ. 573 (2016) 1648–1659. https://doi.org/10.1016/j.scitotenv.2016.09.179.

[33] F. Been, C. Schneider, F. Zobel, O. Delémont, P. Esseiva, Integrating environmental and self-report data to refine cannabis prevalence estimates in a major urban area of Switzerland, Int. J. Drug Policy 36 (2016) 33–42. https://doi.org/10.1016/j.drugpo.2016.06.008.

[34] A. Causanilles, J.A. Baz-Lomba, D.A. Burgard, E. Emke, I. Gonzalez-Marino, I. Krizman-Matasic, A. Li, A.S.C. Love, A.K. McCall, R. Montes, A.L.N. van Nuijs, C. Ort, J.E.B. Quintana, I. Senta, S. Terzic, F. Hernandez, P. de Voogt, L. Bijlsma, Improving wastewater-based epidemiology to estimate cannabis use: focus on the initial aspects of the analytical procedure, Anal. Chim. Acta 988 (2017) 27–33.

[35] S. Castiglioni, E. Zuccato, E. Crisci, C. Chiabrando, R. Fanelli, R. Bagnati, Identification and measurement of illicit drugs and their metabolites in urban wastewater by liquid chromatography-tandem mass spectrometry, Anal. Chem. 78 (2006) 8421–8429. https://doi.org/10.1021/ac061095b.

[36] M.R. Boleda, M.T. Galceran, F. Ventura, Trace determination of cannabinoids and opiates in wastewater and surface waters by ultra-performance liquid chromatography-tandem mass spectrometry, J. Chromatogr. A 1175 (2007) 38–48. https://doi.org/10.1016/j.chroma.2007.10.029.

[37] C. Postigo, M.J. Lopez De Alda, D. Barceló, Fully automated determination in the low nanogram per liter level of different classes of drugs of abuse in sewage water by on-line solid-phase extraction-liquid chromatography-electrospray-tandem mass spectrometry, Anal. Chem. 80 (2008) 3123–3134. https://doi.org/10.1021/ac702060j.

[38] L. Bijlsma, J.V. Sancho, E. Pitarch, M. Ibanez, F. Hernandez, Simultaneous ultra-high-pressure liquid chromatography-tandem mass spectrometry determination of amphetamine and amphetamine-like stimulants, cocaine and its metabolites, and a cannabis metabolite in surface water and urban wastewater, J. Chromatogr. A. 1216 (2009) 3078–3089.

[39] J.D. Berset, R. Brenneisen, C. Mathieu, Analysis of llicit and illicit drugs in waste, surface and lake water samples using large volume direct injection high performance liquid chromatography—electrospray tandem mass spectrometry (HPLC-MS/MS), Chemosphere 81 (2010) 859–866. https://doi.org/10.1016/j.chemosphere.2010.08.011.

[40] I. González-Mariño, J.B. Quintana, I. Rodríguez, R. Cela, Determination of drugs of abuse in water by solid-phase extraction, derivatisation and gas chromatography-ion trap-tandem mass spectrometry, J. Chromatogr. A 1217 (2010) 1748–1760. https://doi.org/10.1016/j.chroma.2010.01.046.

[41] F.Y. Lai, C. Ort, C. Gartner, S. Carter, J. Prichard, P. Kirkbride, R. Bruno, W. Hall, G. Eaglesham, J.F. Mueller, Refining the estimation of illicit drug consumptions from wastewater analysis: co-analysis of prescription pharmaceuticals and uncertainty assessment, Water Res. 45 (2011) 4437–4448. https://doi.org/10.1016/j.watres.2011.05.042.

[42] I. González-Mariño, J.B. Quintana, I. Rodríguez, M. Gonzáez-Díez, R. Cela, Screening and selective quantification of illicit drugs in wastewater by mixed-mode solid-phase extraction and quadrupole-time-of-flight liquid chromatography-mass spectrometry, Anal. Chem. 84 (2012) 1708–1717. https://doi.org/10.1021/ac202989e.

[43] I. Racamonde, E. Villaverde-de-Sáa, R. Rodil, J.B. Quintana, R. Cela, Determination of Δ9-tetrahydrocannabinol and 11-nor-9-carboxy-Δ9-tetrahydrocannabinol in water samples by solid-phase microextraction with on-fiber derivatization and gas chromatography-mass spectrometry, J. Chromatogr. A 1245 (2012) 167–174. https://doi.org/10.1016/j.chroma.2012.05.017.

[44] L. Bijlsma, E. Emke, F. Hernández, P. De Voogt, Performance of the linear ion trap Orbitrap mass analyzer for qualitative and quantitative analysis of drugs of abuse and relevant metabolites in sewage water, Anal. Chim. Acta 768 (2013) 102–110.

[45] G. Fedorova, T. Randak, R.H. Lindberg, R. Grabic, Comparison of the quantitative performance of a Q-Exactive high-resolution mass spectrometer with that of a triple quadrupole tandem mass spectrometer for the analysis of illicit drugs in wastewater, Rapid Commun. Mass Spectrom. 27 (2013) 1751–1762. https://doi.org/10.1002/rcm.6628.

[46] I. Senta, I. Krizman, M. Ahel, S. Terzic, Integrated procedure for multiresidue analysis of dissolved and particulate drugs in municipal wastewater by liquid chromatography-tandem mass spectrometry, Anal. Bioanal. Chem. 405 (2013) 3255–3268. https://doi.org/10.1007/s00216-013-6720-9.

[47] M.J. Andrés-Costa, N. Rubio-López, M. Morales Suárez-Varela, Y. Pico, Occurrence and removal of drugs of abuse in wastewater treatment plants of Valencia (Spain), Environ. Pollut. 194 (2014) 152–162. https://doi.org/10.1016/j.envpol.2014.07.019.

[48] L. Bijlsma, E. Beltran, C. Boix, J.V. Sancho, F. Hernandez, Improvements in analytical methodology for the determination of frequently consumed illicit drugs in urban wastewater, Anal. Bioanal. Chem. 406 (2014) 4261–4272.

[49] A.L.N. van Nuijs, A. Gheorghe, P.G. Jorens, K. Maudens, H. Neels, A. Covaci, Optimization, validation, and the application of liquid chromatography-tandem mass spectrometry for the analysis of new drugs of abuse in wastewater, Drug Test. Anal. 6 (2014) 861–867. https://doi.org/10.1002/dta.1460.

[50] B.J. Tscharke, C. Chen, J.P. Gerber, J.M. White, Temporal trends in drug use in Adelaide, South Australia by wastewater analysis, Sci. Total Environ. 565 (2016) 384–391. https://doi.org/10.1016/j.scitotenv.2016.04.183.

[51] T. Thiebault, L. Fougère, E. Destandau, M. Réty, J. Jacob, Temporal dynamics of human-excreted pollutants in wastewater treatment plant influents: toward a better knowledge of mass load fluctuations, Sci. Total Environ. 596–597 (2017) 246–255. https://doi.org/10.1016/j.scitotenv.2017.04.130.

[52] K.S. Foppe, D.R. Hammond-Weinberger, B. Subedi, Estimation of the consumption of illicit drugs during special events in two communities in Western Kentucky, USA using sewage epidemiology, Sci. Total Environ. 633 (2018) 249–256. https://doi.org/10.1016/j.scitotenv.2018.03.175.

[53] I. González-Mariño, V. Castro, R. Montes, R. Rodil, A. Lores, R. Cela, J.B. Quintana, Multi-residue determination of psychoactive pharmaceuticals, illicit drugs and related metabolites in wastewater by ultra-high performance liquid chromatography-tandem mass spectrometry, J. Chromatogr. A 1569 (2018) 91–100. https://doi.org/10.1016/j.chroma.2018.07.045.

[54] I. González-Mariño, K.V. Thomas, M.J. Reid, Determination of cannabinoid and synthetic cannabinoid metabolites in wastewater by liquid–liquid extraction and ultra-high performance supercritical fluid chromatography-tandem mass spectrometry, Drug Test. Anal. 10 (2018) 222–228. https://doi.org/10.1002/dta.2199.

[55] A.J. Skees, K.S. Foppe, B. Loganathan, B. Subedi, Contamination profiles, mass loadings, and sewage epidemiology of neuropsychiatric and illicit drugs in wastewater and river waters from a community in the Midwestern United States, Sci. Total Environ. 631–632 (2018) 1457–1464. https://doi.org/10.1016/j.scitotenv.2018.03.060.

[56] N. Centazzo, B.M. Frederick, A. Jacox, S.Y. Cheng, M. Concheiro-Guisan, Wastewater analysis for nicotine, cocaine, amphetamines, opioids and cannabis in New York City, Forensic Sci. Res. 4 (2019) 152–167. https://doi.org/10.1080/20961790.2019.1609388.

[57] N. Daglioglu, E.Y. Guzel, S. Kilercioglu, Assessment of illicit drugs in wastewater and estimation of drugs of abuse in Adana Province, Turkey, Forensic Sci. Int. 294 (2019) 132–139. https://doi.org/10.1016/j.forsciint.2018.11.012.

[58] S. Mercan, M. Kuloglu, T. Tekin, Z. Turkmen, A.O. Dogru, A.N. Safran, M. Acikkol, F. Asicioglu, Wastewater-based monitoring of illicit drug consumption in Istanbul: preliminary results from two districts, Sci. Total Environ. 656 (2019) 231–238. https://doi.org/10.1016/j.scitotenv.2018.11.345.

[59] D. Savas, Public Opinion and Illicit Drugs Canadian Attitudes towards Decriminalizing the Use of Marijuana, Fraser Inst. Digit. Publ., 2001, .pp. 1–12. http://oldfraser.lexi.net/publications/books/drug_papers/UDSavas.pdf.

[60] J. Tahirali, 7 in 10 Canadians Support Marijuana Legalization : Nanos Poll, CTV News, 2016. http://www.ctvnews.ca/canada/7-in-10-canadians-support-marijuana-legalization-nanos-poll-1.2968953.

[61] A. Swift, For First Time, Americans Favor Legalizing Marijuana, Gall. Polit., 2013. http://www.gallup.com/poll/165539/first-time-americansfavor-legalizing-marijuana.aspx.

[62] V. Yargeau, B. Taylor, H. Li, A. Rodayan, C.D. Metcalfe, Analysis of drugs of abuse in wastewater from two Canadian cities, Sci. Total Environ. 487 (2014) 722–730. https://doi.org/10.1016/j.scitotenv.2013.11.094.

[63] A. Jacox, J. Wetzel, S.-Y. Cheng, M. Concheiro, Quantitative analysis of opioids and cannabinoids in wastewater samples, Forensic Sci. Res. 2 (2017) 18–25. https://doi.org/10.1080/20961790.2016.1270812.

[64] A. Rodayan, S. Afana, P.A. Segura, T. Sultana, C.D. Metcalfe, V. Yargeau, Linking drugs of abuse in wastewater to contamination of surface and drinking water, Environ. Toxicol. Chem. 35 (2016) 843–849. https://doi.org/10.1002/etc.3085.

[65] T. Werschler, A. Brennan, Wastewater-Based Estimates of Cannabis and Drug Use in Canada: Pilot test Detailed Results, Publication 11-621-M, Stat. Canada Anal. Br., 2019. https://www150.statcan.gc.ca/n1/pub/11-621-m/11-621-m2019004-eng.htm. accessed February 26, 2020.

[66] Statistics-Canada, Prevalence of Cannabis Use in the Past Three Months, Self-Reported. 2019https://doi.org/10.25318/1310038301-eng. Table 13-10-0383-01.

[67] A.K. McCall, A. Scheidegger, M.M. Madry, A.E. Steuer, D.G. Weissbrodt, P.A. Vanrolleghem, T. Kraemer, E. Morgenroth, C. Ort, Influence of different sewer biofilms on transformation rates of drugs, Environ. Sci. Technol. 50 (2016) 13351–13360. https://doi.org/10.1021/acs.est.6b04200.

[68] P. Ramin, A.L. Brock, A. Causanilles, B. Valverde-Pérez, E. Emke, P. De Voogt, F. Polesel, B.G. Plósz, Transformation and sorption of illicit drug biomarkers in sewer biofilms, Environ. Sci. Technol. 51 (2017) 10572–10584. https://doi.org/10.1021/acs.est.6b06277.

[69] P. Ramin, A.L. Brock, F. Polesel, A. Causanilles, E. Emke, P. De Voogt, B.G. Plosz, Transformation and sorption of illicit drug biomarkers in sewer systems: understanding the role of suspended solids in raw wastewater, Environ. Sci. Technol. 50 (2016) 13397–13408. https://doi.org/10.1021/acs.est.6b03049.

Index

Note: Page numbers followed by "f" indicate figures and "t" indicate tables.

A

Acequinocyl, 299, 303f
Aerosols, 132–134, 133f
Agriculture Improvement Act, 4–5
Alkylbenzene-terpene, 223, 223f
Alpha pinene, 199t, 200
   electron ionization spectra, 205, 208f
   GCxGC-TOFMS, 175–176, 176f
Ambient ionization techniques, 370
Americans for Safe Access, 316
Anandamide (AEA), 35–36, 317–319
APCI. *See* Atmospheric pressure chemical
   ionization (APCI)
Appetite loss, cannabis, 341
2-Arachidonyl glyceryl ether (2-AGE), 37
Area under the curve (AUC), 373
Atmospheric pressure chemical ionization
   (APCI), 20, 281–285, 284f
Automated analysis, 390–391
   data analysis, 378
   materials and equipment, 378
   rationale, 377–378
   sample preparation and analysis, 378

B

Ball milling, 407
Beer's Law, 10
Beta pinene, 199t, 200, 205, 208f
Bicyclic monoterpene, 222
Binomial probability equation, 358
Biochemistry analyzers, 320–321
Biological fluid analysis, 78–95, 79–91t
Bisabolol, 226–227, 227f
Blank spike sample. *See* Laboratory control
   sample (LCS)
Blood, cannabinoids analysis in, 92–93
Butylated hydroxytoluene (BHT), 380–381
Butyrolactone, 179f

C

California's Proposition 215 in 1996, 316
Camphene, 198, 221–222, 221f

Cannabichromene (CBC), 440, 442f
Cannabichromene monomethyl ether
   (CBCM), 262, 262t, 263f
Cannabichromenic acid (CBCA), 440, 443f
Cannabichromenolic acid (CBCA), 40f
Cannabichromevarinic acid (CBCVA), 40f
Cannabicitran (CBCT), 445, 446f
Cannabicyclol (CBL), 440–445, 444f
Cannabidiol (CBD), 204–205
   calibration curves, 311f
   cannabis (*see* Cannabis)
   effects on blood pressure and heart
     rate, 342
   entourage effect, 231
   GC/EI-MS, phytocannabinoids,
     235–239
     characteristics, 242–246, 243f, 245f,
      255t, 257t, 259t
     extraction, 241–242, 246–272, 247f,
      249f, 254f, 256f, 261f, 262t, 263f,
      268–269f, 271f
     maintenance issues, 242
     sample preparation, 240
     standards and reagents, 239–240
     ThermoScientific AI/AS3000
      Auto-Sampler, 241
   hemp in (*see* Hemp (*Cannabis sativa*))
   LC-MS/MS, 308–312, 310f, 312f
   mass spectral identification, 228–231, 229f
   MS-MS spectrum for, 423–425, 424f
   total CBD quantitation, 7–18
   wastewater-based epidemiology, 461
Cannabidiolic acid (CBDA), 40f
   LC-MS/MS, 308–312, 310f
   LC/Q-TOF-MS, 425, 427f
Cannabidiol monomethyl ether (CBDM),
   262, 262t, 263f
Cannabidiol (CBD) oils
   sample preparation for, 8
   solid phase microextraction evaluation of,
     124–125
Cannabidivarin (CBDV), 425, 426f

483

Cannabidivarinic acid (CBDVA), 40*f*, 425–428, 429*f*
Cannabielsoin (CBE), 262, 264*f*
Cannabielsolic acid A (CBEA-A), 5–7
Cannabigerol (CBG), 67, 437, 439*f*
Cannabigerolic acid (CBGA), 5, 437–440, 441*f*
Cannabigerovarinic acid (CBGVA), 5
Cannabinodiol (CBND), 5–7
Cannabinoid profiling
  analyses using HPLC-UV and LC-MS, 15–18
  analyses with HPLC-UV, 8–14
  analyses with liquid chromatography, 14–15
  practices, 18
  sample preparation, 7–8
Cannabinoid receptor (CBR), 32–35
Cannabinoids, 72–78, 73–77*t*, 357–363. *See also* Phytocannabinoids
  accuracy %, 329–330, 330*t*
  accurate mass detection of, 420–423
  bioavailability of, 44
  biological samples, 79–91*t*
  biosynthetic pathways for, 38
  chromatogram, 399*f*
  chromatographic separation of, 420, 421*f*
  distribution volume, 46
  in edible chocolates (*see* Chocolates, cannabinoids)
  electron ionization mass spectra, 242–246, 243*f*, 245*f*
  fragmentation patterns of, 423–445
  frequency of detection, 450*f*
  LC-MS/MS, 308–312
  in marijuana products, 72–78, 73–77*t*
  mass spectral identification, 228–231
  metabolism of, 46
  metabolites in organic matrices, 47–51
  multireaction monitoring, 346, 352*t*
  natural and synthetic, 369*f*
  pharmaceutical form, 41–45
  in plants, 68–72, 69–71*t*
  profile for 3 hemp oils, 449*f*
  recoveries for, 331, 332*f*
  separation of, 360–363, 362*f*
  spiked and diluted, 360, 361*t*
  standards, 326, 334, 334*f*

  structures, 47*f*, 399*f*
  testing, suitability for, 391–392
Cannabinol (CBN), 5–7
  LC-MS/MS, 308–312, 310*f*
  LC/Q-TOF-MS, 432–437, 436*f*
Cannabinolic acid (CBNA), 437, 438*f*
Cannabis, 3–4, 340, 356*f*
  alternative methods/procedures, 355
  analyzers, 320–323, 322*f*, 324*f*, 331–332
  biomarkers, 457–461
  calibration, 353–354
  cannabinoid (*see* Cannabinoids)
  classifications of, 169–172
  data analysis and statistics, 354
  development, 320–323
  dynamic range comparisons, 316–317, 318*f*
  and electronic delivery devices, 132–134
  full spectrum extract, 334–335
  history, 340
  LC/MS/MS, 346–352
  legalization, 277
  materials, equipment and reagents, 344–345
  medical application, 341–342
  mycotoxins, 357
  patches, 125–127
  performance data, 325
  pesticides, 355–357
  pharmacokinetics of, 45–47
  phytocannabinoids in (*see* Phytocannabinoids)
  potency testing, 164–165, 317, 320
  products, 72–78, 73–77*t*
  protocol, 345–352
  quality assurance and quality control, 353–354
  quality control testing, 316–319, 317*t*, 321–323
  quantitative analysis, 344
  recreational use today, 342–343
  regulations, 343
  safety considerations and standards, 354
  sample filtration, 325–326
  sample homogenization, 319–320
  sample preparation, 325, 327–328
  solid phase microextraction evaluation of, 122–124

spiked and diluted, 360–363, 361t
taxonomy, 340–341
terpenoids in, 40t
tinctures, 334–335
Title 21 CFR Part 11, 324–325
Cannabis Analyzer for Potency, 322f, 333
capabilities, 335
chromatograms for flower sample, 331f
determination, 326–327
functionality to, 333–334
sample preparation methods, 331
Title 21 CFR Part 11, 325
user interface, 324, 324f
Cannabis indica, 67–68
Cannabis-light industry, 68
Cannabis oil, 42–43
Cannabispiran, 265, 268f
Cannabis ruderalis, 67–68
Cannabis sativa (hemp). *See* Hemp
(*Cannabis sativa*)
Cannflavins, 39
Carbon (C) isotopes transitions, THC,
358–360, 359f, 361f
Caryophyllene, 199t, 202
fragmentation pathway for, 212, 213f
mass spectra of, 211, 211f, 213
Caryophyllene oxide, 199t, 202
mass spectrum of, 214, 214f
proposed fragmentation of, 214, 215f
CBD. *See* Cannabidiol (CBD)
CBDA. *See* Cannabidiolic acid (CBDA)
Certificate of Analysis (COA), 278, 295,
307–308
Certified Reference Material (CRM)
standards, 326–327
Charlotte's Web oils, 42–43
Chemovar classifications
GCxGC-TOFMS application for,
184–193
nontargeted chemical profiling for,
172–183
Chocolates, cannabinoids, 397–400
accuracy, 409
apparatus, 400–401
chemicals and solvents, 401–402
chromatography, 406–407, 407f
cold stabilisation, 408–409, 408f
data processing, 404–405

gravimetric analysis, 405
heated ultrasonic extraction, 407
method detection limit, 411
method reporting limit, 411
method validation, 409, 410–412t
MS and source parameters, 403, 404t
preparation, 403
relative standard deviation, 410–411
repeatability and intermediate precision,
410–411
single-laboratory validation, fortified
samples, 405–406
solutions and calibration standards
preparation, 402–403
ultra-high-pressure liquid
chromatography, 403, 404t
Chordane, 281–283
ChromaTOF Software, 195
ChromaTOF® 5.5 software, 194–195
Chromatography, 173–175, 418
of cannabinoids, 420, 421f
chocolates, cannabinoids in, 399f,
406–407, 407f
nontargeted chemical profiling, 173–175
Chromeleon™ Chromatography Data
System, 145
Cinerin, calibration curve for, 296, 297f
Cold stabilisation, 408–409, 408f
Collision-induced dissociation (CID),
238–239
Concentrates, sample preparation for, 8
Continuing calibration verification (CCV),
305
Controlled Substances Act, 169–170
Convulsions, cannabis, 342
Cryomilling, 407
Cultivar classification, hemp, 171–172
cannabidiol, 228–229, 229f
terpenes, 217, 218–219f, 219–220
tetrahyrocannabinol, 228–229, 229f

D

Daminozide, 278
Decarboxylation, 17, 317
Delta-3-carene, 222, 222f
Δ^9-tetrahyrocannabinolic acid (Δ^9-THCA),
3–4, 308–312, 310f

Δ^9-tetrahyrocannabinol (Δ^9-THC), 3–4, 428, 430f
  analogues and metabolites, 45
  cannabinoid receptors, 32–35
  cannabis biomarkers, 457
  distribution in blood and tissues, 46f
  immunomodulatory activity of, 38
  total THC, 7–18
Delta-9-trans-tetrahydrocannabidiol (d9-trans-THC), 235–236, 244, 245f, 248–251, 250f
Deprenyl-cannabinoids, 265, 269f
Desorption electrospray ionization (DESI), 370
Desorption electrospray ionization (DESI)-mass spectrometry, 400
Dichloromethane (DCM), 17–18
Diffusive head-space extraction, 132
Dilute and shoot approach, 19–20
Dimethylacetamide (DMA), 113
N,N-Dimethylacetamide (DMA)-α,α,α-trifluorotoluene internal standard sample matrix, 25
Dimethylallyl diphosphate (DMAPP), 40f
Dimethylformamide (DMF), 153
Dimethyl sulfoxide (DMSO), 153
Diode-array detector (DAD), 10, 11f, 68–72
Direct analysis in real time (DART), 370
Direct analysis in real time mass spectrometry (DART-MS), 51
DISA Global Solutions, 4
Dispersive solid-phase extraction (dSPE), 19
Distribution volume (DV), 46
Diterpenes, 198
Dravet syndrome, 42–43
Driving under the influence of drugs (DUID), 391
Dronabinol, 371, 379
Drug detection, 50–51
Dynamic head space (DHS), 108, 127–128

E

EI. *See* Electron ionization (EI)
E-juices, method validation, 158–164, 163–164t
Electronic delivery devices, cannabis and, 132–134
Electron ionization (EI)

  alpha and beta pinene, 205, 208f
  terpenes separation, 204–205
  tetrahydrocannabinol, 230, 230f
Electrospray ionization (ESI), 14–15, 49–51
  pesticides analysis, 281–285, 283f
  positive, 470–472
Eluants, 280–281
Endocannabinoid (eCB) system, 32–37, 36f
Endovanilloid systems, 34
Entourage compounds, 35–36
Entourage effect, 171–172, 231
Epidiolex®, 42–43, 342
ESI. *See* Electrospray ionization (ESI)
Essential oils, 241–242
  humulene, 201–202
  myrcene, 200–201
  nerolidol, 225
  terpenes, 198
Ethyl-pyrazine, 179f
Eucalyptol, 223–224, 224f
Evidence-based medicine (EBM) approach, 42
Extracted ion chromatograms (XICs), 176–178
Extraction solvent, 68

F

Farm Bill, 4–5
Farnesyl diphosphate (FPP), 40f
Figures of merit, 292
Fisher Scientific, 239–240
Flavourants, 142
Flower, sample preparation, 8
Fragmentation patterns, cannabinoids, 423–445
Full evaporative transfer (FET), 142–143
Full Extract Cannabis Oil (FECO), 334–335
Full Spectrum Cannabis Oil (FSCO), 334–335

G

Gas chromatography (GC), 68–72
Gas chromatography/mass spectrometry (GC/MS), 142
  electron ionization, phytocannabinoids, 235–239
    characteristics, 242–246, 243f, 245f, 255t, 257t, 259t

extraction, 241–242, 246–272, 247*f*, 249*f*, 254*f*, 256*f*, 261*f*, 262*t*, 263*f*, 268–269*f*, 271*f*
  maintenance issues, 242
  sample preparation, 240
  standards and reagents, 239–240
  ThermoScientific AI/AS3000 Auto-Sampler, 241
headspace, 145–147
  analysis, 143–147
  chemicals, 143–145
  concurrent cannabis potency test, 164–165
  dry hop tissue, 161*f*
  e-juices and related vaping products, 158–164
  liquid-injection GC–MS/MS cross-validation, 147–149
  method development and optimisation, 150–154
  method validation, 149–150
  quantitating terpenoids in, 156–157*t*
  reagents, 143–145
monoterpenoids, 205, 206*f*
sesquiterpenes, 205, 206*f*
terpenes, 202, 202–203*t*, 205, 215–217, 216*f*
  bisabolol, 226–227, 227*f*
  camphene, 221–222, 221*f*
  delta-3-carene, 222, 222*f*
  eucalyptol, 223–224, 224*f*
  extracted ion chromatogram, 217, 218–219*f*
  fragmentation, 205–215
  geraniol, 224, 225*f*
  guaiol, 228, 228*f*
  identification, 215–228, 216*f*, 218–219*f*
  nerolidol, 225–226, 226*f*
  ocimene, 220, 220*f*
  p-cymene, 223, 223*f*
  total ion chromatogram, 215–217, 216*f*
Gas chromatography-tandem mass spectrometry (GC-MS/MS), 23–24
Gas chromatography with flame ionization detectors (GC/FID), 78, 236–238
Gas phase chromatography (GC), 4–5
Geno/Grinder®, 319, 325, 344–345

Geraniol, 224, 225*f*
Geranyl diphosphate (GPP), 40*f*
Geranyl pyrophosphate, 5
Glycerol, 153, 162
G protein-coupled receptor (GPCR), 32–33, 35
G protein receptor kinases (GRKs), 33–34
Graphitized carbon black (GCB), 19
Gravimetric analysis, 405
Guaiol, 228, 228*f*
Gummy bear candies, 122–124, 124*t*

H

Hair, cannabinoids analysis in, 94–95
Half range for the prediction interval of results (HR_{PIR}), 405–406
Headaches, cannabis, 341
Headspace (HS)
  cannabis testing, 107–115
    loop based, 110
    pressure balanced sampling, 110–111, 111*f*
    syringe based, 109–110
  diffusive, 132
  dynamic, 127–128
  gas chromatography–mass spectrometry, 25–26, 145–147
    analysis, 143–147
    chemicals, 143–145
    concurrent cannabis potency test, 164–165
    dry hop tissue, 161*f*
    e-juices and related vaping products, 158–164
    liquid-injection GC–MS/MS cross-validation, 147–149
    method development and optimisation, 150–154
    method validation, 149–150
    quantitating terpenoids in, 156–157*t*
    reagents, 143–145
  sampling techniques, 142–143
  solid-phase microextraction methods, 142–143
  static (*see* Static head-space (SHS))
Headspace Sorbent Pen (HSP), 131
Helium, 145
Hematoencephalic barrier (HEB), 38

Hemiterpenes, 198

Hemp *(Cannabis sativa)*, 4–5, 31–32, 67–68, 105–106, 203–205
  analyzers, 320–323, 333–334
  constituents, 37–40
  cultivar of, 217, 218–219*f*, 219–220, 228–229, 229*f*
  phytocannabinoids in (*see* Phytocannabinoids)
  plant material, 8
  terpenes in (*see* Terpenes)
  2018 Farm Bill, 333

Hemp *(Cannabis sativa)* oils
  analyses of, 445–449, 448*t*
  frequency of detection, 450*f*
  homogenized, sample preparation for, 8
  profile for, 449*f*
  sample preparation for, 418

High-boiling solvent tetradecane, 154

High performance liquid chromatography (HPLC), 4–5, 320–321, 321*f*, 328

High performance liquid chromatography (HPLC)-UV
  cannabinoid analyses with, 8–18
  mobile phase gradient, 13*t*
  parameters, 10–12, 13*t*
  target cannabinoids, 13*t*

High-resolution accurate mass based systems (HRAM), 107, 123–124

High resolution mass spectrometry (HRMS), 48

High-temperature headspace gas chromatography–mass spectrometry, 145–147
  analysis samples, 143–145
  chemicals, 143–145
  concurrent cannabis potency test, 164–165
  e-juices and related vaping products, 158–164
  liquid-injection GC–MS/MS cross-validation, 147–149
  method development and optimisation, 150–154
  method validation, 149–150
  reagents, 143–145

Homogenized hemp oils, 8

Hops *(Humulus lupulus)*, 201–202

Humulene, 199*t*, 201–202, 211, 211*f*
  fragmentation pathway for, 211, 212*f*
  mass spectra of, 211, 211*f*

Humulus lupulus (Hops), 201–202

11-Hydroxy-tetrahydrocannabinol (HTHC), 432, 435*f*

I

Imazalil, 278

Imidacloprid, 280–281, 305–307, 306*f*

Inflammation, cannabis, 341

Intuvo 9000 GC, 202

Ionization suppression, 283–284

Ionotropic cannabinoid receptors, 34

Isopentenyl diphosphate (IPP), 40*f*

Isopulegol, 209–210, 210*f*

Isotope-labelled internal standards (ILIS), 456–457

L

Laboratory control sample (LCS), 304

LC coupled to tandem MS (LC-MS/MS), 456–457

LC/Q-TOF-MS. *See* Liquid chromatography coupled with quadrupole time-of flight mass spectrometry (LC/Q-TOF-MS)

Lennox-Gastaut syndrome, 42–43

Limit of detection (LOD), 18–19, 50–51
  pesticides using LC-MS/MS, 298–303, 300–302*t*, 303*f*
  of THC-COOH, 472

Limit of quantitation (LOQ), 18–19, 21, 50–51, 456–457
  pesticides using LC-MS/MS, 298–303, 300–302*t*, 303*f*
  of THC-COOH, 472

δ-Limonene, 199*t*, 200, 200*f*, 208, 209*f*

Linalool, 199*t*, 201, 209–210, 210*f*

Liposomes, 44

Liquid chromatography (LC), 68–72
  cannabinoid analyses with, 14–15
  LC coupled to mass spectrometry, 15–18, 453–454

Liquid chromatography coupled with quadrupole time-of flight mass spectrometry (LC/Q-TOF-MS), 22–23, 417 accurate mass detection of cannabinoids, 420–423

alternative methods/procedures, 420

analysis and statistics, 419

chromatographic separation, 418, 420

equipment, 417–418

fragmentation patterns of cannabinoids, 423–445

hemp oils, analyses of, 445–449

materials, 417–418

MS-MS conditions, 418

precursor techniques, 419

pros and cons, 419

protocols, 418

rationale, 417

reagents, 417–418

safety considerations and standards, 419

sample preparation for hemp oils, 418

TOF mass detection, 418

troubleshooting and optimization, 420

Liquid chromatography-tandem mass spectrometry (LC-MS/MS), 21–23

analytical challenges to residual pesticide testing using, 22–23

cannabis, 346–352, 355, 357

method commonalties Canada and U.S. States, 22

pesticides

atmospheric pressure chemical ionization process, 281–284, 284*f*

calibration, 292–298

cannabinoid analysis, 308–312, 310*f*, 312*f*

clean up steps procedures, 291

electrospray process, 283, 283*f*

instrumental procedure, 286–290

instrumentation, 279–286, 280*f*, 282*f*

ion ratio and retention time, 280–281, 305–307, 306*f*

limit of quantitation and detection, 298–303, 300–302*t*, 303*f*

preparation procedure, 291–292

procedure, 286–290

pyrethrin, 289, 290*f*

quality assurance/quality control, 304–305

results, 305–308

thiacloprid, 285–286, 285*f*, 287–288*f*

Liquid extraction, 72

Liquid-injection GC–MS/MS cross-validation, 147–149

Liquid-liquid extraction (LLE), 72, 78–92, 469

Liquid-phase micro-extraction (LPME), 78–92

Loop based head-space sampling, 110

Low-boiling hexane, 154

Low-boiling solvent hexane, 154

M

Marijuana. *See* Cannabis

Marijuana Policy Project, 316

MassHunter Quantitative Analysis software, 354

Mass spectrometry (MS), 107. *See also specific mass spectrometry*

acquisition and source parameters, 14*t*

cannabinoid analyses with, 14–15

gas chromatography and (*see* Gas chromatography/mass spectrometry (GC/MS))

monoterpenoids, 205, 206*f*

nontargeted chemical profiling, 175–178

sesquiterpenes, 205, 206*f*

SIM/Scan segments, 15*t*

and source parameters, 403, 404*t*

terpenes, 202, 202–203*t*, 205, 215–217, 216*f*

bisabolol, 226–227, 227*f*

camphene, 221–222, 221*f*

delta-3-carene, 222, 222*f*

eucalyptol, 223–224, 224*f*

extracted ion chromatogram, 217, 218–219*f*

fragmentation, 205–215

geraniol, 224, 225*f*

guaiol, 228, 228*f*

identification, 215–228, 216*f*, 218–219*f*

nerolidol, 225–226, 226*f*

ocimene, 220, 220*f*

p-cymene, 223, 223*f*

total ion chromatogram, 215–217, 216*f*

Matrix-assisted laser desorption/ionization mass spectrometry (MALDI–MS) technique, 51

Matrix matching, 296–298
Matrix spike (MS), 305
Matrix spike duplicate (MSD), 305
Maximum residual limit (MRL)
   acequinocyl, 299
   pesticide detection, 279, 292–293,
      295–296, 299
MCTs. *See* Mid-chain triglycerides (MCTs)
Medical cannabis, 65, 316–319, 341–342
Medic® vaporizer, 43
Method blank (MB), 304
Method detection limit (MDL), 25
   chocolates, cannabinoids in, 411
   high-temperature headspace gas
      chromatography–mass
      spectrometry, 149–150
Method reporting limit (MRL)
   chocolates, cannabinoids in, 411
   high-temperature headspace gas
      chromatography–mass
      spectrometry, 149–150
Microwave-Assisted Extraction and
      Cannabis Activation (MAECA), 194
Mid-chain triglycerides (MCTs),
      241–242
Millipore Sigma, 344–345
Mind-body medicine, 32
Molecularly imprinted solid phase
      extraction (MISPE), 93
Monoterpenes, 198, 199t
   base peak and diagnostic ions for,
      214–215, 216t
   mass spectrum, 205–209, 209f
      camphene, 198, 221–222, 221f
      delta-3-carene, 222, 222f
      ocimene, 220, 220f
Monoterpenoids, 199t
   base peak and diagnostic ions for,
      214–215, 216t
   gas chromatography/mass
      spectrometry, 205, 206f
      eucalyptol, 223–224, 224f
      fragmentation, 209–211, 210f
      geraniol, 224, 225f
MRL. *See* Maximum residual limit (MRL)
MS. *See* Mass spectrometry (MS)
Multidimensional chromatography,
      178–183

Multiple reaction monitoring (MRM), 49,
      285
   cannabinoids, 346, 352t
   mycotoxin transitions, 346, 352t
   pesticides transitions, 346, 347–351t
Multiple-wavelength detector (MWD), 10
Myclobutanil, 278
Mycotoxin
   analysis of, 18–24
   cannabis, 357
   transitions and MRM conditions, 346,
      352t
β-Myrcene, 199t, 200–201

N

Nanostructured lipid carrier particles, 44
N-arachidonoylethanolamine, 317–319
Nausea, cannabis, 342
Nerolidol, 225–226, 226f
Neutral phytocannabinoid, 237f, 260
Nicotine quantitation method, 144t
Nontargeted chemical analyses
   for chemovar classifications, 172–183
   chromatography, 173–175
   classifications, 169–172
   GCxGC-TOFMS application for,
      184–193
   mass spectrometry, 175–178
   multidimensional chromatography,
      178–183
Novel delivery systems, 41–45

O

Ocimene, 220, 220f
Odd electron (OE) ion, 205–208, 212, 213f
Oils
   cannabidiol
      sample preparation for, 8
      solid phase microextraction
      evaluation of, 124–125
   essential (*see* Essential oils)
   hemp
      analyses of, 445–449, 448t
      frequency of detection, 450f
      profile for, 449f
      sample preparation for, 418
   sample preparation procedures for, 8
   sesame (*see* Sesame oil preservation)

Index

Olivetolate geranyltransferase, 5
Olivetolic acid, 265–272, 271*f*
Oral fluid (OF), 93–94
Organic matrices, cannabinoid metabolites in, 47–51
Oromucosal administration, 46

P

Paper spray mass spectrometry, 370–371, 370*f*, 390
Paper strip extraction
  data analysis, 375
  evaluating preconcentration by, 374–375, 381–383
  lowest detectable urine concentration (ng/mL) after, 387–389*t*
  materials, 374
  rationale, 374
  sample preparation and analysis, 374–375
  with and without sesame oil, 385–386*t*
Para-cymene, 223, 223*f*
Parallel reaction monitoring (PRM), 48, 373
Pegasus® BT 4D mass spectrometer, 194–195
Pentachoronitrobenze, 281–283
Pesticides, 277–279
  cannabis, 355–357
  LC-MS/MS, 279–286, 280*f*, 282*f*
    atmospheric pressure chemical ionization process, 281–284, 284*f*
    calibration, 292–298
    cannabinoid analysis, 308–312, 310*f*, 312*f*
    electrospray process, 283, 283*f*
    ion ratio and retention time, 280–281, 305–307, 306*f*
    limit of quantitation and detection, 298–303, 300–302*t*, 303*f*
    procedure, 286–290
    pyrethrin, 289, 290*f*
    quality assurance/quality control, 304–305
    thiacloprid, 285–286, 285*f*, 287–288*f*
    transitions and MRM conditions, 346, 347–351*t*
    triple quadrupole instrument, 285, 286*f*
Pharmaceutical form (PF), 41–45
Photo-oxidation, 5–7

Photosensitizer (PS), 45
Phytocannabinoids, 235–236, 237*f*, 398. *See also* Cannabinoids
  biosynthesis of, 5–7, 6*f*
  chocolate-based edibles (*see* Chocolates, cannabinoids)
  classification given by ElSohly and Slade, 39*f*
  data processing, 404–405
  GC/EI-MS, 236–239
    characteristics, 242–246, 243*f*, 245*f*, 255*t*, 257*t*, 259*t*
    extraction, 241–242, 246–272, 247*f*, 249*f*, 254*f*, 256*f*, 261*f*, 262*t*, 263*f*, 268–269*f*, 271*f*
    maintenance issues, 242
    sample preparation, 240
    standards and reagents, 239–240
    ThermoScientific AI/AS3000 Auto-Sampler, 241
  GC/FID, 236–238
  HPLC/MS, 238–239
  HPLC/UV, 238–239
  liquid chromatography, 238
  neutral, 237*f*, 260
Pinene, 199*t*, 200
  electron ionization spectra, 205, 208*f*
  GCxGC-TOFMS, 175–176, 176*f*
Pipe smoking, 45
Plants, cannabinoids in, 68–72, 69–71*t*
Plant tissue analysis, 154–158
Plasma, cannabinoids analysis in, 92–93
Polydimethylsiloxane (PDMS), 116–117
Poly-ε-caprolactone (PCL), 45
Polyketide pathway, 5
Poly(lactic-*co*-glycolic acid) (PLGA), 45
Polymeric carriers, 45
Polyterpenes, 198
Preconcentration
  data analysis, 377
  materials and equipment, 374
  measuring improvements to detection limits, 387–390
  rationale, 376
  sample preparation and analysis, 376–377
Precursor techniques, 419
Prediction interval of results (PIR), 406
Pressure balanced sampling, 110–111, 111*f*

Principal component analysis (PCA), 188–189, 189*f*, 191*f*, 193*f*
Pro-nano liposphere (PNL), 44
Propylene glycol, 162
Psychoneuroimmunology, 32
Push-button approach, 320–321
Pyrethrin, 289, 290*f*

Q

Quality assurance (QA), 304–305, 353–354
Quality by design (QbD), 42
Quality control (QC), 304–305, 353–354, 456–457
QuEChERS (Quick, Easy, Cheap, Effective, Rugged & Safe), 19, 72, 281, 291–292, 400
Quintozene. *See* Pentachoronitrobenze

R

Relative standard deviation (RSD), 152–153*t*
  chocolates, cannabinoids in, 405, 410–411
  pesticides, LC-MS/MS, 299
Residual pesticides, 18–24
Residual solvent analysis (RSA), 24–26
  headspace GC-MS analysis, 25–26
  reagent and sample preparation, 24–25
  testing, 111–113
Resins, sample preparation for, 8
Restek Corporation, 204–205
Retention Index (RI) values, 174–175
Retention time (RT), pesticide, 280–281, 305–307, 306*f*
Reversed phase (RP), 469

S

Sample Preparation Rail (SPR), 132
San Francisco Cannabis Buyers Club, 316
Sativex®, 43
Schedule I drug, 105–106
Scientific Working Group for Forensic Toxicology (SWGTOX) guidelines, 49–50
Seizures, cannabis, 342
Selected ion monitoring (SIM), 145, 146–147*t*, 151*f*

Selected reaction monitoring (SRM), 148–149*t*
Self-emulsifying drug delivery (SEDD) technology, 44
Sesame oil preservation, 374
  data analysis, 373, 377
  equipment, 372
  materials, 372
  materials and equipment, 374
  measuring improvements to detection limits, 383–386
  methods, 371–374
  rationale, 371, 376
  results, 379–381
  sample preparation and analysis, 372–373, 376
Sesquiterpenes, 198, 199*t*
  base peak and diagnostic ions for, 214–215, 216*t*
  chromatographic separation, 205
  gas chromatography/mass spectrometry, 206*f*
  mass spectral fragmentation, 211–215, 211*f*, 214*f*
Sesquiterpenoids, 199*t*
  base peak and diagnostic ions for, 214–215, 216*t*
  bisabolol, 226–227, 227*f*
  guaiol, 228, 228*f*
  nerolidol, 225–226, 226*f*
Sesterterpenes, 198
Sewage treatment plants (STPs), 453
  sampling of, 455
Shen-Nung, 317
Shimadzu Scientific Instruments USA Innovation Center, 322, 322*f*, 324
Sigma Aldrich, Pennsylvania, 202
Solid-liquid extraction (SLE), 72
Solid phase extraction (SPE), 78–92
Solid phase microextraction (SPME), 78–92, 108, 116–127, 469
  cannabis infused gummy bear candies, 122–124, 124*t*
  cannabis patches, 125–127
  CBD oil, 124–125
  fibres, 72
Solvent extraction, 291
Sorbent pens (SPs), 130–131

Index 493

Sorbent Pen Thermal Conditioner (SPTC), 132
Sorbent Pen Thermal Desorption Unit (SPDU), 132
SPE cleanup and dilution, 20–21
SPEX Freezer/Mill®, 319
Spex 2010 Geno/Grinder®, 319, 325, 344–345
Stable isotopic label (SIL), 374, 379
Standard operating procedure (SOP), 353
Static head-space (SHS), 109
  based cannabis testing, 107–115
  extraction, 114f
  residual solvents for cannabis products using, 111–113
  sampling, 116f
  terpenes and terpenoid analysis by, 113–115
Stationary phase, 281
Stinging nettle (*Urtica dioica* L.), 143, 155–158
Stir bar sorptive extraction (SBSE), 129
Substance Abuse and Mental Health Services Administration (SAMHSA), 391–392
Supercritical fluid chromatography (SFC), 68–72, 469–470
Supercritical fluid extraction (SFE), 41, 58
Superficially porous particle (SPP) column technologies, 9
Synthetic cannabinoids, 78, 367–371, 369f
Syringe based head-space sampling, 109–110
Syringe filters, 325–326

T

Terpenes, 15–17, 198, 199t, 203–205
  analysis by static head space, 113–115
  β-caryophyllene, 202
  β-caryophyllene-oxide, 202
  biosynthetic pathways for, 38
  β-myrcene, 200–201
  chemical characteristics, 206–207t
  chromatographic separation, 205
  δ-limonene, 200, 200f
  in e-cigarette vaping liquids, 142
  extracted ion chromatogram for, 165f
  gas chromatography/mass spectrometry, 202, 202–203t, 205, 215–217, 216f

  extracted ion chromatogram, 217, 218–219f
  total ion chromatogram, 215–217, 216f
  α-humulene, 201–202
  linalool, 201
  mass spectral fragmentation, 205–215
  mass spectral identification, 215–217, 216f, 218–219f
  bisabolol, 226–227, 227f
  camphene, 221–222, 221f
  delta-3-carene, 222, 222f
  eucalyptol, 223–224, 224f
  geraniol, 224, 225f
  guaiol, 228, 228f
  nerolidol, 225–226, 226f
  ocimene, 220, 220f
  p-cymene, 223, 223f
  naturally occurring, 141f
  and nicotine quantitation method, 144t
  ocimene, 220, 220f
  α-pinene and β-pinene, 200
  profiles, 140
  quantitation method, 144t, 159–160t
  terpineol, 201
  terpinolene, 201
Terpenoids. *See* Terpenes
δ-Terpinene. *See* Terpinolene
Terpineol, 199t, 201
Terpinolene, 199t, 201
  fragment ions, 208–209
  mass spectra, 208, 209f
Tetrahydrocannabinol (THC), 204–205, 367–371
  calibration curves, carbon transition, 361f
  cannabis (*see* Cannabis)
  carbon isotopes separation, 358–360, 359f, 361f
  chocolate-based edibles (*see* Chocolates, cannabinoids)
  determination, 317
  entourage effect, 231
  fragmentation with electron ionization, 230, 230f
  GC/EI-MS, phytocannabinoids, 235–239
    characteristics, 242–246, 243f, 245f, 255t, 257t, 259t

Tetrahydrocannabinol (THC) (*Continued*)
  extraction, 241–242, 246–272, 247*f*,
    249*f*, 254*f*, 256*f*, 261*f*, 262*t*, 263*f*,
    268–269*f*, 271*f*
  maintenance issues, 242
  sample preparation, 240
  standards and reagents, 239–240
  ThermoScientific AI/AS3000
    Auto-Sampler, 241
  hemp (*see* Hemp *(Cannabis sativa)*)
  LC-MS/MS, 308–312, 310*f*
  mass spectral identification, 228–231, 229*f*
  potency formula, 330–331
  screening for, 369–370
Tetrahydrocannabinol (THC)-COOH
  analytical methodology, 461–472
  average weekly loads of, 474*f*
  cannabis biomarkers, 457–461
  limits of detection, 472
  limits of quantification, 472
  quantification of, 470–472
  quarterly per capita estimates, 475*f*
  selectivity of, 471*f*
Tetrahydrocannabinolic acid (THCA), 40*f*,
    432, 433*f*
  potency formula, 330–331
  standard curves for, 329, 329*f*
Tetrahydrocannabivarin (THCV), 428–431,
    431*f*
Tetrahydrocannabivarinic acid (THCVA),
    40*f*, 432, 434*f*
Tetraterpenes, 198
THC. *See* Tetrahydrocannabinol (THC)
Thermo Fisher Scientific Q-Exactive Focus
    orbitrap mass spectrometer, 373
Thiacloprid, 285, 285*f*
  MS/MS spectrum, 285, 287*f*
  quantitation transition, 285–286, 288*f*
Time-of-flight mass spectrometry
    (TOFMS), 173–174, 176, 177*f*, 418
  GCxGC-TOFMS application for
    (*see* Two-dimensional GC-time-of-
    flight mass spectrometry
    (GCxGC-TOFMS))
  LC/Q-TOF-MS, 418
    accurate mass detection of
      cannabinoids, 420–423
    alternative methods/procedures, 420

analysis and statistics, 419
chromatographic separation, 418
chromatographic separation of
  cannabinoids, 420
definition, 417
equipment, 417–418
fragmentation patterns of cannabinoids,
  423–445
hemp oils, analyses of, 445–449
materials, 417–418
MS-MS conditions, 418
precursor techniques, 419
pros and cons, 419
protocols, 418
rationale, 417
reagents, 417–418
safety considerations and standards, 419
sample preparation for hemp oils, 418
troubleshooting and optimization, 420
two-dimensional GC (*see* Two-
  dimensional GC-time-of-flight mass
  spectrometry (GCxGC-TOFMS))
Tinctures
  cannabis, 334–335, 335–336*f*
  sample preparation for, 8
Tissues, cannabinoid analysis in, 78–95,
  79–91*t*
Total ion chromatogram (TIC), 114–115,
  114*f*, 173–174
Transdermal administration, 46
Transdermal systems, 126–127
(–)-*Trans*-Δ^9-tetrahyrocannabinol (THC),
  379
Trifloxystrobin
  calibration curve for, 293–295, 293–294*f*
  maximum residual limit, 299
α,α,α-Trifluorotoluene internal standard
  (TFT ISTD), 24
Triple quadrupole (QqQ) mass
  spectrometer, 456–457
  cannabis, 344, 357–358
  pesticide analysis, 285, 286*f*
Triterpenes, 198
Tropylium ion, 223
Turn-key approach, 320–321
Turpentine, 198
2018 Farm Bill, 333
Twister, 129–130

Two-dimensional GC (GCxGC), 178–183, 180–181f
Two-dimensional GC-time-of-flight mass spectrometry (GCxGC-TOFMS)
analysis and statistics, 195
application for nontargeted chemical analysis, 184–193
definition, 194
equipment, 194–195
materials, 194–195
protocols, 195
rationale, 194
reagents, 194–195

U

Ultra-high-performance liquid chromatography (UHPLC) analyses, 9
Ultra-high-performance liquid chromatography coupled with tandem mass spectrometry (UPLC-MS/MS), 49
Ultra-high-pressure liquid chromatography (uHPLC)
chocolates, cannabinoids in, 403
phytocannabinoids, 404t
Ultra-performance liquid chromatography-mass spectrometry (UPLC-MS), 48
Umbilical cord, 49–50
United States Federal Law, 105–106

United States Pharmacopoeia, 317
Urine, cannabinoids analysis in, 94
Urtica dioica L. (stinging nettle), 143, 155–158

V

Vacuum-assisted sorbent extraction (VASE), 108, 130–132, 131f
Vaping liquid (e-juice) method validation, 150
Variable-wavelength detector (VWD), 10

W

Wastewater-based epidemiology (WBE), 453–457
analytical methodology, 456, 461–472, 466–468t
biomarkers, selection, 456
for cannabis estimates in Canada and US, 472–476, 474t
population size, estimation, 456
research and current asset, 476–477
sample preparation steps, 463–465t
sampling of, 455
Water management and storage, 132
Winterization, 19–20
Workplace drug screening, 391

Z

Zero-cannabis matrix, 353